N & N Science Series

Chemistry

**A Comprehensive Review of Chemistry
with a Special Section on**

The College Board Achievement Test in Chemistry ©

Author:

Nicholas R. Romano

Illustrations and Graphics:

Wayne H. Garnse

Cover Design:

D0907888

Eugene B. Fairbanks

N & N Publishing Company, Inc.
18 Montgomery Street   Middletown, New York 10940
(914) 342 - 1677

Special Credits

Thanks to

Ms. Lynn Boruchowitz
Sr. Kathleen McKinney
Dr. Wayne Moreau
Mr. William Schoen
Ms. Gloria Tonkinson

for their assistance in the preparation and proofing of this manuscript

N & N Publishing expresses our appreciation to the College Board and the Educational Testing Service for providing their permission to use materials from *The College Board Achievement Tests in Science*. For students who need a more "in depth" explanation of the Achievement Test in Chemistry, we recommend purchasing the official publication of The College Board: *The College Board Achievement Tests* (14 Tests in 13 Subjects) ISBN: 0-87447-162-1.

It is our sincere hope that the addition of these materials in *N & N's Science Series – Chemistry* will further enhance the education of our students.

College Board Achievement test questions and related material selected from *The College Board Achievement Tests in Science*, College Entrance Examination Board, 1986, 1983. Reprinted by permission of the College Board and Educational Testing Service, the copyright owner of the test questions. Permission to reprint the College Board Test questions and related material does not constitute review or endorsement by Educational Testing Service or the College Board of this publication as a whole or of any other testing information it may contain.

This book has been produced on the Macintosh Quadra 850AV and LaserWriter Pro printer. MacWrite II by Claris and Canvas by Deneba were used to produce the text and graphics. Original line drawings were reproduced with VersaScan on a Microtek MS-300A and modified with Photoshop by Adobe. Formatting, special designs, graphic incorporation, and page layout were accomplished with Ready Set Go by Manhattan Graphics. Special technical assistance was provided by Frank Valenza and Len Genesee of Computer Productions Unlimited, Newburgh, New York. To all, thank you for your excellent software, hardware, and technical support.

© Copyright 1985, 2000
N & N Publishing Company, INC.

SAN # - 216-4221   ISBN # - 0935487 41 7
Printed in the United States of America
6 7 8 9 0   BMP   2000

Table
Of Contents

UNIT I - Matter and Energy

UNIT II - Atomic Structure

UNIT III - Bonding

UNIT IV - Periodic Table

Unit V - Mathematics of Chemistry

Additional Materials in Mathematics of Chemistry

Unit VI - Kinetics and Equilibrium

Additional Materials in Kinetics and Equilibrium

Unit VII - Acids and Bases

Unit VIII - Redox and Electrochemistry

Additional Materials in Redox and Electrochemistry

Unit IX - Organic Chemistry

Please Note:
     Bold Face Type indicates important Topics, Concepts, and Vocabulary
     Underlined Topics indicate **Additional Materials** (AM) within the Units

Review Overview

N & N's Science Series – Chemistry is a condensed, complete review of high school chemistry material.

N & N's Science Series – Chemistry is written in a concise outline–styled narrative to best help students read and comprehend the chemistry content in a relatively short time. It may be necessaery for the teacher to provide additional examples and explanations to further amplify any of the topics for which greater depth of information or understanding is desired or necessary.

Each of the more than fifty concept areas contain many sample problems and questions, followed by over six hundred practice questions. Also included in the body of the text are additional questions. The specific answers are not given. These questions are asked so that the student may review more actively, rather than just "reading over" the material and questions.

The text is followed by an explanation of each of the tables in the Reference Tables for Chemistry. *Note: This section is periodically updated or revised in order to include those chemical value changes proposed by the American Chemical Society and IUPAC (International Union of Pure and Applied Chemists).*

Finally, the text material is followed by a review of Math Skills and Laboratory Activities. Supplimentary Materials, including a Glossary with Page Index, Answers to the Self–Help Questions, the Collge Board Achievement Test materials and questions with answers, and several complete course Practice Examinations in Chemistry complete the text.

Self-Help Questions

At the end of each Unit are a special group of Self-Help Questions. These questions are in addition to the Unit Section Questions found after each major conceptual area within each Unit. Answers to the Unit Section Questions and the Practice Exam Questions may be obtained from the teacher.

The **Self-Help Questions** are intended as a review of each Unit. Since these questions are intended to be answered by the student as "self-help," the answer for each question and page references where additional explanations can be found for each question are included at the end of the Glossary/Index.

The Achievement Test

There is also a special section for those students who plan to take *The Chemistry Achievement Test,* produced by The College Board. The answer to each question and page references where additional explanations can be found for each question are included at the end of the section.

Unit 1
Matter and Energy

Important Terms and Concepts

Homogeneous Matter
Elements
Mixtures - Solutions
   Homogeneous, Heterogeneous
Potential & Kinetic Energy
Law - Conservation of Energy
Exothermic Reactions
Endothermic Reactions
Calories and Joules
Specific Heat
Thermometry
   Celsius and Kelvin
   Absolute Zero
Phases of Matter
   Gas, Liquid, Solid

Boyle's & Charles' Laws
Stand. Temp. & Pres. (STP)
Compounds - Binary, Ternary
Partial Pressures
Kinetic Theory of Gases
Ideal Gas Model
Avogadro's Hypothesis
Vapor Pressure
Boiling Point
Heat of Vaporization
Crystals
Melting Point
Heat of Fusion
Sublimation

I. Definition Of Chemistry

Chemistry is the study of the composition, structure, and properties of matter, the changes that matter undergoes, and the energy accompanying these changes.

II. Matter

A. Substances

We define a substance as **homogeneous matter having identical properties and composition**. For example, all samples of a particular substance have the same heat of vaporization, melting point, boiling point, and other properties related to composition. These properties can be used for the identification of the substance.

Question: Which of the following is *not* classified as a substance?

        A. Oxygen   B. Water   C. Concrete   D. An iron slab

Substances include:

1. Elements. All samples of an element are composed of atoms of the same atomic number and are considered substances that cannot be decomposed by chemical means. Although the periodic table of elements names just one hundred and nine elements, more have recently been discovered or made.

Question: What are some elements which are metals?

        Answer: 1. _____   2. _____   3. _____

Question: Can you name some nonmetallic elements?

Answer: 1. _____ 2. _____ 3. _____

2. Compounds are two or more different elements chemically combined in a definite ratio by weight. Therefore, all samples of a compound have identical composition and can only be decomposed by chemical change. Those compounds, which are made up of just two elements, are called **binary compounds**. In addition, those compounds that are made up of three elements are called **ternary compounds**.

The properties of a compound are quite different from the separate elements which make them up. For example, sodium is a soft, very active, metal that must be stored under benzene so that it does not react with air. Chlorine is a green, deadly gas. They combine in a definite weight ratio of 23g of sodium to 35.5g of chlorine. When combined they form the compound sodium chloride or common table salt which is stable in air and is required, to a limited extent, for the body to function normally.

Question: Can you name a compound with three elements in it?

Answer: _____

Question: If a compound formed by two elements is called a binary compound, what would you call a compound with three elements in the formula?

Answer: _____

B. Mixtures

Mixtures are **combinations** of varying amounts of two or more distinct substances (either elements or compounds) that differ in properties and composition. Mixtures may be **homogeneous** or **heterogeneous**.

Homogeneous mixtures have a uniform intermixture of particles and are called **solutions**. A solution is produced when one substance dissolves or dissociates in another. Examples include:

- Gas in Gas — such as air
- Solid in Liquid — such as salt dissolved in water [NaCl(aq)]
- Solid in Solid — an alloy, such as brass, a combination of copper and zinc
- Liquid in Liquid — such as alcohol and water, in which the components are considered **miscible** [$C_3H_5OH(aq)$]

Heterogeneous mixtures have uniformly dispersed ingredients. Examples include:

- Iron and Sulfur
- Oil and Water
- Concrete
- Sand and Water

Mixtures differ from compounds in that the amounts of the different substances which make up mixtures are not in a fixed ratio by weight, whereas, in compounds they are. For example, a sand and water mixture can contain various quantities of sand and water and still be considered a mixture of sand and water. Also, the components of a mixture can be separated by physical means and do not lose their identity. We would separate the sand and water mixture above by either boiling off the water or by filtering out the sand. These are both physical means of separation.

Finally, **compounds** are made up of elements, but mixtures can be made up of either elements or compounds.

Questions I-1

1 Which formula represents a mixture?
   (1) NaCl(aq)   (2) NaCl(s)   (3) $H_2O(l)$ (4) $H_2O(s)$

2 An example of a substance that can be decomposed by a chemical change is:
   (1) ammonia   (3) argon
   (2) antimony   (4) arsenic

3 Sugar dissolves in water to form a solution. This solution is a:
   (1) heterogeneous compound   (3) heterogeneous mixture
   (2) homogeneous compound   (4) homogeneous mixture

4 Which is the formula of a binary compound?
   (1) KOH   (2) $NaClO_3$ (3) Al_2S_3 (4) $Bi(NO_3)_3$

5 Which substance cannot be decomposed by a chemical change?
   (1) mercury (II) oxide   (3) water
   (2) potassium chlorate   (4) copper

III. Energy
(The Ability To Do Work)
A. Forms Of Energy

Mechanical energy, heat energy, radiant energy (such as light, radio waves, and all other forms of electromagnetic radiation), chemical energy (derived from movements of electrons in forming bonds), and nuclear energy (as in fission or fusion reactions) are examples of energy.

Energy is broken down into two types:

1. Potential energy. The energy of position (such as, dammed up water).

2. Kinetic energy. The energy of motion (such as, a moving truck).

The **Law of the Conservation of Energy** states:

Energy May Be Converted From One Form To Another But Is Never Created Or Destroyed.

The classical example is the process of starting up a car. In this process, energy is converted from electrical energy from the battery, to chemical energy from combustion, and then to mechanical energy.

B. Energy And Chemical Change

Energy is absorbed by molecules when chemical bonds are broken and liberated when bonds are formed. Usually, this energy occurs in the form of heat. During a chemical reaction, when the bonds of molecules are broken and new ones are formed, the net energy absorbed or given off depends on the strength of the bonds broken, as compared to the strength of the bonds formed. Reactions involving heat energy are classified as:

1. Exothermic Reactions. When the energy required to break the existing bonds is less than the energy given off as new bonds are formed, the **excess heat energy is given off**. This reaction is called an exothermic reaction, and the container in which the exothermic reaction is taking place becomes warm. For example:

$$C + O_2 \longrightarrow CO_2 + 100 \text{ kcal}$$

2. Endothermic Reactions. When the energy required to break the existing bonds is greater than the energy given off in the formation of new ones, **heat energy is absorbed** into the reaction. This reaction is called an endothermic reaction, and the container in which the reaction is taking place becomes cool. For example:

$$12 \text{ kcal} + H_2 + I_2 \longrightarrow 2 \text{ HI}$$

C. Measurement Of Energy

Since energy in various forms may be converted to heat, the chemist uses calories or kilocalories as heat units to measure the energies involved in chemical reactions.

1. Calories. One calorie is the amount of heat required to raise the temperature of one gram of water one Celsius degree.

From the above definition, therefore, in order to find the number of calories given off or absorbed by the reaction, one requires two values:

• The mass of the water in grams.

• The temperature change, whose symbol is Δt, Δ is the sign for the Greek letter **delta**, and we use it to represent change.

$$\text{calories} = \text{grams of water} \times \Delta t$$

Sample Problem:
How many calories are required to heat 150g. of water from 30°C to 40°C?

Solution:
First —	Determine the change in the temperature	= 10°C.
Second —	Note the mass of the water	= 150g
Third —	Substitute the two values in the formula	
	Calories	= 150g x 10°C
		= 1500

Since 1000 calories = 1 kilocalorie (kcal),
the above answer could be stated as 1.5 kcal.

2. Specific Heat. The specific heat of a material is the amount of heat energy required to raise the temperature of 1 gram of the material, one degree Celsius.

For determining reaction heats in calories, the value of the specific heat of water is considered a standard. It is 1 calorie per gram of water, per degree Celsius or cal/g°C. One calorie is equivalent in energy to 4.18 joules. When determining the reaction heat in joules, the value of the specific heat of water is 4.18 joules per gram of water, per degree Celsius or 4.18 J/g°C.

Sample Problem:
A 3.0×10^3 gram mass of water in a calorimeter has its temperature raised 5.0°C. How much heat energy was transferred to the water in calories and in joules?

Solution in Calories:
Heat transferred to the water is the product of three factors:

$$\text{Heat} \;=\; \left(\begin{array}{c}\text{Mass of}\\\text{Water}\end{array}\right) \times \left(\begin{array}{c}\text{Specific Heat}\\\text{of Water}\end{array}\right) \times \left(\begin{array}{c}\text{Change in Temperature}\\\text{of Water}\end{array}\right)$$

$$=\; 3.0 \times 10^3 \, g \;\times\; 1.0 \, \text{cal/g°C} \;\times\; 5.0°C$$

$$=\; 15{,}000 \text{ calories or } 15 \text{ kilocalories}$$

To the correct number of significant digits, the answer is: 1.5×10^4 calories

Solution in Joules:
To calculate the heat energy in joules, use the specific heat of water as 4.18 J/g°C.

$$\text{Heat} \;=\; 3.0 \times 10^3 \, g \;\times\; 5.0°C \;\times\; 4.18 \, \text{J/g°C}$$

$$=\; 6.27 \times 10^4 \text{ joules}$$

To the correct number of significant digits, the answer is: 6.3×10^4 joules

Questions I-2

1 If 4 grams of water at 1°C absorbs 8 calories of heat, the temperature of the water will change by:
(1) 1°C   (2) 2°C   (3) 3°C   (4) 4°C

2 How many kilocalories of heat energy are absorbed when 100 grams of water is heated from 20°C to 30°C?
(1) 1 kcal   (2) 10 kcal   (3) 100 kcal   (4) 0.1 kcal

3 Which is the equivalent of 750. calories?
(1) 0.750 kcal   (2) 7.50 kcal   (3) 75.0 kcal   (4) 750. kcal

4 When 5 grams of water at 20°C absorbs 10 calories of heat, the temperature of the water will be increased by a total of:
(1) 0.5°C   (2) 2°C   (3) 10°C   (4) 50°C

5 The number of calories needed to raise the temperature of 10 grams of water from 20°C to 30°C is:
(1) 10   (2) 20   (3) 100   (4) 40

6   A 5-gram sample of water is heated and the temperature rises from 10°C to 15°C. The total amount of heat energy absorbed by the water is:
(1) 25 cal   (2) 20 cal   (3) 15 cal   (4) 5 cal

7   Which of the following best describes exothermic chemical reactions?
(1) They never release heat.
(2) They always release heat.
(3) They never occur spontaneously.
(4) They always occur spontaneously.

3. Thermometry. Temperature is a measure of the average kinetic energy of the particles in a system. The average kinetic energy of the particles in a system is the same, when the temperature remains constant. Also, temperature indicates the direction of heat flow. **Heat flows spontaneously from a system at higher temperature to a system of lower temperature.**

Thermometers are instruments used to measure temperature. Most contain liquid mercury, which has the advantage of remaining liquid over a wide range of temperatures, and also has the advantage of expanding and contracting evenly. The two measurement scales frequently used by scientists in calibrating thermometers are:

The **Celsius** (**°C**) temperature scale has fixed points. They are: 0°C at the ice-water equilibrium temperature and 1 atmosphere pressure, and 100°C at the steam—water equilibrium temperature and 1 atmosphere pressure.

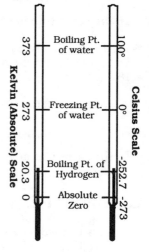

The **Kelvin** (**K**) or **Absolute temperature scale** has fixed points. They are: 273 K at the ice-water equilibrium temperature and 1 atmosphere pressure, and 373 K at the steam—water equilibrium temperature and 1 atmosphere pressure. **Absolute zero or zero K (Kelvin) is equivalent to -273°C.**

When the temperature is given in degrees Celsius, but one needs to find the temperature in Kelvin, one should use the following equation:

Kelvin = 273 + degrees Celsius

Note: All measurements on the Kelvin scale are positive, but the difference between one Kelvin (1K) and one Celsius degree (1°C) is the same.

Figure I-1
Kelvin (left)
Celsius (right)
Scales

A rise in the Celsius Scale from 100°C to 101°C would be stated in the Kelvin Scale as 373 to 374. The Kelvin (Absolute) temperature scale has its zero point at -273°C, with the size of the degrees the same as on the Celsius scale, as illustrated above.

It should be pointed out, however, that all gases liquefy before they reach absolute zero, and that this point of absolute zero has not been reached yet.

Questions I-3

1 At which temperature would the molecules in a one gram sample of water have the lowest average kinetic energy?
   (1) 5°C   (2) -100°C   (3) 5 K   (4) 100 K
2 At standard pressure, what is the number of Kelvin degrees between the melting point of ice and the boiling point of water?
   (1) 100 K   (2) 180 K   (3) 273 K   (4) 373 K
3 Which Kelvin temperature is equal to -33°C?
   (1) -33 K   (2) 33 K   (3) 240 K   (4) 306 K
4 Which temperature is the same as 260 K?
   (1) -333 °C   (2) -13°C   (3) 286°C   (4) 533°C

IV. Phases of Matter

The term "**phase**" is used to refer to the **gas (g)**, **liquid (*l*)**, or **solid (s)** form of matter. *Note that the word "phase" is now used instead of "state" to avoid confusion with other conditions, such as the "state of equilibrium."* The change of phase is accompanied by the absorption (called an **endothermic process**) or the release (called an **exothermic process**) of heat energy. The phase that a substance is in is dependent on temperature and pressure. Also, the phases of matter are characterized by the type of motion that the particles are undergoing.

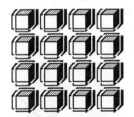

Figure I-2
Solid
(Vibrating Particles)

For example, in a **solid phase**, particles (atoms or molecules) are in a fixed position with little space between them. Although to our eyes, the particles appear stationary, they are **vibrating**.

In a **gaseous phase**, the molecules are all **translating**, as well as **rotating** and **vibrating**. When translating in the gaseous state, the molecules have broken the intermolecular bonds between themselves and other molecules, and the distance between molecules is great. In this state, the molecules have attained a greater amount of randomness. The degree of the randomness of a substance is defined as **entropy**.

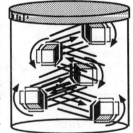

Figure I-4
Gas
(Vibrating,
Rotating, and
Translating Particles)

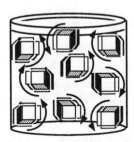

Figure I-3
Liquid
(Vibrating and
Rotating Particles)

In the **liquid phase** the particles are characterized as having acquired two types of motion, **vibrating** and **rotating**. However, this phase is also considered an intermediate state having all three types of motion found in the gaseous state, but to a much more limited degree of motion with limited space between the particles. Liquid water is a good example of this phase.

Reading the diagram below from left to right, the substance is a solid, and, as heat energy is added to it, the temperature of the substance increases until it reaches its melting point. **While a phase change takes place, the temperature remains the same** until all of the sample has melted. The temperature begins to rise until the boiling point is reached. During this second phase change, the temperature remains constant until all of the sample has changed from a liquid phase to a gaseous phase.

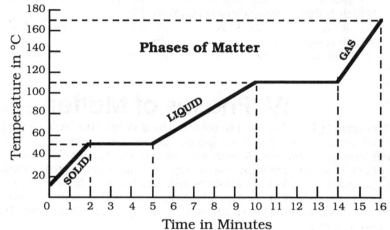

Figure I-5
Phases
of Matter

Time in Minutes

Note: Both plateaus on the graph (Figure I-5) **indicate a change of phase is taking place, and the temperature does not change during a phase change**. The energy added is being used to change the potential energy of the molecules, therefore, the kinetic energy (temperature) remains constant.

Multiplying the time factor by the constant or average rate of heat energy supplied allows us to determine the calories of heat energy used during any portion of the graph. This graph could read from right to left as a cooling curve.

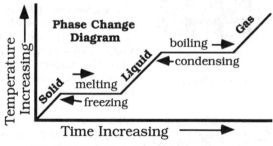

Figure I-6 Phase Change Diagram

Sample Problem:
   If heat energy is being applied at the rate of 50 cal./minute, how many calories does it take for the sample to melt? (Refer to Fig. I-5 Phases of Matter, top of page)

   Record the time that it takes for the sample to melt. Multiply the time by the melting rate, which is 50 cal./min., by the recorded time.

Solution: **3 min x 50 cal/min = 150 cal**

Questions I-4

Base your answers to questions 1 & 2 on the diagram (right) which represents a substance being heated from a solid to a gas, the pressure remaining constant.

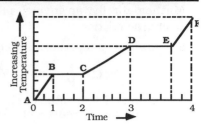

1 The substance begins to boil at point
   (1) *E*　　　　(2) *B*　　　　(3) *C*　　　　(4) *D*
2 Between points *B* and *C* (on the diagram above) the substance exists in:
   (1) the solid state only　　　(3) both solid and liquid states
   (2) the liquid state only　　　(4) neither solid nor liquid state

3 The graph (right) represents changes of state for an unknown substance. What is the boiling temperature of the substance?
   (1) 0°C
   (2) 20°C
   (3) 70°C
   (4) 40°C

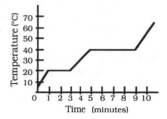

4 The graph at the right represents the uniform cooling of a substance, starting with the substance as a gas above its boiling point. How much time passes between the first appearance of the liquid phase of the substance and the presence of the substance completely in its solid phase
   (1) 5 minutes　　(2) 2 minutes　　(3) 7 minutes　　(4) 4 minutes

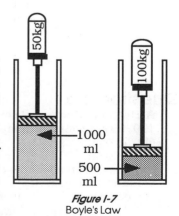

5 Which change of phase is exothermic?
   (1) gas to liquid　　　　(3) solid to gas
   (2) solid to liquid　　　(4) liquid to gas

A. Gases

In this phase, molecules possess vibrational, rotational, and translational movements. **This allows them to fill the volume of the container in which they are located.** The laws that describe their behavior are:

1. Boyle's Law. When the temperature remains a constant, the volume of a given mass of gas varies inversely with the pressure. If one increases the pressure on a definite mass of gas molecules, the volume it occupies will decrease proportionally.

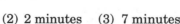

Figure I-7
Boyle's Law

Mathematically, an inverse proportion concerning just two items has definite characteristics:

a. Multiplied together they equal a constant. For example:

$$V \; \alpha \; \frac{1}{P} \quad \text{or} \quad VP = k, \quad \text{where k is a constant.}$$

b. Using the two values as graphing coordinates, the graph should represent a hyperbola.

This can also be represented in a mathematical formula:

$$\frac{V}{V'} = \frac{P'}{P}$$

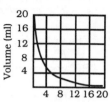

Pressure (mm Hg)

Where V and P represent the initial volume and pressure, and V' and P' represent a new volume and pressure. Note: One atmosphere (atm) is equal to 760 mm Hg.

Sample Problem:

A storage container holds 400 liters of a gas at 2 atm. If the pressure is increased to 5 atmospheres at constant temperature, what will the volume of the gas be?

Solution:

Since the temperature is a constant, the two variables will be volume and pressure.

Knowing that they are inversely proportional, we can determine whether the volume will increase or decrease by comparing the second pressure of the gas with the first pressure. Since the second pressure is higher than the first pressure, the volume will decrease.

Simply write the amounts for each of the values and key them into the formula:

$$\frac{V}{V'} = \frac{P'}{P}$$

Where: V = 400 L
           P = 2 atm
           P' = 5 atm
           Solve for V'

$$\frac{400 \text{ L.}}{V'} = \frac{5 \text{ atm}}{2 \text{ atm}}$$

$$\frac{\overset{80}{\cancel{400} \text{ L.} \times \cancel{2 \text{ atm}}}}{\cancel{5 \text{ atm.}}} = V'$$

$$160 \text{ L.} = V'$$

Questions 1-5

1   A 100. milliliter sample of a gas at a pressure of 380. torr is reduced to 190. torr at constant temperature. What is the new volume of the gas?
    (1) 50.0 mL   (3) 200.0 mL
    (2) 90.0 mL   (4) 290.0 mL

2   As pressure on a sample of gas increases at constant temperature, the volume of the gas:
    (1) decreases   (2) increases   (3) remains the same

3   As the pressure on a given sample of a gas increases at constant temperature, the mass of the sample:
    (1) decreases   (2) increases   (3) remains the same

4   A 100 milliliter sample of gas is enclosed in a cylinder under a pressure of 760 torr. What volume would the gas sample occupy at a pressure of 1,520 torr, temperature remaining constant?
    (1) 50 mL   (3) 200 mL
    (2) 100 mL   (4) 380 mL

5   As a given volume of gas is compressed, the number of molecules:
    (1) decreases   (2) increases   (3) remains the same

6   At constant temperature the pressure on 8.0 liters of a gas is increased from 1 atmosphere to 4 atmospheres. What will be the new volume of the gas?
    (1) 1.0 liter   (3) 32 liters
    (2) 2.0 liters   (4) 4.0 liters

7   At constant temperature, which line best shows the relationship between the volume of an ideal gas and its pressure?
    (1) A   (3) C
    (2) B   (4) D

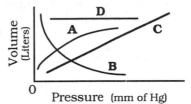

2. Charles' Law. At constant pressure, the volume of a given mass of gas varies directly with the Kelvin (Absolute) temperature.

A good example would be illustrated by the difference in size of two hot air balloons. Both are at the same altitude (constant pressure), but the air in the second balloon is heated to double the temperature (in Kelvin) of the first balloon. The volume of the higher air temperature balloon increases to double that of the first balloon.

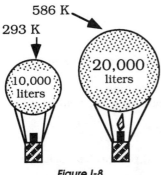

Figure I-8
Charles' Law

Mathematically, Charles' Law can be represented by the formula:

$$\frac{V}{V'} = \frac{T}{T'}$$

V and T represent the initial volume and temperature, and V' and T' represent the new volume and temperature. Graphically, it is represented by a straight line.

Starting at 0°C, with each decrease of 1°C, the volume of a gas decreases by $\frac{1}{273}$ of its original volume.

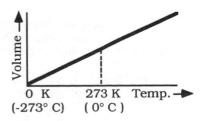

Sample Problem:
At constant pressure and 27°C, a gas has a volume of 150 mL. If the temperature is increased to 327°C, what will be the new volume?

Solution:
First: Since the pressure is constant, the only two variables are temperature and pressure.

Knowing that there is a direct relationship between volume and temperature, we can estimate the following: Since the temperature is rising, the volume will increase.

Second: The next step is to convert the temperature from degrees Celsius to Kelvin and list the known items along with the unknown. Then key them into the formula and solve for V'.

$$\frac{V}{V'} = \frac{T}{T'}$$

$$\frac{150 \text{ mL}}{V'} = \frac{300 \text{ K}}{600 \text{ K}}$$

Where: V   =   150 L
              T   =   300 K
              T' =   600 K
              Solve for V'

$$\frac{150 \text{ mL} \times \overset{2}{\cancel{600 \text{ K}}}}{\underset{1}{\cancel{300 \text{ K}}}} = V'$$

$$300 \text{ mL} = V'$$

Questions 1-6

1   As the temperature of a sample of an ideal gas increases at constant pressure, the volume occupied by the sample:
    (1) decreases   (2) increases   (3) remains the same

2   A gas sample is at l0.0°C. If pressure remains constant, the volume will increase when the temperature is changed to:
    (1) 263 K   (3) 283 K
    (2) 273 K   (4) 293 K

3   As the temperature of a sample of gas decreases at constant pressure, the volume of the gas:
    (1) decreases   (2) increases   (3) remains the same

4   A gas occupies a volume of 30 milliliters at 273 K. If the temperature is increased to 364 K while the pressure remains constant, what will be the volume of the gas?
    (1) 60 mL   (3) 30 mL
    (2) 40 mL   (4) 20 mL

3. Relationship between temperature and pressure in a rigid container (volume is a constant). At constant volume as the Kelvin temperature is increased, the pressure exerted by a given mass of a gas is also increased.

The relationship is a direct proportion and can be shown mathematically as:

$$\frac{T}{T'} = \frac{P}{P'}$$

Sample problem:
In a rigid container, a gas exerts a pressure of 500 mmHg at 77 degrees C. What would the pressure be at -98 degrees C?

Solution:
Convert the temperature values from Celsius to Kelvin degrees, then place these temperatures in the the above equation and solve:

$$\frac{350\ K}{175\ K} = \frac{500\ mmHg}{P'}$$

$$P' = \frac{500\ mmHg \ x \ 175\ K}{350\ K}$$

$$P' = 250\ mmHg$$

4. Standard Temperature and Pressure (STP). As noted above, there are three variables which we must consider when we study gases. They are volume, temperature, and pressure. Since the volume of a given mass of gas varies with changes in temperature and pressure, we cannot give the volume of a gas without specifying its temperature and pressure. Therefore, gas volumes are usually calculated to an arbitrary standard, which we abbreviate as STP. **Standard temperature and pressure (STP)** of a gas are defined as 0°C (273 K) and 760 mm of mercury (760 torr.) or 1 atmosphere pressure. Standard temperature for those phases other than a gas is 25°C (298 K).

5. Partial Pressures. The pressure exerted by each of the gases in a gas mixture is called the partial pressure of that gas. Therefore, the total pressure of a gas mixture is equal to the sum of the individual partial pressures of each of the gases comprising the mixture. The partial pressure of each gas is determined by its **molecular ratio** (or mole ratio).

Sample Problem:
A container holds three gases, each of which exert pressure on the side of the container. Gas A exerts a pressure of 1.0 atm, gas B exerts a pressure of 1.5 atm, and gas C exerts a pressure of 2.5 atm. What is the total pressure of the gases on the sides of the container?

Solution:
Total pressure = Pressure of gas A + Pressure of gas B + Pressure of gas C
= 1.0 atm + 1.5 atm + 2.5 atm
= 5.0 atm

6. Kinetic Theory of Gases. Studies of gas behavior have led to a model based on the following assumptions about perfect or "ideal" gases:

a. A gas is composed of individual particles which are in continuous, random straight line motion.
b. Not all of the particles of a gas have the same kinetic energy, but the average kinetic energy is proportional to the measured Kelvin (Absolute) temperature of the gas
c. Collisions between gas particles may result in a transfer of energy between particles and also may result in perfect elastic collisions with each other and the sides of the container. However, the system's net total energy remains the same.

d. The volume of the gas particles themselves is ignored in comparison with the volume of the space in which they are contained.
e. Gas particles are considered as having no attraction to each other.

The model based on the assumptions listed above is referred to as the **"Ideal Gas" Model**. Such models can be useful in the study of the behavior of gases. *It should be emphasized that a model is only an approximation and is only as good as its ability to predict behavior under new conditions.*

7. Deviations from the Gas Laws. The "Ideal Gas" model does not exactly represent any gas under all conditions. Hydrogen (H_2) and helium (He) are the two most ideal gases. No real gas is ideal under all conditions of temperature and pressure. Deviations from the gas laws occur because the model is not perfect. That is, gas particles:

a. Do have volume.
b. Do exert some attraction for each other.

These factors become significant under conditions of relatively **high pressure, low temperature,** and **decreased velocity** due to increased molecular mass.

Questions I-7

1   A 1-liter flask contains two gases at a total pressure of 3.0 atmospheres. If the partial pressure of one of the gases is 0.5 atmosphere, then the partial pressure of the other gas must be:
    (1) 1.0 atm   (2) 2.5 atm   (3) 1.5 atm   (4) 0.50 atm
2   Which change must result in an increase in the kinetic energy of the molecules of a sample of $N_2(g)$?
    (1) The pressure changes from 0.5 atmosphere to 1 atmosphere
    (2) The volume changes from 1 liter to 2 liters
    (3) The temperature changes from 20°C to 30°C
    (4) The density changes from 2.0 g/L to 2.5 g/L
3   Under which conditions does a real gas behave most like an ideal gas?
    (1) at high temperatures and low pressures
    (2) at high temperatures and high pressures
    (3) at low temperatures and low pressures
    (4) at low temperatures and high pressures
4   What is the total pressure exerted by a mixture containing two gases if the partial pressure of one gas is 70 torr and the partial pressure of the other gas is 30 torr?
    (1) 30 torr   (2) 40 torr   (3) 70 torr   (4) 100 torr
5   A sealed flask contains 1 mole of hydrogen and 3 moles of helium at 20°C. If the total pressure is 400 torr, the partial pressure of the hydrogen is
    (1) 100 torr   (3) 300 torr
    (2) 200 torr   (4) 400 torr
6   At STP, which gas would most likely behave as an ideal gas?
    (1) H_2 (2) CO_2 (3) Cl_2 (4) SO_2

8. Avogadro's hypothesis. Avogadro's hypothesis states that equal volumes of all gases under the same conditions of temperature and pressure contain equal numbers of particles. For example, at the same temperature and pressure, the number of particles in 1 liter of hydrogen is the same as the number of particles in 1 liter of oxygen although the individual particles of oxygen are heavier (by a ratio of 16 to 1) and larger than the individual particles of hydrogen.

The amount of matter than contains **6.02×10^{23} (Avogadro's number)** particles is called a **mole of matter.** One mole of any substance contains as many particles (molecules, atoms, ions, etc.) as there are atoms of carbon-12 in 12.000g of carbon-12 isotope.

Since it is inconvenient to work with individual particles (atoms, molecules, ions, electrons, etc.), chemists have chosen a unit containing many particles for comparing amounts of different materials. The mole is a unit which contains 6.02×10^{23} particles. **A mole of particles of any gas occupies a volume of 22.4 liters at STP and is called a molar volume.**

Questions 1-8

1 Which sample of methane gas contains the greatest number of molecules at standard temperature?
   (1) 22.4 liters at 1 atmosphere   (3) 11.2 liters at 1 atmosphere
   (2) 22.4 liters at 2 atmospheres   (4) 11.2 liters at 2 atmospheres

2 Which quantity represents the total amount of $N_2(g)$ in a 22.4 liter sample at STP?
   (1) 1.00 mole   (3) 3.01×10^{23} molecules
   (2) 14.0 grams   (4) 6.02×10^{23} atoms

3 At STP, 44.8 liters of CO_2 contains the same number of molecules as:
   (1) 1.00 mole of He   (3) 0.500 mole of H_2
   (2) 2.00 moles of Ne   (4) 4.00 moles of N_2

4 A sample of nitrogen containing 1.5×10^{23} molecules has the same number of molecules as a sample containing:
   (1) 1.0 mole of H_2 (3) 0.25 mole of O_2
   (2) 2.0 moles of He   (4) 0.50 mole of Ne

B. Liquids

Each particle in a liquid has attained vibrational, translational, and rotational movement, which allows them to move about freely. With no regular arrangement, the particles take the shape of the container into which they are placed. However a liquid does have a definite volume because the particles are bonded stronger and, therefore, closer together than a gas.

Liquids exhibit certain characteristics which include:

1. Vapor Pressure. As a result of the free movement of particles in a liquid, a certain quantity of collisions occurs between them, releasing energy in the process. This energy, absorbed by individual particles, allows them to acquire the translational energy with which they break those intermolecular bonds holding them to one another, and they become gaseous particles. When a liquid substance changes to a gas the process is called **evaporation**. Evaporation tends to take place at the surface of a liquid and at all temperatures.

The term "vapor" is frequently used to refer to the gas phase of a substance that is normally a liquid or solid at room temperature.

The vapor (gas) produced exerts a pressure, known as **vapor pressure**, which increases as the temperature of the liquid is raised and is specific for each substance and temperature.

2. Boiling Point. A liquid will boil at the temperature at which the vapor pressure equals the pressure on the liquid. At this temperature, **all** of the particles in a liquid have acquired the translational movement, which transforms them from a liquid phase to a gas phase.

The "normal boiling point" is the temperature at which the vapor pressure of the liquid equals one atmosphere. (H_2O: 100°C or 373 K)

Usually when reference is made to the "boiling point" of a substance, it is the normal boiling point that is indicated.

3. Heat of Vaporization. The energy required to vaporize a unit mass of a liquid to a gas at constant temperature is called its heat of vaporization.

The energy which is involved in the change of phase is needed to overcome the forces of attraction between particles (potential energy) and does not increase their average kinetic energy. Since temperature is defined as the measure of average kinetic energy in a system, there is no increase in temperature during this phase change.

As found in Table A, at 100°C (373 K) and 1 atmosphere pressure, the heat of vaporization of water is 540 calories per gram of water to be vaporized.

Note 1: When a gas is changed to a liquid by condensation, the same quantity of heat is given off.

Note 2: The boiling points of gases can be found on the Chemistry Reference Table *C*. The vapor pressure of water can be found on the Chemistry Reference Table *O*.

Questions 1-9

1   Which substance has a normal boiling point of 263 K?
    (1) NO(g)   (2) CO(g)   (3) CO_2(g)   (4) SO_2(g)

2   A sample of pure water is boiling at 90.0°C. The vapor pressure of the water is closest to:
    (1) 90.0 torr   (2) 363 torr   (3) 526 torr   (4) 760 torr

3   The normal boiling point of water is equal to:
    (1) 173 K   (2) 273 K   (3) 373 K   (4) 473 K

4   When the vapor pressure of a liquid in an open container equals the atmospheric pressure, the liquid will
    (1) freeze   (3) melt
    (2) crystallize   (4) boil

5   As a 1-gram sample of $H_2O(l)$ changes to $H_2O(g)$ at 100°C, the kinetic
     energy of the molecules:
     (1) decreases   (2) increases   (3) remains the same

C. Solids

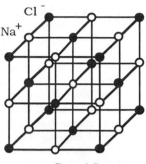

Figure I-9
NaCl Crystal Lattice

In order to distinguish one solid from another,
their respective **densities** must be compared.

Solids have crystalline structures with definite
shape and volume. Their characteristics are explained below.

1. Crystals. Although the particles in a crystal
are constantly vibrating, they do not change their
regular positions and are arranged in a regular
geometric pattern (called the crystal lattice structure).

The particles in glass and certain plastics are not arranged in a regular
geometric pattern and behave as highly viscous liquids. Since all true solids
have crystalline structures, these materials are often considered solids, but
are really **super-cooled liquids**. Therefore, they are not considered true
solids.

2. Melting point. The "normal melting point" is the temperature at which
a solid substance will change to a liquid at 1 atmosphere pressure. Melting
point may also be defined as the temperature at which the solid and liquid
phases can exist in equilibrium. At the melting point temperature, all of the
particles once in a definite geometric pattern have acquired rotational energy
and are no longer a part of a lattice structure.

Melting points can be determined from cooling curves which are obtained
experimentally.

3. Heat of fusion. The energy required to change a unit mass of a solid to
a liquid at constant temperature is called its **heat of fusion. Keep in mind
that fusion is another word for melting**.

As found in Table A, the heat of fusion of water is 80 cal. per gram of water.

**Note: When a gram of water is frozen into ice, the same amount of
heat is given off**.

4. Sublimation. Sublimation is a change from the solid phase directly to
the gas phase without passing through an apparent liquid phase.

Solids which sublime have high vapor pressures and low intermolecular
attractions. **Examples of solids that sublime at room temperature are solid
carbon dioxide (dry ice), iodine crystals, and naphthalene.**

Questions 1-10

1 Which substance readily sublimes at room temperature?
   (1) $H_2O(l)$ (3) $Fe(s)$
   (2) $O_2(g)$ (4) $CO_2(s)$

2 Solid substances are most likely to sublime if they have:
   (1) high vapor pressures and strong intermolecular attractions
   (2) high vapor pressures and weak intermolecular attractions
   (3) low vapor pressures and strong intermolecular attractions
   (4) low vapor pressures and weak intermolecular attractions

3 The heat of fusion for ice is 80 calories per gram. Adding 80 calories of heat at STP - will cause the ice to:
   (1) increase in temperature
   (2) decrease in temperature
   (3) change to water at a higher temperature
   (4) change to water at the same temperature

4 Which change of phase represents sublimation?
   (1) $H_2O(g) \rightarrow H_2O(l)$ (3) $CO_2(s) \rightarrow CO_2(g)$
   (2) $H_2O(l) \rightarrow H_2O(s)$ (4) $CO_2(s) \rightarrow CO_2(l)$

5 Which change of phase represents sublimation?
   (1) solid to liquid   (3) liquid to gas
   (2) solid to gas   (4) liquid to solid

6 Which term represents the change of a substance from the solid phase to the liquid phase?
   (1) condensation   (3) evaporation
   (2) vaporization   (4) fusion

7 At 1 atmosphere, which substance will sublime when heated?
   (1) $CO_2(s)$ (3) $CH_4(g)$
   (2) $H_2O(l)$ (4) $HCl(aq)$

Self-Help Questions I

Unit 1 - Matter And Energy

Matter - substances, elements, compounds and mixtures

1 In an equation, which symbol would indicate a mixture?
   (1) $NH_3(s)$ (2) $NH_3(l)$ (3) $NH_3(aq)$ (4) $NH_3(g)$

2 Which is a formula of a binary compound?
   (1) $BaSO_4$ (2) $BiPO_4$ (3) $MgCl_2$ (4) $Mg(ClO)_2$

3 Which substance can be decomposed by a chemical change?
   (1) ammonia   (3) iron
   (2) argon   (4) helium

4 Which sample represents a homogeneous mixture?
   (1) $C_2H_5OH(l)$ (3) $C_2H_5OH(g)$
   (2) $C_2H_5OH(aq)$ (4) $C_2H_5OH(s)$

5   A sample is prepared by adding 10. grams of oxygen gas to 10. grams of nitrogen gas. No energy change is observed. Which best describes this sample?
    (1) a mixture that is 50.% oxygen by mass
    (2) a compound that is 50.% oxygen by mass
    (3) a mixture in which 50.% of the atoms are oxygen
    (4) a compound in which 50.% of the atoms are oxygen

6   Which of the following substances can *not* be decomposed by chemical change?
    (1) sulfuric acid   (3) water
    (2) ammonia   (4) argon

7   Which formula represents a binary compound?
    (1) NH_4NO_3 (2) CH_4 (3) CH_3COCH_3 (4) $CaCO_3$

Forms of energy, energy and chemical change, measurement of energy

8   How many calories of heat energy are released when 50 grams of water are cooled from 70°C to 60°C?
    (1) 10 calories   (3) 500 calories
    (2) 50 calories   (4) 1,000 calories

9   Which unit is used to express the amount of energy absorbed or released during a chemical reaction?
    (1) degree   (2) torr   (3) gram   (4) calorie

10 The number of calories per gram required to melt ice at its melting point is called
    (1) sublimation   (3) heat of vaporization
    (2) vapor pressure   (4) heat of fusion

11 When a quantity of electricity is converted to heat, the heat energy produced is measured in
    (1) volts   (2) amperes   (3) calories   (4) degrees

Thermometry

12 A flask containing molecules of gas *A* and a separate flask containing molecules of gas *B* are both at the same temperature. Gases *A* and *B* must have equal
    (1) volumes   (3) pressures
    (2) masses   (4) average kinetic energies

13 Which Kelvin temperatures represent, respectively, the normal freezing point and the normal boiling point of water?
    (1) 0 K and 273 K   (3) 100 K and 273 K
    (2) 0 K and 100 K   (4) 273 K and 373 K

14 When the average kinetic energy of a gaseous system is increased, the average molecular velocity of the system
    (1) increases and the molecular mass increases
    (2) decreases and the molecular mass increases
    (3) increases and the molecular mass remains the same
    (4) decreases and the molecular mass remains the same

15 At which temperature does a water sample have the highest average kinetic energy?
    (1) 0°C   (2) 100°C   (3) 0 K   (4) 100 K

Phases of matter

16 The diagram represents the uniform heating of a substance that is a solid at t_0. What is the freezing point of the substance?

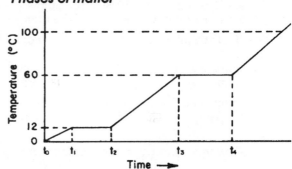

    (1) 1°C
    (2) 12°C
    (3) 60°C
    (4) 100°C

17 Given the equilibrium $H_2O(s) \leftrightarrow H_2O(l)$ at a pressure of 1 atmosphere. The temperature of the ice water mixture must be
    (1) 0°C   (2) 32°C   (3) 100°C   (4) 212°C

18 The graph at the right represents the relationship between temperature and time as heat was added uniformly to a substance, starting as a solid below its melting point.
During the *BC* portion of the curve, the average kinetic energy of the molecules of the substance

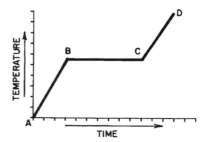

    (1) increases and the potential energy increases
    (2) decreases and the potential energy increases
    (3) remains the same and the potential energy increases
    (4) remains the same and the potential energy decreases

19 Which sample contains particles arranged in a regular geometric pattern?
    (1) $CO_2(l)$ (2) $CO_2(s)$ (3) $CO_2(g)$ (4) $CO_2(aq)$

Boyle's Law

20 At constant temperature, which 10-milliliter sample, measured at STP, will uniformly take the shape and volume of a 100-milliliter container into which it is placed?
    (1) water   (3) sodium chloride
    (2) mercury   (4) carbon dioxide

21 A 0.500-mole sample of a gas has a volume of 11.2 liters at 273 K. What is the pressure of the gas?
    (1) 11.2 torr   (2) 273 torr   (3) 380. torr   (4) 760. torr

Charles' Law

22 A sample of oxygen gas has a volume of 150. milliliters at 300 K. If the pressure of the sample is held constant and the temperature is raised to 600 K, the new volume of the sample will be
(1) 75.0 mL   (2) 150. mL   (3) 300. mL   (4) 600. mL

23 The volume of 50.0 milliliters of an ideal gas at STP increases to 100. milliliters. If the pressure remains constant, the new temperature must be
(1) 0 K   (2) 100. K   (3) 273 K   (4) 546 K

24 A 100. milliliter sample of helium gas is placed in a sealed container of fixed volume. As the temperature of the confined gas increases from 10.°C to 30.°C, the internal pressure
(1) decreases   (2) increases   (3) remains the same

STP, Partial Pressures, Kinetic Theory of Gases, Ideal Gas Model and Deviations from Ideal Gases.

25 Under the same conditions of temperature and pressure, which of the following gases would behave most like an ideal gas?
(1) $He(g)$ (2) $NH_3(g)$ (3) $Cl_2(g)$ (4) $CO_2(g)$

26 Which gas under high pressure and low temperature has a behavior closest to that of an ideal gas?
(1) $H_2(g)$ (2) $O_2(g)$ (3) $NH_3(g)$ (4) $CO_2(g)$

27 One reason that a real gas deviates from an ideal gas is that the molecules of the real gas have
(1) a straight-line motion
(2) no net loss of energy on collision
(3) a negligible volume
(4) forces of attraction for each other

28 Which temperature represents absolute zero?
(1) 0 K   (2) 0°C   (3) 273 K   (4) 273°C

Avogadro's number, the mole, and molar volume

29 A sample of $H_2(g)$ and a sample of $N_2(g)$ at STP contain the same number of molecules. Each sample must have
(1) the same volume, but a different mass
(2) the same mass, but a different volume
(3) both the same volume and the same mass
(4) neither the same volume nor the same mass

30 At STP, equal volumes of $N_2(g)$ and $CO_2(g)$ contain equal numbers of
(1) atoms   (2) electrons   (3) molecules   (4) protons

31 What is the volume of 4.40 grams of CO_2 at STP?
(1) 2.24 L   (3) 22.4 L
(2) 4.48 L   (4) 44.8 L

32 What is the total number of molecules in 11.2 liters of N_2 gas at STP?
    (1) 3.01×10^{23} (3) 14.0
    (2) 6.02×10^{23} (4) 28.0

33 Which gas has properties that are most similar to those of an ideal gas?
    (1) N_2 (2) O_2 (3) He   (4) Xe

Liquids, Vapor pressure, boiling point, relationship of vapor pressure to boiling point, Heat of vaporization

34 When the vapor pressure of a liquid in an open container equals the atmospheric pressure, the liquid will
    (1) freeze   (3) melt
    (2) crystallize   (4) boil

35 In a closed system, as the temperature of a liquid increases, the vapor pressure of the liquid
    (1) decreases   (2) increases   (3) remains the same

36 As the temperature of a liquid increases, its vapor pressure
    (1) decreases   (2) increases   (3) remains the same

37 As the temperature of liquid water decreases, its vapor pressure
    (1) decreases   (2) increases   (3) remains the same

38 The boiling point of water at standard pressure is
    (1) 0.000 K   (2) 100. K   (3) 273 K   (4) 373 K

39 Which sample of water has the greatest vapor pressure?
    (1) 100 mL at 20°C   (3) 20 mL at 30°C
    (2) 200 mL at 25°C   (4) 40 mL at 35°C

Solids, crystals, melting point, heat of fusion, sublimation

40 The melting of sodium is accompanied by the
    (1) destruction of energy   (3) absorption of energy
    (2) creation of energy   (4) release of energy

41 Which process occurs when dry ice, CO_2(s) is changed into CO_2(g)?
    (1) crystallization   (3) sublimation
    (2) condensation   (4) solidification

42 The heat of fusion of a substance is the energy measured during a
    (1) phase change   (3) chemical change
    (2) temperature change   (4) pressure change

Important Terms and Concepts

Subatomic Particles
  Electrons
  Nucleons
  Protons and Neutrons
Atomic Nucleus
  Atomic and Mass Numbers
  Isotopes
Atomic Structural Models
Energy Levels
  Ground and Excited States

Quantum of Energy
Spectral Lines – Bohr Model
Atomic Orbital Model
Orbital Notation – Valence Electrons
  Kernel, Electron Dot Symbols
  Ionization Energy
Natural Radioactivity
  Alpha and Beta Decay
  Gamma Radiation
Half-Life

I. Atoms

A. Introduction To Atomic Structure

Our concept of the nature of the atom has undergone change and probably will continue to do so. All atomic concepts, however, must be based on the structure of the atom and its various fundamental or subatomic particles.

Experimentally, it has been shown that for all practical purposes, the volume of the atom is made up of whirling electrons. Its mass is located in a central core particle called the nucleus, which contains all the atom's positive charge.

B. Important Subatomic Particles

1. Electrons ($_{-1}^{0}e$). An electron has a mass of approximately $1/1836$ of a proton and a unit negative charge (**-1**). Although its mass is considered negligible, because of the velocity with which it moves around the nucleus, *it accounts for the volume of the atom*.

2. Nucleons. Those particles that compose the nucleus are called nucleons and include:

Protons ($_{+1}^{1}p$). A proton has a mass of approximately **one atomic mass unit** and a unit positive charge (**+1**). Although protons and neutrons are the only nuclear particles that have been identified in an intact nucleus, other particles have been identified among the breakdown products of certain nuclear disintegrations. The relationship of these particles to the structure and stability of the nucleus is the subject of much current research.

Neutrons ($_{0}^{1}n$). A neutron also has a mass of approximately **one atomic mass unit**. It has a unit charge of zero (**0**), therefore, the name neutron. It is found in the nucleus.

C. Structure Of Atoms

Atoms differ in the number of protons and neutrons in the nucleus and in the configuration of electrons surrounding the nucleus.

1. "Empty space" concept. Most of the atom consists of empty space. **Rutherford's** gold foil experiments showed an atom to be mostly empty space with the size of the nucleus very small compared to the atom's size.

2. Nucleus. The mass of the atom is concentrated almost entirely in the nucleus, which is the basis for determining:

Atomic Number. The atomic number indicates the number of protons in the nucleus. It is the number of protons, called the atomic number, that identifies the element.

Isotopes. Isotopes are atoms with the same atomic number (number of protons) but a different number of neutrons. This difference in the number of neutrons affects the mass of an atom but does not affect its chemical identity.

Note: For a given element the number of protons in the nucleus remains constant, but the number of neutrons may vary.

Mass number. The mass number indicates the total number of protons and neutrons. Since the masses of the protons and neutrons are approximately one, the mass number approximates the total mass of the isotope. The number of neutrons in an atom can be calculated by subtracting the atomic number from the mass number.

Atomic mass. The mass of a neutral atom, called its atomic mass, is measured in atomic mass units (amu), which are standardized on the isotope carbon-12 ($^{12}_{6}C$). This isotope of carbon is equal to 12.000 atomic mass units. Therefore, the definition for an atomic mass unit can be stated as $\frac{1}{12}$th of a carbon-12 ($^{12}_{6}C$) atom.

The atomic mass of an element, *given in the Reference Tables for Chemistry,* is the weighted average mass of the naturally occurring isotopes of that element. Since most elements occur naturally as mixtures of isotopes, this average is weighted according to the proportions in which the isotopes occur. This accounts for fractional atomic masses found in reference tables. For example, the **element hydrogen** exists in three different **isotopes**:

a. **protium** ($^{1}_{1}H$) occurs about 99% of the time in nature.

b. **deuterium** ($^{2}_{1}H$) occurs about 0.6% of the time in nature.

c. **tritium** ($^{3}_{1}H$) occurs about 0.4% of the time in nature.

Their masses, along with their occurrence, average out to be 1.00797 amu. In general, we determine the mass number by rounding off the atomic mass of the element to the nearest whole number. For example, the **atomic mass** of a single atom (isotope) such as neon-20 is 19.992 amu, while the atomic mass of the element neon (which is the weighted average of all its natural isotopes) is 20.183 amu.

The **gram atomic mass** (the mass of one mole of atoms) of an element is the mass in grams of Avogadro's number of atoms of that element as it occurs naturally. It is numerically equal to the atomic mass. Examples:

    a.   12 grams of carbon is one gram atomic mass of carbon.

    b.   23 grams of sodium is one gram atomic mass of sodium.

    3. Electrons. The electrons are outside the nucleus at various energy levels. In a neutral atom the total number of electrons is equal to the number of protons in the nucleus. They have a mass of $1/1836$ of a proton and have a negative charge.

Questions II-1

1   Which atom has an equal number of protons and neutrons?

    (1) $^{1}_{1}H$ (2) $^{12}_{6}C$ (3) $^{19}_{9}F$ (4) $^{39}_{19}K$

2   The atomic mass unit is defined as exactly one-twelfth of the mass of an atom of:

    (1) $^{12}_{6}C$ (2) $^{11}_{5}B$ (3) $^{23}_{11}Na$ (4) $^{24}_{12}Mg$

3   What is the total number of nucleons (protons and neutrons) in an atom of Se?

    (1) 34   (2) 45   (3) 79   (4) 93

4   The atomic number of an atom is always equal to the total number of:

    (1) neutrons in the nucleus

    (2) protons in the nucleus

    (3) neutrons plus protons in the atom

    (4) protons plus electrons in the atom

5   Which of the following nuclei is an isotope of $\left(\begin{array}{c}10\ p\\11\ n\end{array}\right)$?

    (1) $\left(\begin{array}{c}10\ p\\9\ n\end{array}\right)$ (2) $\left(\begin{array}{c}11\ p\\10\ n\end{array}\right)$ (3) $\left(\begin{array}{c}9\ p\\11\ n\end{array}\right)$ (4) $\left(\begin{array}{c}11\ p\\12\ n\end{array}\right)$

6   Compared to an atom of $^{14}_{6}C$, an atom of $^{12}_{6}C$ has:

    (1) more protons   (3) more neutrons

    (2) fewer protons   (4) fewer neutrons

7   Which of the following atoms has the greatest nuclear charge?

    (1) N   (2) C   (3) H   (4) He

8   Isotopes of the same element must differ in their:

    (1) atomic number   (3) number of electrons

    (2) mass number   (4) number of protons

9   An atom of potassium containing 23 neutrons has a mass number of:

    (1) 19   (2) 20   (3) 23   (4) 42

10 Which particle has a negative charge?

    (1) alpha particle   (3) proton

    (2) electron   (4) neutron

11 Which two particles have approximately the same mass?

    (1) neutron and electron   (3) proton and neutron

    (2) neutron and deuteron   (4) proton and electron

D. Atomic Structure Models

The model for atomic structure of the elements has passed through many stages of development. Sixty years ago, Danish physicist, Niels Bohr, made one proposal for the model.

Although not currently used by chemists to describe atomic structure, the significance of the Bohr model of the atom is that it is concerned with the first applications of the quantum mechanical concepts to atomic structure.

Figure II-1
Example of Bohr Model

In the Bohr model (Figure II-1), electrons were considered to revolve around the nucleus in one of several concentric circular orbits, similar to the solar system.

1. Principal energy levels. The principal energy level approximates how far the electron is from the nucleus and can be denoted by the **letters K, L, M, N, O, P, Q**, or by the **numbers 1, 2, 3, 4, 5, 6, 7**.

Electrons in orbits near the nucleus are at lower energy levels than those in orbits farther from the nucleus. When the electrons are in the lowest available energy levels, the atom is said to be in the **"ground state."**

When atoms absorb energy, electrons may shift to a higher energy level. At this higher energy level, the atom is said to be in an **"excited state."** The excited state is unstable, and the electrons fall back to lower energy levels. In the process, they release energy equal to the energy difference of the two energy levels involved.

2. Quanta. Electrons can absorb energy only in discrete amounts called **quanta**. When electrons return to a lower energy level, energy is emitted in quanta. A single quanta is called a **quantum of energy**.

Note: A quantum is a distinct, discrete amount of energy, and fractions of that quantum are not allowed. So, energy is not given off nor absorbed in a continuous flow, but in small packets or quanta.

3. Spectral lines. When electrons in an atom in the excited state return to lower energy levels, the quanta of energy emitted is called radiant energy. This radiant energy, made up of quanta (also called **photons**), has a wave characteristic with a specific frequency. We identify elements by measuring the wave length of the radiant energy being emitted. These waves are called **spectral lines**.

The study of spectral lines has provided much of the evidence regarding energy levels within the atom. This study is carried on with an instrument called the **spectroscope,** which investigates the wavelength of one type of radiant energy called **light**. A spectroscope separates light into specific bright line bands of color called a **spectrum**. Each element gives off a characteristic bright line spectra when investigated by a spectroscope and can therefore be identified by this method.

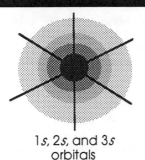

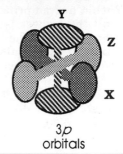

Figure II-2
Orbital Models of Atom

1*s*, 2*s*, and 3*s*
orbitals

3*p*
orbitals

E. Atomic Orbital Model

Although Bohr's model accounted for the lines of the hydrogen spectrum, it did not account for the spectra of heavier and more complicated atoms.

Electrons occupy orbitals that may differ in size, shape, or orientation in space. The term **orbital** refers to the *average region of the most probable electron location*. The orbital model (Figure II-2) differs from the Bohr model in that it does not represent electrons as moving in planetary orbits around the nucleus. Instead, it is defined so that no two electrons will have the same four (4) quantum numbers.

1. Energy levels. The energy levels of electrons within an atom are represented by four (4) quantum numbers. They include:

a. Principal quantum number. The principal quantum number (n) represents the ***principal energy level*** (also referred to as shells). This number, (n), is equal to the number of the principal energy level (as referred to under the Bohr atom) and is the same as the period number in the periodic table. For example, in the Bohr model, if the electron was described as being in the second shell, then its orbital model description would include a principal quantum number of 2.

b. 2nd quantum number. The second quantum number (l) represents *sublevels*. Additional spectral lines appearing in the spectrum of atoms heavier than hydrogen can be explained only by assuming that the principal energy levels are divided into sublevels.

The total number of possible sublevels for each principal energy level is equal to the number of the principal energy level (n). These sublevels are designated by the letters *s*, *p*, *d*, and *f*.

1) Principal energy level number 1 is made up of one energy sublevel called 1*s*.

2) Principal energy level 2 is made up of two sublevels called 2*s* and 2*p*.

3) Principal energy level 3 is made up of three sublevels called 3*s*, 3*p*, and 3*d*.

4) Principal energy level 4 is made up of four sublevels called 4*s*, 4*p*, 4*d*, and 4*f*.

Within a given principal energy level, the lowest in energy is the "*s*" sublevel and the highest is the "*f* " sublevel. The number of the principal energy levels and one of the letters *s*, *p*, *d*, or *f* are used to describe the energy of an electron in a particular sublevel. (The number of occupied sublevels does not exceed four even when n is greater than 4.)

c. 3rd quantum number. The third quantum number (m_l) represents *orbitals*. Each sublevel may consist of one or more orbitals with each orbital having a different spatial orientation. The maximum number of electrons possible in the various energy levels and their distribution are shown in the following table:

Principal Quantum Number (n)	Number of Orbitals (n^2)	s orbitals / p orbitals / d orbitals / f orbitals	Maximum Number of Electrons ($2n^2$)
1	1	1	2
2	4	1 3	8
3	9	1 3 5	18
4	16	1 3 5 7	32

Each electron occupies an orbital, which can hold no more than two electrons. The number of orbitals within the same principal energy level (n) is equal to n^2. When n is less than or equal to 4, the following sequence occurs:

1) The *s* sublevel consists of **1** orbital.
2) The *p* sublevel consists of **3** orbitals.
3) The *d* sublevel consists of **5** orbitals.
4) The *f* sublevel consists of **7** orbitals.

d. 4th quantum number. The fourth quantum number (m_s) represents the *spin of the electron*. In order for two electrons to occupy the same orbital, they must have opposite spins.

Questions II-2

1   When an atom goes from the excited state to the ground state, the total energy of the atom:
    (1) decreases   (2) increases   (3) remains the same
2   What is the maximum number of electrons that can occupy the second principal energy level?
    (1) 6   (2) 8   (3) 18   (4) 32
3   A neutral atom of an element has an electron configuration of 2-8-2. What is the total number of p electrons in this atom?
    (1) 6   (2) 2   (3) 10   (4) 12
4   Which principal energy level can hold a maximum of 18 electrons?
    (1) 5   (2) 2   (3) 3   (4) 4

5   Which electron transition is accompanied by the emission of energy?
    (1) $1s$ to $2s$ (2) $2s$ to $2p$ (3) $3p$ to $3s$ (4) $3p$ to $4p$
6   The total number of orbitals in the 4f sublevel is
    (1) 1   (2) 5   (3) 3   (4) 7
7   The total number of d orbitals in the third principal energy level is:
    (1) 1   (2) 5   (3) 3   (4) 7
8   What is the total number of electrons in an atom with an atomic number
    of 13 and a mass number of 27?
    (1) 13   (2) 14   (3) 27   (4) 40
9   What is the maximum number of sublevels in the third principal energy
    level?
    (1) 1   (2) 2   (3) 3   (4) 4

2. Electron configurations. Electron configurations of the atoms in order
of their atomic numbers starting with hydrogen can be built up by adding one
electron at a time according to the following rules:

a.   No more than two electrons can be accommodated in any orbital.

b.   The added electron is placed in the unfilled orbital of lowest energy.

c.   The two electrons in an orbital have opposite spins.

d.   In a given sublevel, according to Hund's rule, a 2nd electron is not add-
     ed to an orbital until each orbital in the sublevel contains one electron.

e.   No more than four orbitals are occupied in the outermost principal
     energy level of any atom except for palladium.

In an electron configuration, the number of electrons in a sublevel is shown
by a superscript following the designation of the sublevel. For example:

Calcium is represented as: $1s^2\ 2s^2\ 2p^6\ 3s^2\ 3p^6\ 4s^2$ or (Ar) $4s^2$

Sulfur is represented as: $1s^2\ 2s^2\ 2p^6\ 3s^2\ 3p^4$ or (Ne) $3s^2\ 3p^4$

Missing in electron configuration notation is the manner in which the
electrons are distributed in the orbitals. This is shown with diagrams called
orbital notation.

In each instance, an orbital is shown as either:

a. an empty orbital with no electrons

b. a half-filled orbital with one electron

c. a full orbital with a maximum number
   of two electrons

Below, note the electron notation and orbital notation of a chlorine atom:

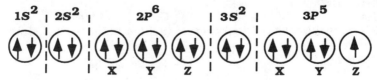

3. Valence electrons. The electrons in the outermost principal energy level of an atom are called the valence electrons. The chemical properties of an atom are related to the valence electrons.

The term **"kernel"** is sometimes used to refer to the atom exclusive of the valence electrons. The kernel consists of the nucleus and all electrons except the valence electrons.

The valence electrons may be represented by **electron dot symbols** in which the kernel of the atom is represented by the letter symbol for the element and the valence electrons are represented by dots. For example, to the right is the electron dot diagram for a chlorine molecule.

F. Ionization Energy

Ionization energy is the amount of energy required to remove the most loosely bound electron from an atom in the gaseous phase. The *Reference Tables for Chemistry* lists the ionization energy in kcal./mole required for the removal of the first (outermost) electron. The second ionization energy refers to the removal of the second most loosely bound electron. Each successive ionization energy is greater than the previous one.

Questions II-3

1 What is the electron configuration for Be^{2+} ions?
    (1) $1s^1$ (2) $1s^2$ (3) $1s^2\, 2s^1$ (4) $1s^2\, 2s^2$

2 What is the electron configuration of an O^{2-} in the ground state?
    (1) 2-4   (2) 2-8   (3) 2-8-4   (4) 2-8-8

3 The number of completely filled orbitals in a fluorine atom in the ground state is:
    (1) 5   (2) 6   (3) 9   (4) 4

4 Which orbital notation represents the outermost principal energy level of a phosphorus atom in the ground state?

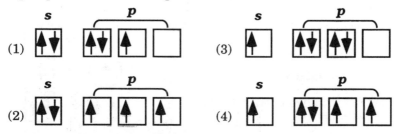

5 Which 1-mole sample of atoms requires the least energy to form a mole of positive ions?
    (1) Ge   (2) Ca   (3) Ga   (4) K

6 Which is the electron configuration of an atom in the excited state?
    (1) $1s^2\, 2s^1$ (3) $1s^2\, 2s^2\, 2p^5$
    (2) $1s^2\, 2s^2\, 2p^1$ (4) $1s^2\, 2s^2\, 2p^5\, 3s^1$

7   If X is the symbol of a noble gas atom in the ground state, its electron-dot symbol could be:

(1) $X \bullet$ (2) $\overset{\bullet}{X}\overset{\bullet}{}$ (this is X with dots)   (3) $\bullet \overset{\bullet}{X} \bullet$ (4) $\overset{\bullet}{\underset{\bullet}{{:}X{:}}}$

8   Usually the term "kernel" includes all parts of the atom except the:
(1) neutrons   (3) valence electrons
(2) protons   (4) orbital electrons

9   Which electron configuration represents an atom in an excited state?
(1) $1s^2\, 2s^2\, 2p^6\, 3p^1$ (3) $1s^2\, 2s^2\, 2p^6\, 3s^2\, 3p^2$
(2) $1s^2\, 2s^2\, 2p^6\, 3s^2\, 3p^1$ (4) $1s^2\, 2s^2\, 2p^6\, 3s^2$

10 Which element forms the ion represented in the following orbital notation?

(1) boron   (3) magnesium
(2) bromine   (4) potassium

11 Which is the electron configuration of a neutral atom in the ground state with a total of six valence electrons?
(1) $1s^2\, 2s^2\, 2p^2$ (3) $1s^2\, 2s^2\, 2\, p^6$
(2) $1s^2\, 2s^2\, 2p^4$ (4) $1s^2\, 2s^2\, 2p^6\, 3s^2\, 3p^6$

12 The amount of energy required to remove the most loosely bound electron from an atom in the gaseous phase is called:
(1) kinetic energy   (3) ionization energy
(2) potential energy   (4) electron affinity

13 Which electron dot symbol could represent a noble gas?

(1) $X \bullet$ (2) $X{:}$ (3) $\bullet \overset{\bullet\bullet}{X} \bullet$ (4) $X{:}$

14 The number of valence electrons in an atom with an electron configuration of $1s^2\, 2s^2\, 2p^6\, 3s^2\, 3p^4$ is:
(1) 6   (2) 2   (3) 16   (4) 4

15 Which diagram correctly represents a fluorine atom in an excited state?

16 A Ca^{2+} ion differs from a Ca atom in that the Ca^{2+} ion has:
(1) more protons   (3) more electron
(2) fewer protons   (4) fewer electrons

17 What is the total number of valence electrons in an atom of phosphorus in the ground state?
(1) 5   (2) 2   (3) 3   (4) 7

18 What is the total number of principal energy levels that are completely filled in an atom of magnesium in the ground state?
(1) 1   (2) 2   (3) 3   (4) 4

19 What is the total number of occupied principal energy levels in a sodium atom in the ground state?
(1) 1   (2) 2   (3) 3   (4) 4

20 Which atom in the ground state has three unpaired electrons in its outermost principal energy level?
(1) Li   (2) B   (3) N   (4) Ne

21 Which atom in the ground state contains only one orbital that is partially occupied?
(1) Si   (2) Ne   (3) Ca   (4) Na

22 What is the total number of electrons in a Mg^{2+} ion?
(1) 10   (2) 2   (3) 12   (4) 24

23 In the ground state, the atoms of which element have an incomplete 3rd principal energy level?
(1) Zn   (2) Mn   (3) Kr   (4) Sr

II. Natural Radioactivity

Radioactivity is the spontaneous disintegration of the nucleus of an atom with the emission of particles and/or radiant energy. Some naturally occurring elements are radioactive. Therefore, the term **natural radioactivity** is used.

When the nucleus of an element contains a disproportionate amount of neutrons, as compared to protons, the nucleus starts to emit particles. It is then classified as **radioactive**. Elements with an atomic number of higher than 82 (Lead) fall into this category. When one element is changed to another element because of change in the nucleus, the change is called **transmutation**.

A. Difference In Emanations

Emanations differ from each other in mass, charge, penetrating power, and ionizing power. The nuclear disintegration of naturally radioactive atoms produces alpha particles, beta particles and gamma radiation.

1. Alpha decay. When an alpha particle is given off as the result of nuclear disintegration, the reaction is called alpha decay. Alpha particles can be considered helium nuclei because they consist of 2 protons and 2 neutrons; therefore, they have a mass of four amu.

An atom that emits an alpha particle is called an **alpha emitter**. When an atom emits an alpha particle, the atomic number is reduced by 2 and the mass number is reduced by 4.

Example: $$^{226}_{88}Ra \longrightarrow {}^{4}_{2}He + {}^{222}_{86}Rn$$

2. Beta decay. Beta particles are high-speed electrons. When a beta particle is given off as the result of neutron disintegration, the reaction is called beta decay. The atom that emits a beta particle is called a **beta emitter**.

When an atom emits a beta particle, the atomic number is increased by 1 and the mass number remains the same.

Example:

$$^{234}_{90}\text{Th} \longrightarrow {}^{234}_{91}\text{Pa} + {}^{0}_{-1}\text{e}$$

Question: If a beta particle is a fast moving electron and electrons are found outside of the nucleus, what is the beta particle doing in the nucleus, in the first place?

Answer: _____

3. Gamma radiation. Gamma rays are not particles and do not have mass or charge. They are similar to high energy X-rays. Most nuclear changes involve the emission of gamma rays, reducing the energy content of the nucleus without affecting its charge or mass.

B. Separating Emanations

Separation of emanations is possible by an electric field or magnetic field. In such an electric field, alpha particles are deflected toward the negative electrode, beta particles toward the positive electrode, but gamma rays are not affected by the field.

C. Detection Of Radioactivity

The study of radioactivity is made possible by its ionizing, fluorescent, and photographic effects.

D. Half—Life

The half-life of a radioactive isotope is the time required for one-half of the nuclei of a given sample of that isotope to disintegrate. The decay of radioactive elements occurs erratically, so that a small amount might decay one day and a large amount might decay the following month. The scientist, knowing that the mass of an atom is essentially derived from its nucleus, has picked a span of time in which half of the mass of the sample has decayed.

Sample Problem:

If the half-life of the isotope ^{99}Tc is 60 hours, and if you had 100 grams of the isotope, how much would you have left after 180 hours?

Solution:

First - divide the total time by $\dfrac{180 \text{ hours}}{60 \text{ hours}}$ = 3 Half-Life Periods

During the first half-life period, the following would occur: 100 g. $\longrightarrow$ 50 g.

During the second half-life period, the following would occur: 50 g. $\longrightarrow$ 25 g.

During the third half-life period, the following would occur: 25 g. $\longrightarrow$ 12.5 g.

Answer: 12.5 grams

Note: It is strongly suggested that you should know the definition for half-life and then use your own good reasoning power, instead of using equations or formulas, to solve such problems.

Questions II-4

1   At the end of 12 days, $\frac{1}{4}$ of an original sample of a radioactive element remains. What is the half life of the element?
    (1) 24 days   (3) 3 days
    (2) 48 days   (4) 6 days

2   In the reaction $^{238}_{92}U \rightarrow X + {}^{4}_{2}He$, the particle represented by X is:
    (1) $^{234}_{90}Th$ (2) $^{234}_{92}U$ (3) $^{238}_{93}Np$ (4) $^{242}_{94}Pu$

3   What is represented by X in the equation? $X \rightarrow {}^{14}_{7}N + {}^{0}_{-1}e$
    (1) $^{13}_{6}C$ (2) $^{14}_{6}C$ (3) $^{13}_{7}N$ (4) $^{14}_{7}N$

4   A sample of element X contains 90. % ^{35}X atoms, 8.0 % ^{37}X atoms, and 2.0 % ^{38}X atoms. The average isotopic mass is closest to:
    (1) 32   (2) 35   (3) 37   (4) 38

5   An original sample of a radioisotope had a mass of 10 grams. After 2 days, 5 grams of the radioisotope remains unchanged. What is the half-life of this radioisotope?
    (1) 1 day   (2) 2 days   (3) 5 days   (4) 4 days

6   In the equation $^{32}_{15}P \rightarrow {}^{32}_{16}S + X$, the particle represented by X is:
    (1) $^{4}_{2}He$ (2) $^{1}_{0}n$ (3) $^{1}_{1}H$ (4) $^{0}_{-1}e$

7   Which nuclear emission moving through an electric field would be deflected toward the positive electrode?
    (1) alpha particle   (3) gamma radiation
    (2) beta particle   (4) proton

8   What is the number of hours required for potassium-42 to undergo 3 half-life periods?
    (1) 6.2 hours   (3) 24.8 hours
    (2) 12.4 hours   (4) 37.2 hours

9   After 3 half-life periods, 12.5 grams of an original sample of radioisotope remains unchanged. What was the mass of the original sample?
    (1) 250.0 g   (3) 100. g
    (2) 50.0 g   (4) 200. g

10 If X is the symbol of an element, which pair correctly represents isotopes of X?

    (1) $\left(\begin{array}{c}1p\\2n\end{array}\right)$ & $\left(\begin{array}{c}2p\\1n\end{array}\right)$ (3) $\left(\begin{array}{c}5p\\6n\end{array}\right)$ & $\left(\begin{array}{c}7p\\6n\end{array}\right)$

    (2) $\left(\begin{array}{c}3p\\3n\end{array}\right)$ & $\left(\begin{array}{c}3p\\4n\end{array}\right)$ (4) $\left(\begin{array}{c}10p\\10n\end{array}\right)$ & $\left(\begin{array}{c}11p\\11n\end{array}\right)$

11 How many grams of a 32-gram sample of ^{32}P will remain after 71.5 days?
    (1) 1   (2) 2   (3) 8   (4) 4

12 Which pair of nuclei represent isotopes of the same element correctly?

(1) $^{158}_{64}X$ and $^{158}_{64}X$ (3) $^{158}_{64}X$ and $^{159}_{64}X$

(2) $^{64}_{158}X$ and $^{158}_{64}X$ (4) $^{158}_{64}X$ and $^{158}_{65}X$

13 A radioactive isotope has a half-life of 10 years. What fraction of the original mass will remain unchanged after 50 years?
(1) $\frac{1}{2}$ (2) $\frac{1}{8}$ (3) $\frac{1}{16}$ (4) $\frac{1}{32}$

14 Which emission from a radioactive source is *not* affected by an electric field?
(1) alpha particles   (3) positrons
(2) beta particles   (4) gamma rays

15 Which of the following particles has the *least* mass?
(1) alpha particle   (3) proton
(2) beta particle   (4) neutron

Self-Help Questions II

Unit II - Atomic Structure

Atoms - Sub-atomic particles, structure-including atomic number, isotopes, mass number, atomic mass, gram atomic mass, electrons.

1 Which pair of atoms are isotopes of element X?

(1) $^{226}_{90}X$ and $^{226}_{91}X$ (3) $^{227}_{91}X$ and $^{227}_{90}X$

(2) $^{226}_{91}X$ and $^{227}_{91}X$ (4) $^{226}_{90}X$ and $^{227}_{91}X$

2 In which pair of atoms do both nuclei contain the same number of neutrons?

(1) $^{7}_{3}Li$ and $^{9}_{4}Be$ (3) $^{40}_{20}Ca$ and $^{38}_{18}Ar$

(2) $^{40}_{19}K$ and $^{40}_{17}Cl$ (4) $^{14}_{7}N$ and $^{16}_{8}O$

3 Which of the following particles has the *least* mass?
(1) alpha particle   (3) proton
(2) beta particle   (4) neutron

4 Which atoms represent different isotopes of the same element?

(1) $^{39}_{18}Ar$ and $^{39}_{19}K$ (3) $^{12}_{6}C$ and $^{13}_{6}C$

(2) $^{58}_{27}Co$ and $^{59}_{28}Ni$ (4) $^{35}_{17}Cl$ and $^{35}_{17}Cl$

5 The nucleus of an atom consists of 8 protons and 6 neutrons. The total number of electrons present in a neutral atom of this element is
(1) 6   (2) 2   (3) 8   (4) 14

6 The total number of protons and neutrons in the nuclide $^{37}_{17}Cl$ is

(1) 54   (2) 37   (3) 20   (4) 17

7   Which particle has the greatest mass?
    (1) an alpha particle   (3) an electron
    (2) a beta particle   (4) a neutron

8   All of the atoms of argon have the same
    (1) mass number   (3) number of neutrons
    (2) atomic number   (4) number of nucleons

9   Which particle has a negative charge and a mass that is
    approximately $1/1836$ the mass of a proton?

    (1) a neutron   (3) an electron
    (2) an alpha particle   (4) a positron

10 Which pair must represent isotopes of the same element?

    (1) $^{120}_{51}X$ and $^{120}_{52}X$ (3) $_{21}X^{2+}$ and $_{19}X^{2+}$

    (2) $^{38}_{18}X$ and $^{39}_{18}X$ (4) $_{26}X^{2+}$ and $_{26}X^{3+}$

11 A sodium atom and a sodium ion must have the same number of
    (1) neutrons   (3) occupied principal energy levels
    (2) protons   (4) outermost electrons

12 What is the mass number of an atom which contains 21 electrons,
    21 protons, and 24 neutrons?
    (1) 21   (2) 42   (3) 45   (4) 66

13 As the number of neutrons in the nucleus of an atom increases, the
    nuclear charge of the atom
    (1) decreases   (2) increases   (3) remains the same

14 Which nuclide contains the greatest number of neutrons?
    (1) ^{37}Cl (2) ^{39}K (3) ^{40}Ar (4) ^{41}Ca

*Atomic structure models, principal energy levels, spectural lines,
ground state-excited state, quantum numbers,
atomic orbital model including energy levels, 1,2,3, and 4*

15 If n represents the principal energy level, the maximum number of
    electrons possible in that principal energy level is equal to
    (1) n (2) $2n$ (3) n^2 (4) $2n^2$

16 The greatest absorption of energy occurs as an electron moves from
    (1) $1s$ to $3s$ (2) $3p$ to $3s$ (3) $4d$ to $4s$ (4) $4s$ to $3p$

17 The total number of completely filled orbitals in an atom of nitrogen in
    the ground state is
    (1) 1   (2) 2   (3) 3   (4) 5

18 Which electron transition results in the emission of energy?
    (1) $2s$ to $2p$ (2) $2p$ to $3s$ (3) $3d$ to $2p$ (4) $3p$ to $4d$

19 The total number of orbitals in an f sublevel is
    (1) 1   (2) 5   (3) 3   (4) 7

20 When the electrons of an excited atom fall back to lower energy levels, there is an emission of energy that produces
   (1) beta particles   (3) spectral lines
   (2) alpha particles   (4) gamma radiation

21 An atom has 8 electrons in a d sublevel. How many d orbitals in this sublevel are half filled?
   (1) 1   (2) 2   (3) 3   (4) 4

22 Which principal energy level has a maximum of three sublevels?
   (1) 1   (2) 2   (3) 3   (4) 4

23 The maximum number of electrons that a single orbital of the $3d$ sublevel may contain is
   (1) 5   (2) 2   (3) 3   (4) 4

Electron and orbital configurations, valence electrons, electron dot symbols, ionization energy

24 Which is the correct electron dot symbol for an aluminum atom in the ground state?
   (1) Al :   (2) Al :   (3) • Al :   (4) • Al :

25 Which electron notation represents the valence electrons of a phosphorus atom in the ground state?

26 According to Reference Table K, the element in Period 2 with the highest first ionization energy is
   (1) a noble gas   (3) an alkali metal
   (2) a halogen   (4) an alkaline earth metal

27 Which electron configuration represents an atom in the excited state?
   (1) $1s^2 2s^1$ (2) $1s^2 2s^2 2p^4$ (3) $1s^2 2s^2 3s^1$ (4) $1s^2 2s^2 2p^6 3s^2$

28 Based on Reference Table K, the atom of which of the following elements requires the *least* amount of energy to remove the most loosely bound electrons?
   (1) lithium   (2) sodium   (3) potassium   (4) rubidium

29 Which could be the first ionization energy, in kilocalories per mole of atoms, of a nonmetal?
   (1) 96   (2) 141   (3) 169   (4) 273

30 Given an atom with the electron configuration $1s^2 2s^2 2p^3$, how many orbitals are completely filled?
   (1) 1   (2) 2   (3) 3   (4) 4

31 Which is the electron dot symbol of an atom of boron in the ground state?
   (1) • B :   (2) B •   (3) • B :   (4) B :

32 Which is an electron configuration of a fluorine atom in the excited state?
   (1) $1s^22s^22p^4$ (3) $1s^22s^22p^43s^1$
   (2) $1s^22s^22p^5$ (4) $1s^22s^22p^53s^1$

33 If the nucleus of an atom is represented as $^{24}_{11}X$, the atom is

   (1) Na   (2) Al   (3) Mg   (4) Br

Natural radioactivity - Emanations: their separations and detection, Half-life

34 A radioactive isotope has a half-life of 10 years. What fraction of the original mass will remain unchanged after 50 years?
   (1) $\frac{1}{2}$ (2) $\frac{1}{8}$ (3) $\frac{1}{16}$ (4) $\frac{1}{32}$

35 Which emission from a radioactive source is *not* affected by an electric field?
   (1) alpha particles   (3) positrons
   (2) beta particles   (4) gamma rays

36 Given the equation: $X \rightarrow {}^4_2He + {}^{222}_{86}Rn$ The nucleus represented by X is

   (1) $^{218}_{84}Po$ (2) $^{218}_{88}Po$ (3) $^{218}_{84}Ra$ (4) $^{226}_{88}Ra$

37 After 62.0 hours, 1.0 gram remains unchanged from a sample of ^{42}K. How much ^{42}K was in the original sample?
   (1) 8.0 g   (2) 16 g   (3) 32 g   (4) 64 g

38 In 6.20 hours, a 100.-gram sample of $^{112}_{47}Ag$ decays to 25.0 grams. What is the half-life of $^{112}_{47}Ag$?

   (1) 1.60 hours   (2) 3.10 hours   (3) 6.20 hours   (4) 12.4 hours

39 The diagram at the right represents radiation passing through an electric field. The arrow marked 2 would most likely represent

   (1) an alpha particle
   (2) a beta particle
   (3) gamma radiation
   (4) a proton

lead block
radioactive sources

40 In the equation $X \rightarrow {}^4_2He + {}^{216}_{85}At$,   the element represented by X is

   (1) Fr   (2) Bi   (3) Rn   (4) Ra

41 In the equation: $^{232}_{90}Th \rightarrow {}^{228}_{88}Ra + X$, which particle is represented by the letter X?
   (1) an alpha particle   (3) a positron
   (2) a beta particle   (4) a deuteron

42 Which element has no known stable isotope?

   (1) carbon   (2) potassium   (3) polonium   (4) phosphorus

3 Bonding

Important Terms and Concepts

Chemical Energy	Molecular Attraction
Bonding and Stability	Dipoles
Exothermic Reaction	Hydrogen Bonding
Endothermic Reaction	Van der Waals Forces
Electronegativity	Molecular Ion Attraction
Ionic and CovalentBonds	Electron Density Formulas
Nonpolar and Polar Bonds	Chemical Formula and Symbols
Coordinate Covalent Bonds	Binary Compounds
Molecular and Network Solids	Stock System and Metal Ions
Metallic Bonding	Binary and Ternary Acids
Mono– and Polyatomic Ions	Chemical Equations and Balancing

I. The Nature Of Chemical Bonding

A chemical bond results from the simultaneous attraction of electrons (either single or paired) to two nuclei.

A. Chemical Energy

Potential energy is stored in molecules, and the transfer or release of some of this energy manifests itself in the form of chemical energy.

Substances possess energy because of their composition and structure. Factors such as mass, types of bonding, and types of motion influence the absorption and storage of energy by molecules.

B. Energy Changes In Bonding

When two atoms are held together by a chemical bond, they are generally at a lower energy condition than when they are separated. Therefore, we state that when a chemical bond is formed, energy is released, and when a chemical bond is broken, energy is absorbed.

C. Bonding And Stability

Because there is a release of energy when bonds are formed, systems at lower energy levels are more stable than systems at higher energy levels. So, it follows that bonding will more often occur among atoms if the changes lead to a lower energy condition and, therefore, a more stable structure. The more energy given off when a bond is formed, the stronger and more stable the bond will be. Also, the less energy given off in the formation of a bond, the weaker and less stable it will be.

For example, when an element such as fluorine, with a nearly filled outer shell, reacts with another element, it usually releases energy in the process, mainly because the resulting state of the compound is in a lower energy level. Such a change in condition would be an example of an **exothermic reaction**. However, when the compound thus formed is decomposed, energy must be put into the compound to decompose or break it up. Therefore, this would be an example of an **endothermic reaction**.

Note: the exact same amount of energy would be required to force this endothermic reaction to proceed as was released when the compound was formed.

D. Electronegativity

Electronegativity is a measure of the ability of an atom to attract the electrons that form a bond between it and another atom. The values designated are based on an arbitrary scale on which fluorine, the most electronegative element, is assigned a value of 4.0.

Keep in mind that this electronegativity value does NOT necessarily measure the reactivity of the element. However, the scale can be used to predict the type of intramolecular (attractive forces inside the molecule) bond formed.

The ionic or covalent character of a bond can be approximated from the differences in electronegativity of the resulting species. Electronegativity differences of 1.7 or more indicate a bond that is predominately ionic in character. Differences of less than 1.7 indicate that the bond is predominately covalent. Some exceptions to this may be found. (For example, the metal hydrides, with an electronegativity difference of less than 1.7, are predominately ionic.)

II. Bonds Between Atoms

The electrons involved in bond formation may be transferred from one atom to another or may be shared equally or unequally between two atoms. When atoms of the elements enter into a chemical reaction, they do so in a manner that results in their becoming more like "inert" gas atoms. In this state they contain their maximum complement of valance electrons, and they are in a condition of maximum stability.

A. Ionic Bonds

An ionic bond is formed by the **transfer** of one or more electrons from metals to nonmetals. This transfer of electrons results in the formation of **ions**. The attraction between a positive ion and a negative ion is called an **ionic bond**.

In ionic bonding the number of electrons transferred is such that the atoms involved achieve an "inert" gas configuration, except for some transition elements. Since the ion has a different electron configuration than the atom, the properties of the ion differ from those of the atom. Also, ionic bonds may form between ions that were formed in a previous reaction. For example:

$$AgNO_3(aq) \; + \; NaCl(aq) \; \longrightarrow \; AgCl(s) \; + \; NaNO_3(aq)$$

Characteristics of Ionic solids

Ionic solids have high melting points and do not conduct electricity. In the geometric structure of the solid ionic crystal, ions form the crystal lattice and are held in relatively fixed positions by electrostatic attraction. When melted or dissolved in water, the crystal lattice is destroyed and the ions move freely. This free movement of ions permits electrical conductivity. Examples of ionic solids are **sodium chloride and magnesium oxide**.

B. Covalent Bonds

A **covalent bond** is a simultaneous attraction of two nuclei for the same electrons resulting in the sharing of those electrons. A covalent bond is formed when two atoms *share* electrons, instead of transferring them. In order to form this type of bond, the electronegativity difference between the two atoms forming the bond must be less than 1.7. Covalent bonds are classified as two types:

1. Nonpolar covalent bond. When electrons are shared between atoms of the same element, they are shared equally, and the resulting bond is a nonpolar bond. An example of a nonpolar covalent bond is found in the fluorine molecule.
Since the electronegativity of both fluorine atoms in the molecule is the same, the difference is zero, and the electron density of the molecule is symmetrical.

2. Polar covalent bonds. When electrons are shared between atoms of different elements, they are usually shared unequally. The resulting bond is polar. An example of a polar covalent bond is found in the hydrogen chloride molecule.

Chlorine, having an electronegativity value of 3.2 will attract the bonding electrons to a greater extent than hydrogen, which has an electronegativity value of 2.2. The difference of 1.0 denotes a covalent bond. Since the chlorine end of the molecule will show a greater electron density probability than the hydrogen end, the molecule will be asymmetrical and therefore polar.

3. Coordinate covalent bonds. When the two shared electrons forming a covalent bond are both donated by one of the atoms, this bond is called a coordinate covalent bond.

A coordinate covalent bond, once formed, is not different from an ordinary covalent bond. The difference lies in the source of the electrons involved in the bond. This type of bond is frequently involved in the bonding within polyatomic ions and is very important in modern acid-base theories.

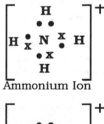
Ammonium Ion

A classic example of a coordinate covalent bond is the ammonium ion. In the ammonium ion, the nitrogen atom has five valence electrons. Three electrons are unpaired, but are shared with the electrons from three hydrogen atoms. The other two form a full pair and are not shared.

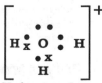
Hydronium Ion

It is at this unshared pair of electrons that the electron density is so great that the molecule may attract a hydrogen ion (proton). When that occurs, the ion is formed and takes on a charge of positive one (+1).

4. Molecular substances. A **molecule** may be defined as a discrete particle formed by covalently bonded atoms. A molecule has also been defined as the smallest particle of an element or compound capable of independent existence. When a stable molecule is born, a covalent bond is formed. The atoms that form the bond *usually* assume electronic structures of inert gases by sharing electrons. Examples of molecules are:

$$H_2 \quad NH_3 \quad H_2O \quad HCl \quad CCl_4 \quad S_8 \quad C_6H_{12}O_6$$

Characteristics of Molecular solids. Molecular substances may exist as gases, liquids, or solids, depending on the attraction that exists between the molecules. Generally, molecular solids are soft, good electrical insulators, poor heat conductors, and have low melting points.

5. Network Solids. Certain solids consist of **co-valently bonded** atoms linked in a network that extends throughout the sample with an absence of simple discrete particles. Such a substance is said to be a network solid.

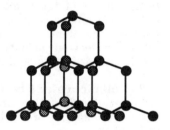

Generally, network solids are hard, are poor conductors of heat and electricity, and have high melting points. Examples of network solids are diamond (C), silicon carbide (SiC), and silicon dioxide (SiO_2).

Figure III-1
Carbon Atom Crystal Lattice

C. Metallic Bonding

Metallic bonding occurs between atoms of metals that have a small number of valence electrons leaving them with many vacant valence orbitals and low ionization energies.

A metallic bond consists of an arrangement of positive ions that are located at the crystal lattice sites and are immersed in a "sea" of mobile electrons. These mobile electrons can be considered as belonging to the whole crystal rather than to individual atoms.

Note: This mobility of electrons distinguishes the metallic bond from an ionic or covalent bond and gives the metal the following characteristics:

*   good conductors of electricity and heat
*   great strength
*   malleability and ductility
*   luster

D. Polyatomic Ions

A single atom with a charge is called a **monatomic ion**. A compound of two or more covalently bonded atoms with a charge is called a **polyatomic ion**.

A polyatomic ion is very stable and behaves like a monoatomic particle, because it contains strong covalent bonds. These bonds are stronger than the bonds that hold it to the rest of the atoms in the compound. Therefore, during reactions, the polyatomic ion usually remains intact as it passes from the reactants to the products.

Some polyatomic ions include: NH_4^+ (ammonium ion), PO_4^{-3} (phosphate ion), and NO_3^- (Nitrate ion). Other examples can be found in Table *F* of the *Reference Tables for Chemistry*.

Although the bonds which keep the atoms in a polyatomic ion are covalent bonds, the polyatomic ions possess a charge. When they attach themselves to a metal atom or another polyatomic ion, they do so by forming an ionic bond. The final compound contains both ionic and covalent bonds. Some examples include: $Na^+NO_3^-$ and $NH_4^+OH^-$.

Questions III-1

1 Which molecule contains a nonpolar covalent bond?

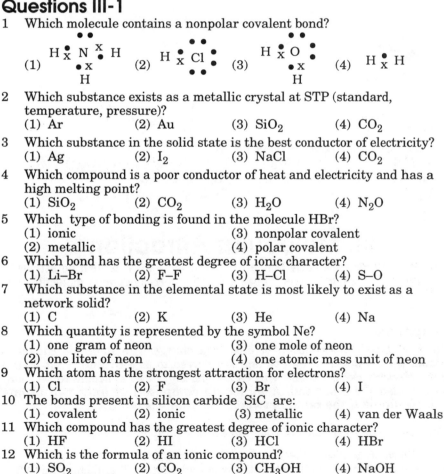

2 Which substance exists as a metallic crystal at STP (standard, temperature, pressure)?
(1) Ar   (2) Au   (3) SiO_2 (4) CO_2

3 Which substance in the solid state is the best conductor of electricity?
(1) Ag   (2) I_2 (3) NaCl   (4) CO_2

4 Which compound is a poor conductor of heat and electricity and has a high melting point?
(1) SiO_2 (2) CO_2 (3) H_2O (4) N_2O

5 Which type of bonding is found in the molecule HBr?
(1) ionic   (3) nonpolar covalent
(2) metallic   (4) polar covalent

6 Which bond has the greatest degree of ionic character?
(1) Li–Br   (2) F–F   (3) H–Cl   (4) S–O

7 Which substance in the elemental state is most likely to exist as a network solid?
(1) C   (2) K   (3) He   (4) Na

8 Which quantity is represented by the symbol Ne?
(1) one gram of neon   (3) one mole of neon
(2) one liter of neon   (4) one atomic mass unit of neon

9 Which atom has the strongest attraction for electrons?
(1) Cl   (2) F   (3) Br   (4) I

10 The bonds present in silicon carbide SiC are:
(1) covalent   (2) ionic   (3) metallic   (4) van der Waals

11 Which compound has the greatest degree of ionic character?
(1) HF   (2) HI   (3) HCl   (4) HBr

12 Which is the formula of an ionic compound?
(1) SO_2 (2) CO_2 (3) CH_3OH (4) NaOH

13 The bonding in NH_3 is most similar to the bonding in:
    (1) H_2O (2) NaCl   (3) MgO   (4) KF

14 Which formula represents an ionic compound?
    (1) $H_2O(l)$ (2) NaCl(s)   (3) $NH_3(g)$ (4) $CCl_4(l)$

15 When hydrogen atoms change to hydrogen molecules according to the
    equation $2H(g) \rightarrow H_2(g)$,   energy is:
    (1) absorbed, a bond is formed   (3) released, a bond is formed
    (2) absorbed, a bond is broken   (4) released, a bond is broken

16 Which electron dot formula represents a molecule that contains a
    nonpolar covalent bond?

17 Which species contains a coordinate covalent bond?

18 Which compound contains ionic bonds?
    (1) NaBr(s)   (2) HBr(g)   (3) $C_6H_{12}O_6(s)$ (4) $CO_2(g)$

19 Which molecule contains a nonpolar covalent bond?
    (1) HCl   (2) F_2 (3) CO_2 (4) NH_3

III. Molecular Attraction

Groups of atoms covalently bonded in a molecule may in turn be attracted to similar molecules or to ions. These attractive forces between molecules are classified as follows:

A. Dipoles

The asymmetric distribution of electrical charge in a molecule causes a molecule that is **polar** in nature and is referred to as a **dipole**. That is, the uneven electron cloud density will cause one end of a molecule to be more negative than the other end. **A molecule composed of only two atoms will be a dipole if the bond between the atoms is polar.**

Example: The hydrogen chloride molecule is a dipole because:

- the chlorine atom is larger than the hydrogen atom
- the difference in electronegativity of hydrogen (2.2) and chlorine (3.2)

These factors allow chlorine to share electrons closer to itself than to hydrogen. When the bond is formed, the electron density around the chlorine atom is greater than around the hydrogen atom, leading to a polar molecule (a dipole). The bond between two hydrogen chloride molecules is a result of dipole—dipole attraction (at right)

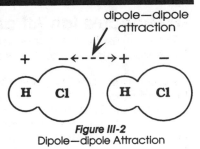

Figure III-2
Dipole—dipole Attraction

B. Hydrogen Bonding

Hydrogen bonds are formed between molecules when hydrogen is covalently bonded to an element of small atomic radius and high electronegativity. When a hydrogen atom is bonded to a highly electronegative atom, the hydrogen has such a small share of the electron pair that it acts like a bare proton.

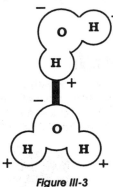

Figure III-3
H —Bonding

As such it can be attracted to the more electronegative atom of an adjacent molecule. The sulfur atom is larger in radius and has an electronegativity of 2.6, compared to the smaller radius of the oxygen atom and its higher electronegativity of 3.5. The greater polarity of the water molecules results in stronger electrostatic bonds between them. This accounts for the lower vapor pressure and the higher boiling point of H_2O, as compared with the boiling point of H_2S.

Hydrogen bonding (shown in a quantity of water) is important in compounds of hydrogen with **fluorine, oxygen**, or **nitrogen**. These compounds represent special cases of dipole—dipole attraction.

C. Van der Waals Forces

In the absence of dipole attraction and hydrogen bonding, as in nonpolar molecules, weak attractive forces exist between molecules. These forces are called **van der Waals forces**. Van der Waals forces make it possible for species of small nonpolar molecules, such as hydrogen, helium, oxygen, etc., to exist in the liquid and solid phases under conditions of low temperature and high pressure.

Van der Waals forces appear to be due to chance distribution of electrons resulting in momentary dipole attractions. Therefore, we can say that these forces are momentary electrostatic forces that increase as the distance between molecules decreases. Also, as we increase the molecular size, the number of bonds between the molecules will increase.

The effect of molecular size on the magnitude of the van der Waals forces accounts for the increasing boiling points of a series of similar compounds (such as the alkane series of hydrocarbons).

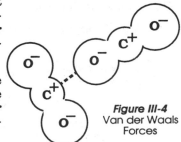

Figure III-4
Van der Waals
Forces

D. Molecule Ion Attraction

Polar solvents, when interacting with ionic compounds, attract ions from these compounds and form a solution.

Ionic compounds are generally soluble in polar solvents such as water, alcohol, and liquid ammonia. The negative ion of the substance being dissolved is attracted to the positive end of the adjacent polar molecules, while the positive ion is attracted to the negative end of the polar molecules. Water is the polar substance most commonly used to dissolve these ionic compounds. When an ionic compound is dissolved in water, its crystal lattice is destroyed and water molecules surround each ion forming hydrated ions. It is because water is a dipole that this attraction between the water molecules and the positive or negative ion exists. The orienting of water molecules around ions is called the **hydration of the ions**. This process is important in aqueous chemistry.

IV. Directional Nature Of Covalent Bonds

Generally the geometric structure of covalent substances which result from the directional nature of the covalent bond helps to explain properties of the resulting molecule. The polarity of a water molecule is explained by the asymmetrical shape of the molecule. The water molecule (H_2O) is shown as follows:

Figure III-5
Electron
Density
Formula
of Water

Electron Dot
Notation

Structural
Formula

104.5°

Oxygen's electronegativity is 3.5, and hydrogen's electronegativity is 2.2. The difference, 1.3, indicates a polar covalent bond. The bond angle is such that there exists an unsymmetrical distribution of electron density. Therefore, a polar molecule results.

Because of the directional nature of the covalent bonds which form them, some molecules composed of more than two atoms may be nonpolar, even though the individual bonds are polar if the shape of the molecule is such that symmetric distribution of charge results. For example, carbon dioxide (CO_2) is a nonpolar molecule.

1. The CO_2 molecule can be shown in many ways. For Example:

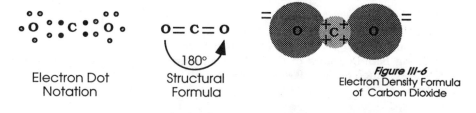

Electron Dot
Notation

Structural
Formula

180°

Figure III-6
Electron Density Formula
of Carbon Dioxide

Because of the greater electronegativity of oxygen, the electron density is greater around the oxygen atoms. The bond with an electronegativity difference of 0.9 is polar, but because the bond angle is 180° and the molecule is symmetrical, it is considered a nonpolar molecule.

2. The carbon tetrafluoride molecule can be shown to be a nonpolar molecule with polar bonds (other examples include: CH_4, CCl_4, CBr_4, CI_4):

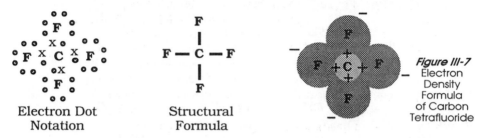

Electron Dot Notation	Structural Formula	Figure III-7 Electron Density Formula of Carbon Tetrafluoride

Fluorine has an electronegativity of 4.0, and Carbon has an electronegativity of 2.6. The difference, 1.4, indicates a polar covalent bond. However, because the tetrahedral molecule is symmetrical, the molecule is nonpolar.

Questions III-2

1   Which formula represents a tetrahedral molecule?
    (1) CH_4 (2) $CaCl_2$ (3) HBr   (4) Br_2

2   Which formula represents a polar molecule containing polar covalent bonds?
    (1) H_2O (2) CO_2 (3) NaCl   (4) Cl_2

3   A molecule of ammonia (NH_3) contains:
    (1) ionic bonds, *only*   (3) *both* covalent *and* ionic bonds
    (2) covalent bonds, *only*   (4) *neither* covalent *nor* ionic bonds

4   When sodium reacts with chlorine to form sodium chloride, electrons are lost by:
    (1) sodium, *only*   (3) *both* sodium *and* chlorine
    (2) chlorine, *only*   (4) *neither* sodium *nor* chlorine

5   The strongest hydrogen bonds are formed between molecules of:
    (1) H_2Te (2) H_2Se (3) H_2O (4) H_2S

6   Which type of bond is formed when a water molecule accepts a proton?
    (1) metallic   (3) ionic
    (2) covalent   (4) coordinate covalent

7   Molecule-ion attractions are found in:
    (1) Cu(s)   (2) CO(g)   (3) KBr(*l*)   (4) NaCl(aq)

8   When a reaction occurs between atoms with ground state electron configurations of $1s^22s^1$ and $1s^22s^22p^5$, the predominant type of bond formed is:
    (1) polar covalent   (3) ionic
    (2) nonpolar covalent   (4) metallic

9   Which pair of atoms would form the most polar bond?
    (1) I and Br   (2) Cl and Br   (3) I and F   (4) Cl and F

10 The attraction between molecules of methane in the liquid state is primarily due to:
    (1) hydrogen bonds   (3) covalent bonds
    (2) van der Waals forces   (4) electron affinities

11 Hydrogen bonds are strongest between the molecules of:
    (1) $HF(l)$ (2) $HCl(l)$ (3) $HBr(l)$ (4) $HI(l)$

12 A pure substance melts at 38°C and does not conduct electricity in either the solid or liquid phase. The substance is classified as:
    (1) ionic   (2) metallic   (3) electrovalent (4) molecular

13 The weakest van der Waals forces exist between molecules of:
    (1) $C_2H_6(l)$ (2) $C_3H_8(l)$ (3) $C_4H_{10}(l)$ (4) $C_5H_{12}(l)$

14 Which of the following compounds has the highest boiling point?
    (1) CH_4 (2) C_2H_6 (3) C_5H_{12} (4) C_7H_{16}

15 Hydrogen bonds are most likely to exist between molecules of:
    (1) H_2 (2) CH_4 (3) HI (4) H_2O

16 Which compound has the lowest normal boiling point?
    (1) HCl (2) H_2S (3) NH_3 (4) CH_4

17 At 298 K, the vapor pressure of H_2O is less than the vapor pressure of CS_2. The best explanation for this is that H_2O has:
    (1) larger molecules   (3) stronger ionic bonds
    (2) a larger molecular mass   (4) stronger intermolecular forces

18 Which type of solid does pure water form when it freezes?
    (1) ionic   (2) network   (3) metallic   (4) molecular

19 Helium may be liquefied at low temperature and high pressure primarily because of:
    (1) hydrogen bonding   (3) van der Waals forces
    (2) covalent bonds   (4) ionic attraction

20 The strongest hydrogen bonds are formed between molecules in which hydrogen is covalently bonded to an element with:
    (1) high electronegativity and large atomic radius
    (2) high electronegativity and small atomic radius
    (3) low electronegativity and large atomic radius
    (4) low electronegativity and small atomic radius

V. Formula Writing

A. Chemical Formula

A chemical formula is both a qualitative and a quantitative expression of the composition of an element or a compound. For example, the formula for phosphoric acid is H_3PO_4. This tells us that a molecule of phosphoric acid contains 3 atoms of hydrogen, 1 atom of phosphorus, and 4 atoms of oxygen. It also states that if we wanted to make a mole of this compound, we would need 3 moles of hydrogen atoms, 1 mole of phosphorus atoms, and 4 moles of oxygen atoms.

B. Symbol

A **symbol** may represent **one atom or one mole of atoms of an element.** One mole of atoms contains Avogadro's number (6.02×10^{23}) of atoms.

C. Formula

A formula is a statement in chemical symbols that represents the composition of a substance. There are two basic types of formulas:

1. Molecular Formula. A molecular formula indicates the total number of atoms of each element needed to form the molecule, given the mole mass of the substance. That is, the molecule C_2H_6 is a representation of the molecular formula for ethane.

2. Empirical Formula. An empirical formula represents the simplest ratio in which the atoms combine to form a compound.

Question: What is the empirical formula of ethane (C_2H_6)?

Answer: _____

Empirical formulas are used to represent ionic compounds that do not exist as discrete molecular entities. For example, CaF_2 is an ionic substance and combines, in definite proportions of 1 atom of calcium to 2 atoms of fluoride into a solid. CaF_2 does not exist as a molecule.

VI. Naming and Writing Chemical Compound Formulas

The discussion of oxidation numbers and their rules will be covered later in this book in greater detail. However, the rules are included here in brevity because the student cannot write a valid formula without knowing the rules for applying oxidation numbers.

The common **oxidation number** (**state**) is found in the Periodic Table of Elements and is given for each of the elements. It represents the charge which an atom has, or appears to have, when the electrons are counted according to certain arbitrary rules. These rules result in the following operational rules for determining oxidation numbers:

- In the free element, each atom has an oxidation number of zero.

- In simple ions (ions containing one atom), the oxidation number is equal to the charge on the ion.

- The algebraic sum of all the oxidation numbers of the atoms of any molecule or compound is zero.

- The oxidation number of oxygen is -2, except in peroxides it is -1.

- The oxidation number of hydrogen is +1, except in hydrides it is -1.

- In compounds formed by nonmetals, the less electronegative element is positive and the more electronegative element is negative.

- The algebraic sum of the charges of the atoms in a polyatomic ion is equal to the charge on the ion.

The chemical name of a compound generally indicates the chemical composition of the substance. The procedures used in naming and writing chemical formulas are as follows:

1. In **binary compounds composed of metals and nonmetals**, the metallic element is usually named and written first. The name of the nonmetal ends in "**-ide**" (for example, sodium chloride, potassium bromide, lithium iodide).

2. In **compounds composed of two nonmetals**, the less electronegative element is usually named and written first. The name of the compound still ends in "**-ide**" (for example, hydrogen chloride, nitrogen bromide).

3. Prefixes are used to indicate the number of atoms of each nonmetal in the compound. Examples include carbon dioxide (CO_2), sulfur trioxide (SO_3), carbon tetrachloride (CCl_4).

4. In the naming of compounds, which include one or more polyatomic ions, the metal is usually named first with polyatomic ions named last (for example, sodium hydroxide - $NaOH$, magnesium sulfate - $MgSO_4$).

The various polyatomic ions and their charges are listed in Table *F* in the Reference Tables for Chemistry. There are two exceptions:

1. The mercury ion is a polyatomic ion of a single element. Although it has a +2 charge, there are 2 atoms in the ion. Therefore, each of the atoms in the mercury ion has a +1 charge, and it is called the mercury (I) ion. For example, $HgCl$ mercurous chloride and $HgCl_2$ mercury (++) chloride.

2. The only polyatomic ion that has a positive charge and will be written first in compounds is the ammonium ion (NH_4^+). For example, ammonium nitrate is written NH_4NO_3 and ammonium chloride is written NH_4Cl.

In naming compounds of metals which may have more than one oxidation number, the **Stock System** should be used. In this system, Roman numerals indicate the oxidation number of the **metal ion**. For example, FeO is named iron (II) oxide, and Fe_2O_3 is named iron (III) oxide.

In naming compounds of two nonmetals with multiple oxidation states, the Stock system should be used to denote the oxidation state of the less electronegative member. For example, N_2O is named nitrogen (I) oxide, NO is named nitrogen (II) oxide, and NO_2 is named nitrogen (IV) oxide.

In naming acids, you should follow the following suggestions:

1. In **binary acids** (acids in which there are only two elements in the formula, hydrogen being one of them), hydrogen takes the place of a metal. Therefore, the prefix for a binary acid will always be **Hydro-**. For example, Hydrochloric acid (HCl) and Hydrosulfuric acid (H_2S).

2. In **ternary acids** (acids which combine hydrogen with a polyatomic ion and contain three elements in their formulas), the prefix **Hydro-** is dropped, and the acid is named after the polyatomic ion as follows:

     a) If the polyatomic ion ends with *ate*, the name of the acid will end with *ic*. (For example, $SO_4{}^{2-}$ is called the sulfate ion. When it combines with hydrogen, its formula becomes H_2SO_4 and is called sulfuric acid.)

     b) If the polyatomic ion ends with *ite*, the name of the acid will end with *ous*. (For example, $SO_3{}^{2-}$ is called the sulfite ion. When it combines with hydrogen, its formula becomes H_2SO_3 and is called sulfurous acid.)

Question: What are the names of the following acids?

HNO_3 _____ HNO_2 _____

H_3PO_4 _____ H_2CO_3 _____

In naming the common bases, all we need do is to follow the name of the metal with the word hydroxide. (Examples, KOH - potassium hydroxide, NaOH - sodium hydroxide, and $Ca(OH)_2$ - calcium hydroxide.)

We name salts as follows:
1. Binary salts always end with *ide*. (Such as KBr - potassium bromide and $CaCl_2$ - calcium chloride.)

2. When a metal is combined with a polyatomic ion, we simply follow the name of the metal with the name of the polyatomic ion. (Such as Na_2CO_3 – sodium carbonate and $K_2Cr_2O_7$ – potassium dichromate.)

Questions III-3

1   What is the correct formula for sodium thiosulfate?
     (1) $Na_2S_2O_4$ (2) Na_2SO_3 (3) Na_2SO_4 (4) $Na_2S_2O_3$

2   Which represents both an empirical and molecular formula?
     (1) P_2O_5 (2) N_2O_4 (3) C_3H_6 (4) $C_6H_{12}O_6$

3   What is the name of the calcium salt of sulfuric acid?
     (1) calcium thiosulfate   (3) calcium sulfide
     (2) calcium sulfate   (4) calcium sulfite

4   What is the formula of nitrogen (I) oxide?
     (1) NO   (2) N_2O (3) NO_2 (4) N_2O_4

5   The correct formula for the thiosulfate ion is:
     (1) $SO_3{}^{2-}$ (2) $SO_4{}^{2-}$ (3) SCN^- (4) $S_2O_3{}^{2-}$

6   Which symbol represents a molecule at room temperature?
     (1) I   (2) N   (3) Cl   (4) Ne

7   What is the correct formula of potassium hydride?
     (1) KH   (2) KH_2 (3) KOH   (4) $K(OH)_2$

8   Which is the correct formula for nitrogen (IV) oxide?
     (1) NO   (2) NO_2 (3) NO_3 (4) NO_4

9   Which is an empirical formula?
     (1) C_2H_2 (2) C_2H_4 (3) Al_2Cl_6 (4) K_2O
10   Sodium chloride will be produced by a reaction between sodium hydroxide
     and:
     (1) chlorous acid   (3) hydrochloric acid
     (2) chloric acid   (4) hypochlorous acid

VII. Chemical Equations

A chemical equation represents the type and the amount of changes in bonding and energy that take place in a chemical reaction. It is made up of reactants (those items that enter into the reaction and are usually found on the left side of the equation) and the products (those items produced and usually found on the right side of the equation). For example:

$$2H_2 + O_2 \longrightarrow 2H_2O + \text{heat}$$

1. Two molecules of hydrogen plus **one molecule** of oxygen will give us two molecules of water and also give off heat in the process.

2. Two moles of hydrogen plus **one mole** of oxygen will give us 2 moles of water molecules.

Equations must conform to the laws of conservation of mass and charge. Usually, energy is omitted from the equation, when we are not concerned with it.

In an equation, it is often desirable to indicate the phase of reactants and products. This may be done by using the symbols: (s) - solid; (*l*) - liquid; (g) - gas; and (aq) - in aqueous solution. The "written equation" could be written:

$$2H_2\,(g) + O_2\,(g) \longrightarrow 2H_2O\,(l) + \text{heat}$$

Balancing an Equation. The number of atoms of each element must be the same on both sides of the equation. This is done by placing numbers in front of the formulas (coefficients) in order to equalize the number of atoms.

Questions III-4

1   When the equation, $H_2 + Fe_3O_4 \rightarrow Fe + H_2O$, is completely
     balanced, using smallest whole numbers, the coefficient of H_2 would be:
     (1) 1   (2) 2   (3) 3   (4) 4
2   When the equation, $NH_3 + O_2 \rightarrow HNO_3 + H_2O$, is completely
     balanced, using smallest whole numbers, the coefficient of O_2 would be:
     (1) 1   (2) 2   (3) 3   (4) 4
3   Given the balanced equation, $2\,Na + 2\,H_2O \rightarrow 2\,X + H_2$, what is the
     correct formula for the product represented by the letter X?
     (1) NaO   (2) Na_2O (3) NaOH   (4) Na_2OH

4 When the equation, $\_C_2H_4 + \_O_2 \rightarrow \_CO_2 + \_H_2O$, is correctly balanced, using smallest whole-numbered coefficients, the sum of all the coefficients is:
   (1) 16   (2) 12   (3) 8   (4) 4

5 When the above (question # 4) equation is balanced, using smallest whole numbers, what is the coefficient of the O_2?
   (1) 1   (2) 2   (3) 3   (4) 4

6 When the equation: $\_Na(s) + \_H_2O(l) \rightarrow \_NaOH(aq) + \_H_2(g)$ is correctly balanced using smallest whole numbers, the coefficient of the water is
   (1) 1   (2) 2   (3) 3   (4) 4

Self-Help Questions III

Unit III -bonding

I. Nature of the chemical bond; including energy changes, stability, electronegativity and ionic or covalent character of the bond.

1 If the electronegativity difference between the elements in compound NaX is 3.1, the element represented by X is
   (1) F   (2) Cl   (3) Br   (4) I

2 Energy is released when the atoms of two elements bond to form a compound. Compared to the total potential energy of the atoms before bonding, the total potential energy of the atoms after bonding is
   (1) lower and the compound formed is stable
   (2) lower and the compound formed is unstable
   (3) higher and the compound formed is stable
   (4) higher and the compound formed is unstable

3 Which of the following pairs of atoms would form the most stable compound?
   (1) Li and F   (3) Li and Cl
   (2) Li and Br   (4) Li and I

4 Which compound contains a bond with the *least* ionic character?
   (1) CO   (2) CaO   (3) K_2O (4) Li_2O

5 Which atom has the greatest attraction for a pair of electrons in a bond between the atom and oxygen?
   (1) Al   (2) Si   (3) S   (4) P

6 Which of the following atoms has the most stable outermost principal energy level?
   (1) H·   (2) He :   (3) Li ·   (4) Be :

7 Which atom has the *least* attraction for the electrons in a bond between that atom and an atom of hydrogen?
   (1) carbon   (2) nitrogen   (3) oxygen   (4) fluorine

II. Bonds between atoms; Including
Ionic, Covalent (both polar and nonpolar), coordinate covalent bonds, molecular substances, network substances, and metallic bonding. Characteristics of ionic and molecular substances

8   An ionic bond forms between atoms of
   (1) I and Cl   (2) K and Cl   (3) P and Cl   (4) H and Cl

9   A characteristic of ionic solids is that they
   (1) have high melting points   (3) conduct electricity
   (2) have low boiling points   (4) are noncrystalline

10   Mobile electrons are a distinguishing characteristic of
    (1) an ionic bond   (3) a metallic bond
    (2) an electrovalent bond   (4) a covalent bond

11   Which type of bond exists between carbon atoms in diamond?
    (1) ionic   (2) covalent   (3) metallic   (4) hydrogen

12   Which substance contains a polar covalent bond?
    (1) Na_3N (2) Mg_3N_2 (3) NH_3 (4) N_2

13   Which molecule can form a coordinate covalent bond?

$$H \overset{\bullet\bullet}{\underset{\underset{\displaystyle H}{\bullet\bullet}}{\overset{}{\bullet}N\bullet}} H \qquad H \bullet H \qquad H \bullet \overset{\bullet\bullet}{\underset{\underset{\displaystyle H}{\bullet\bullet}}{C}} \bullet H \qquad H \bullet C \bullet\bullet\bullet C \bullet H$$

    (1)   (2)   (3)   (4)

14   Which electron dot formula represents a molecule that can combine with a proton (H^+)?

$$H \bullet H \qquad H \bullet C \bullet\bullet\bullet C \bullet H \qquad H \bullet \overset{\overset{\displaystyle H}{\bullet\bullet}}{\underset{\underset{\displaystyle H}{\bullet\bullet}}{C}} \bullet H \qquad H \bullet \overset{\bullet\bullet}{\underset{\underset{\displaystyle H}{\bullet\bullet}}{N}} \bullet H$$

    (1)   (2)   (3)   (4)

15   Which type of bond is contained in a water molecule?
    (1) nonpolar covalent   (3) ionic
    (2) polar covalent   (4) electrovalent

16   Which substance is classified as a network solid?
    (1) SiO_2 (2) Li_2O (3) H_2O (4) CO_2

17   When a calcium atom loses its valence electrons, the ion formed has an electron configuration which is the same as an atom of
    (1) Cl   (2) Ar   (3) K   (4) Sc

18   Which compound in the solid state has a high melting point and conducts electricity when it is liquefied?
    (1) carbon dioxide   (3) hydrogen chloride
    (2) silicon dioxide   (4) potassium chloride

19   Which element consists of positive ions immersed in a "sea" of mobile electrons?
    (1) sulfur   (2) nitrogen   (3) calcium   (4) chlorine

20   Which molecule is nonpolar and contains a nonpolar covalent bond?
    (1) CCl_4 (2) F_2 (3) HF   (4) HCl

III Molecular attraction; Including dipole-dipole, hydrogen bonding, van der Waals forces, and molecule ion attraction. Directional nature of covalent bonds.

21 Hydrogen bonds are strongest between molecules of
   (1) HF   (2) HCl   (3) HBr   (4) HI

22 Which compound contains both covalent and ionic bonds?
   (1) KCl   (2) NH_4Cl (3) $MgCl_2$ (4) CCl_4

23 Van der Waals forces of attraction between molecules decrease with
   (1) increasing number of electrons and increasing distance between the molecules
   (2) increasing number of electrons and decreasing distance between the molecules
   (3) decreasing number of electrons and increasing distance between the molecules
   (4) decreasing number of electrons and decreasing distance between the molecules

24 Which formula represents a nonpolar symmetrical molecule?

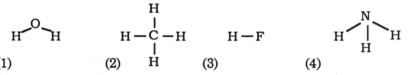

   (1)   (2)   (3)   (4)

25 The table below shows boiling points for the elements listed.

Elements	Normal Boiling Points(°C)
Fluorine	-188.1
Chlorine	-34.6
Bromine	+58.8
Iodine	+184.4

Which statement best explains the pattern of boiling points relative to molecular size?
   (1) Stronger van der Waals forces occur in larger molecules.
   (2) Weaker van der Waals forces occur in larger molecules.
   (3) Stronger hydrogen bonds occur in larger molecules.
   (4) Weaker hydrogen bonds occur in larger molecules.

26 Which kinds of bonds are found in a sample of $H_2O(s)$?
   (1) hydrogen bonds, only   (3) both ionic and hydrogen bonds
   (2) covalent bonds only   (4) both covalent and hydrogen bonds

27 The forces of attraction that exist between nonpolar molecules are called
   (1) van der Waals   (3) covalent
   (2) ionic   (4) electrovalent

28 Which substance is made up of molecules that are dipoles?
   (1) N_2 (2) H_2O (3) CH_4 (4) CO_2

29 In which of the following liquids are hydrogen bonds between molecules the strongest?
   (1) HI(l)   (2) HBr(l)   (3) HCl(l)   (4) HF(l)

30 When a salt dissolves in water, the water molecules are attracted by dissolved salt particles. This attraction is called
(1) atom-atom　　　　　　　　(3) ion-ion
(2) molecule-molecule　　　　　(4) molecule-ion

31 Which of the following liquids has the weakest van der Waals forces of attraction between its molecules?
(1) Xe(l)　　　　(2) Kr(l)　　　　(3) Ne(l)　　　　(4) He(l)

IV. Formula writing; including chemical, molecular and empirical formulas. Naming and writing compound formulas.

32 A binary compound of sodium is
(1) sodium chlorate　　　　　　(3) sodium perchlorate
(2) sodium chlorite　　　　　　(4) sodium chloride

33 Which quantity of helium may be represented by the symbol He?
(1) 1 gram　　　　　　　　　　(3) 6×10^{23} atoms
(2) 2 moles　　　　　　　　　　(4) 4 liters

34 Which formulas could represent the empirical formula and the molecular formula of a given compound?
(1) CH_2O, $C_4H_6O_4$　　　　　(3) CH_4, C_5H_{12}
(2) CHO, $C_6H_{12}O_6$　　　　　(4) CH_2, C_3H_6

35 Which is the formula for the compound that forms when magnesium bonds with phosphorus?
(1) Mg_2P　　　　(2) MgP_2　　　　(3) Mg_2P_3　　　　(4) Mg_3P_2

36 What is the name of the compound whose formula is N_2O_5?
(1) nitrogen (V) oxide　　　　　(3) nitrogen (III) oxide
(2) nitrogen (II) oxide　　　　　(4) nitrogen (IV) oxide

37 What is the total number of moles of atoms represented by one mole of $(CH_3)_2NH$?
(1) 5　　　　(2) 8　　　　(3) 9　　　　(4) 10

38 Which is the correct formula for iron (III) sulfate?
(1) Fe_3SO_4　　(2) Fe_2SO_4　　(3) $Fe(SO_4)_3$　　(4) $Fe_2(SO_4)_3$

39 Which is the correct formula for titanium (III) oxide?
(1) Ti_2O_3　　　(2) TiO　　　(3) Ti_3O_2　　　(4) Ti_2O_4

V. Chemical equations, writing and balancing an equation

40 When the equation: __Na(s) + __$H_2O(l)$ → __NaOH(aq) + __$H_2(g)$ is correctly balanced using smallest whole numbers, the coefficient of the water is
(1) 1　　　　(2) 2　　　　(3) 3　　　　(4) 4

41 When the equation __Al(s) + __$O_2(g)$ → __$Al_2O_3(s)$ is correctly balanced using the smallest whole numbers, the coefficient of Al(s) is
(1) 1　　　　(2) 2　　　　(3) 3　　　　(4) 4

4 Periodic Table

Important Terms and Concepts

Periodic Table
  Periods, Rows, and Series
  Groups and Families
Atomic Radius
  kernel
Metals and Nonmetals
Metalloids (Semimetals)
Periodic Table Blocks, Groups

S Block - Groups
  1 - Alkali Metals
  2 - Alkaline Earth Metals
P Block - Groups 13 – 18
  17 - Halogen Family
  18 - Monoatomic Gases
D Block - Groups 3 – 12
  Transition Elements

I. Development Of Periodic Table

The periodic table of the elements has passed through many stages of development, evolving into the present form. Observed regularities in properties of elements led Mendeleev and others to consider these regularities to be functions of the atomic mass. Moseley established that properties of elements are periodic functions of the atomic number. This is known as the **Periodic Law**.

The atomic number is the basis of the arrangement in the present form of the periodic table. The properties of the elements depend on the structure of the atom and vary with the atomic number in a systematic way.

II. Properties Of Elements

The horizontal rows of the periodic table are called **periods**, **rows**, or **series**. The properties of elements change systemically through a period. Period 2 examples include: Li, Be, B, C, N, O, F, and Ne. The vertical columns of the periodic table are called **groups or families**. The elements of a group exhibit similar or related properties. Group 17 examples include: F, Cl, Br, I, and At.

Most elements are **solids** at room temperature, except Mercury and Bromine which are **liquids**. **Gases** include Hydrogen, Oxygen, Nitrogen, Fluorine, Chlorine, and all of group 18.

A. Atomic Radius

The radius of a solitary atom cannot be measured easily, because the electron probability distribution will not allow a single distance measurement; however, measurements can be made using large numbers of the same atoms (in molecules and crystals). This results in the atomic radius being a property of the combined atoms.

The radius of an atom is the closest distance to which one atom can approach another. Since each atom in a molecule or crystal is affected by the presence of other atoms, the radius of an atom will vary under certain specified circumstances.

For example, the radius of adjacent atoms *within the same molecule* is defined as one-half the distance between the nuclei. However, in adjacent atoms *in different but adjoining molecules*, the radius will be a different distance. Also, different values will be measured if the types of movements that the atoms have attained are characterized as solid, liquid, or gas.

The atomic radius is very useful in determining density, solubility, melting point, and acid strength and is therefore considered a **periodic property**. Valid comparisons of properties can be made, if the atomic radius is specified as Covalent Radius, van der Waals Radius, or Atomic Radius in Metals. High school chemistry problems should be solved using the **covalent radius** method, unless otherwise specified.

1. **Covalent radius** is the effective distance from the center of the nucleus to the outer valence shell of that atom in a typical covalent or coordinate bond.
2. **Van der Waals radius** is half the internuclear distance or radius of the closest approach of an atom with another atom with which it forms no bond.
3. **Atomic radius in metals** is half the internuclear distance or radius in a crystalline metal.

The relationship between atomic radius and atomic number can be interpreted in terms of the arrangement of electrons in the orbitals of atoms and in terms of nuclear charge. Within a single period of the periodic table, the atomic radius generally decreases as the atomic number increases. Within any one period, the electrons in the outer orbitals are arranged around a **kernel** containing the same number of filled levels.

As one proceeds from left to right in the period, the increase in nuclear charge due to the increasing number of protons pulls the electrons more tightly around the nucleus. This increased attraction more than balances the repulsion between the added electron and other electrons, thus the atomic radius is reduced.

The members of any group in the periodic table generally show an increase in atomic radius with an increase in the atomic number.

For a group of elements, the atoms of each successive member have a larger kernel containing more filled levels. Therefore, the electrons in the unfilled orbitals are farther from the nucleus. This results in an increase in atomic radius as the atomic number increases among the elements in a group.

B. Ionic Radius
A loss or gain of electrons by an atom causes a corresponding change in size.

    1. Metal atoms lose one or more electrons when they form ions. Ionic radii of metals are smaller than the corresponding atomic radii.

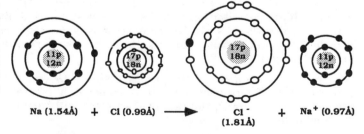

Na (1.54Å)   +   Cl (0.99Å) ⟶ Cl⁻ (1.81Å)   +   Na⁺ (0.97Å)

2. Nonmetal atoms gain one or more electrons when they form ions. Ionic radii of nonmetals are larger than the corresponding atomic radii. Atomic and ionic radii are usually measured in Angstrom (Å) units (1Å= 1 x 10^{-10} meter).

C. Metals

Metal atoms possess relatively low ionization energies and low electronegativity. More than two-thirds of the elements are metals; the rest are metalloids (semimetals), nonmetals, and inert gases.

Metal atoms tend to lose electrons and to form positive ions when combining with other elements. Metallic properties are most pronounced in those elements on the lower left side of the periodic table. Metals usually possess the properties of high thermal and electrical conductivities, metallic luster, malleability (drawn into sheets), and ductility (drawn into wire). The metal mercury is a liquid at room temperature, but the rest are all solids.

D. Nonmetals

Nonmetal atoms possess high ionization energies and have high electronegativities. Nonmetallic properties are most pronounced in those elements in the upper right corner of the periodic table (not including Group 18). Nonmetal atoms tend to gain electrons when in combination with metals or to share electrons when in combination with other elements. Nonmetals tend to be gases, molecular solids, or network solids. The exception is bromine which is a volatile (ability to change from liquid to gas) liquid at room temperature. Nonmetals in the solid phase tend to be brittle, to have low thermal and electrical conductivities, and to lack metallic luster. A good example is the element sulfur.

E. Metalloids (Semimetals)

*Note: IUPAC recommends that the term **metalloid** be abandoned because of language inconsistencies. Using the word **semimetal** may be better.*

Metalloids (semimetals) are those elements that have some properties characteristic of metals and other properties characteristic of nonmetals. Their electronegativity rating is usually about 2.0, and they are used in the manufacture of semiconductors. Examples of metalloids (semimetals) are boron, silicon, arsenic, tellurium, germanium, and antimony.

Questions IV-1

1 Elements that have properties of both metals and nonmetals are called:
   (1) alkali metals   (3) transition elements
   (2) metalloids (semimetals) (4) halogens
2 The group 2 (IIA) element having the largest atomic radius is found in Period:
   (1) 1   (2) 2   (3) 6   (4) 7
3 As the elements in Group 1 (IA) are considered in order of increasing atomic number, the atomic radius of each successive element increases. This is primarily due to an increase in the number of:
   (1) neutrons in the nucleus   (3) unpaired electrons
   (2) electrons in the outermost shell (4) principal energy levels
4 Which of the following ions has the smallest radius?
   (1) Na^+ (2) K^+ (3) Mg^{2+} (4) Ca^{2+}

5   Atoms of metallic elements tend to:
     (1) gain electrons and form negative ions
     (2) gain electrons and form positive ions
     (3) lose electrons and form negative ions
     (4) lose electrons and form positive ions

6   Given the same conditions, which of the following Group 17 elements has the least tendency to gain electrons?
     (1) fluorine   (2) iodine   (3) bromine   (4) chlorine

7   Which of the following Group 17 (VIIA) elements has the highest melting point?
     (1) fluorine   (2) chlorine   (3) bromine   (4) iodine

8   Which is the most active nonmetal in the Periodic Table of the Elements?
     (1) Na   (2) F   (3) I   (4) Cl

9   Which element is considered malleable?
     (1) gold   (2) hydrogen   (3) sulfur   (4) radon

10 Which of the following particles has the smallest radius?
     (1) Na^0 (2) K^0 (3) Na^+ (4) K^+

11 In a chemical reaction, which element forms an ion with a smaller radius than its atom?
     (1) Mg   (2) Br   (3) O   (4) Ne

12 Which are two properties of most nonmetals?
     (1) low ionization energy and good electrical conductivity
     (2) high ionization energy and poor electrical conductivity
     (3) high ionization energy and good electrical conductivity
     (4) low ionization energy and poor electrical conductivity

A Key to the Periodic Table

When referring to any reproduction of the Periodic Table of Elements, the first item the student should look at is the **key**. The key of the Periodic Table is illustrated below.

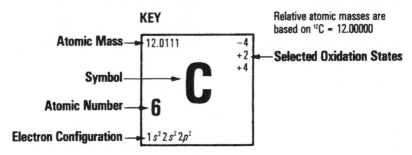

III. Chemistry of a Group (Family)

The elements in the **Periodic Table** are divided into **Blocks *s, p, d,*** and *f* and are further divided into **Groups 1** through **18**.

The chemical properties of the elements in each group are related. Similarities in chemical properties within a group are associated with the similarity in the number of valence electrons.

Related chemical properties are illustrated by the similarity in the type of compound formed by the members of a group. For example, the elements in **Group 1** (formerly known as Group IA) form chlorides having the general formula MCl, where M represents any member of the group. Elements in **Group 2** (formerly known as Group IIA) form chlorides having the general formula MCl_2. In general, the properties of elements in a group change progressively as the atomic number increases.

Properties of the members of the *s* and *p* **Blocks** (formerly known as the "A" group) depend on the following:

a. **As we go down a group**, a new, fully occupied shell is added, giving the atom a **larger atomic radius**.

b. These newly occupied shells allow the **"kernel"** (all the occupied shells except the valence shell) to enlarge. The positive nuclear charge which holds the valence electrons must penetrate a "screening or shielding effect" caused by the electrons in the inner shells. This, coupled with the increased distance of the valence electron(s) from the nucleus, causes the nuclear attraction to diminish. **Generally, we can state that the ionization energy of the valence electrons decreases as we proceed down a group**.

Exceptions in the properties of elements within a group do occur. For example, in **Group 13** (formerly known as Group IIIA), boron does not form an ion as do other members of the group. These anomalies occur most frequently among the elements in Period 2 because of the relative closeness of the valence electrons to the nucleus and the somewhat small screening or shielding effect of the two electrons in the 1s sublevel.

c. **As we go down a group, the electronegativity of the element generally decreases**.

d. **As we go down a group, the elements tend to have more metallic properties**.

Note: The relative tendency of the atoms to form compounds (sometimes called the reactivity of the element) cannot be predicted from the electronegativity of the atom, but should be found from the **Table of Standard Electrode Potentials** given in the Reference Tables for Chemistry.

A. *S*-Block: Groups 1 And 2

Properties of the elements in Groups 1 and 2 include:

a. Groups 1 and 2 include the most reactive metals.

b. Elements in **Group 1** are called the **alkali metals**. *Note: Hydrogen is not an alkali metal.*

s-block 1 IA	
1.00794 +1 −1 **H** 1 1s¹	

s-block **GROUP**

1 IA	2 IIA
6.941 +1 **Li** 3 1s²2s¹	9.01218 +2 **Be** 4 1s²2s²
22.98977 +1 **Na** 11 [Ne]3s¹	24.305 +2 **Mg** 12 [Ne]3s²
39.0983 +1 **K** 19 [Ar]4s¹	40.08 +2 **Ca** 20 [Ar]4s²
85.4678 +1 **Rb** 37 [Kr]5s¹	87.62 +2 **Sr** 38 [Kr]5s²
132.905 +1 **Cs** 55 [Xe]6s¹	137.33 +2 **Ba** 56 [Xe]6s²
(223) +1 **Fr** 87 [Rn]7s¹	226.025 +2 **Ra** 88 [Rn]7s²

c. Elements in **Group 2** are called the **alkaline earth metals**.

d. Because of their reactivity, Groups 1 and 2 elements occur in nature only in compounds.

e. The elements in both groups have relatively **low ionization energies and electronegativities**. Therefore, they lose electrons readily to form ionic compounds that are somewhat stable.

f. Generally, the reactivity within both groups usually increase with an increase in atomic number. Exceptions to this occur in Group 1. The anomaly of lithium is due to the small size of the lithium atom and ion.

g. In the same period, each Group 1 metal is more active than the corresponding Group 2 metal.

h. The elements in both groups are usually reduced to their free state by the electrolysis of their fused compounds.

B. *P* Block - Groups 13 Through 18

Group 13 (formerly known as Group IIIA). Properties of the elements in this group include:

a. All members have oxidation states of +3.

b. Boron is the only Group 13 element that has nonmetallic properties and is classified as a metalloid (semimetal).

c. Aluminum is the best known element of Group 13. It is a shiny easily workable, non—corroding metal.

d. Gallium, indium, and thallium are lustrous moderately reactive metals.

Group 14 (formerly known as Group IVA). Properties of the elements in this group include:

a. The Group 14 elements progress from carbon (nonmetallic) to silicon and germanium (metalloids) to tin and lead (metals). They have the oxidation states of +2, +4, and -4.

b. Carbon exists in **allotropic forms**, such as graphite, diamond, and buckyballs.

c. Silicon is a dull solid and the second most abundant element in the Earth's crust.

d. Germanium is used as a semiconductor in transistors.

e. Tin and lead are relatively inert metals.

Group 15 (formerly known as Group VA). Properties of the elements in this group include:

a. The elements in Group 15 show a marked progression from nonmetallic to metallic properties with increasing atomic number.

b. Nitrogen and phosphorus are typical nonmetals; arsenic is classified as a metalloid (semimetal); bismuth is metallic in both appearance and properties.

c. The element nitrogen is relatively inactive at room temperature.

d. In general, the reactivity of nonmetals in the same group decreases with increasing atomic number.

e. Nitrogen exists as a diatomic molecule with a triple bond between the two atoms. The high energy required to break a triple bond ($N \equiv N$) explains the relative inactivity of nitrogen.

f. Nitrogen compounds are essential constituents of all living matter.

g. Generally, nitrogen compounds are relatively unstable.

h. The element phosphorus is more reactive than nitrogen at room temperature because nitrogen has a strong triple bond.

i. Phosphorus does not exist as a diatomic molecule at room temperature but exists as a tetratomic molecule, P_4.

j. Phosphorus compounds are essential constituents of all living matter.

Group 16 (formerly known as Group VIA). Properties of the elements in this group include:

a. The elements in Group 16 show a marked progression from nonmetallic to metallic properties with an increase in atomic number.

b. Oxygen and sulfur are typical nonmetals; selenium and tellurium are classified as a metalloid (semimetal); polonium shows metallic properties.

c. The element oxygen is an active nonmetal.

d. Oxygen forms compounds with most elements. The existence of oxygen in its free state, in spite of its high reactivity, is explained by the continuous production of oxygen by plants during photosynthesis. Because of its high electronegativity, oxygen in compounds always shows a negative oxidation state unless combined with fluorine.

e. Sulfur is less reactive than oxygen.

f. Sulfur in compounds shows both $\overset{+1 \ -2}{H_2S}$ and $\overset{+1 \ +6 \ -2}{H_2SO_4}$ negative and positive oxidation states.

g. Examples of elements, in group 16, having **allotropic** forms are oxygen and sulfur (i.e. ozone O_3 and rhombic sulfur S_8).

h. Selenium and tellurium are rare elements.

i. Selenium and tellurium show negative oxidation states when combined with hydrogen. In most other compounds, they show positive oxidation states.

j. Polonium is a radioactive element that emits alpha particles. Polonium is a degradation product of uranium.

Group 17. Properties of elements in this group include:

a. The elements in Group 17 (formerly known as Group VIIA) are typical nonmetals.

b. Group 17 is known as the **halogen family**.

c. Although the metallic character increases with increasing atomic number, none of the elements in the group is a metal. (Astatine is radioactive with a short half-life. It has not been found in nature, and its properties are not well known.)

d. The elements in Group 17 have relatively **high electronegativities** and **high ionization energies** as shown in Table K.

e. **Fluorine** has the highest electronegativity of any element and in compounds can show only a **negative oxidation state**.

f. The other elements of the group may exhibit positive oxidation states in combination with more electronegative elements (for example, ClO_2^{-2}). The ease with which positive oxidation states of the halogens are formed increases with increasing atomic number.

g. The physical form of the free element at room temperature varies with increasing atomic number. At room temperature fluorine and chlorine are gases, bromine is a liquid, and iodine a solid. It is the only group that contains all three phases of matter.

h. The change in physical form as the atomic radius increases is due to an increase in the number of van der Waals forces.

i. The elements are usually prepared from the corresponding halide ion by removing one of the electrons from the ion.

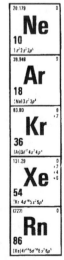

j. Because of their high reactivity, the halogens occur in nature only in compounds. Since fluorine is the most electronegative element, there is no chemical oxidizing agent that can oxidize the fluoride ion to fluorine. Fluorine is prepared by the electrolysis of its fused compounds. Chlorine, bromine, and iodine can be prepared by various chemical methods.

Group 18. Properties of the elements in this group include:

a. Group 18 (formerly known as Group O) elements are **monatomic gases**. The atoms of these elements have complete outer shells, which result in an electron configuration of s^2p^6, and is a stable configuration. Therefore, they possess the highest *first ionization energies* in their Periods.

b. Group 18 elements are referred to by a variety of terms including **rare gases**, "**inert**" **gases**, and **noble gases**.

c. The term "inert" is no longer strictly applicable to this group, since it is possible to form compounds of krypton, xenon, and radon with fluorine and oxygen.

d. However, the term is still in general use and the electron configuration is quite generally referred to as the "inert gas structure."

Questions IV-2

1 Alkali metals, alkaline earth metals, and halogens are elements found respectively in Groups:
   (1) 1, 2, and 18   (3) 1, 2, and 14
   (2) 2, 13, and 18   (4) 1, 2, and 17

2 Which element in Group 16 has no stable isotopes?
   (1) O   (2) S   (3) Po   (4) Te

3 Which element in Group 15 has the most metallic character?
   (1) nitrogen   (3) arsenic
   (2) bismuth   (4) phosphorus

4 In which group do all elements have the same number of electrons in the outermost principal energy level?
   (1) 5   (2) 6   (3) 13   (4) 18

5 In which group do the elements usually form oxides which have the general formula XO_2?
   (1) 1   (2) 2   (3) 13   (4) 14

6 Which of the following noble gases has the lowest normal boiling point?
   (1) Ne   (2) Ar   (3) Kr   (4) Xe

7 Which is the symbol of an alkaline earth element?
   (1) Na   (2) Ne   (3) Ca   (4) Ce

8 Which group contains the most active metals?
   (1) 1   (2) 11   (3) 17   (4) 7

9 Which molecule is relatively inactive and contains a triple bond?
   (1) N_2 (2) O_2 (3) Cl_2 (4) H_2

10 As the elements are considered from the top to the bottom of Group 15, which sequence in properties occurs?
   (1) metal → metalloid → nonmetal
   (2) metal → nonmetal → metalloid
   (3) metalloid → metal → nonmetal
   (4) nonmetal → metalloid → metal

11 If X represents an element of Group 1, the formula of its oxide would be
   (1) XO   (2) X_2O (3) XO_2 (4) X_2O_3

12 Which of the following elements is most likely to form a compound with radon?
   (1) iodine   (2) fluorine   (3) sodium   (4) calcium

13 In which group do the elements usually form chlorides which have the general formula MCl_2?
   (1) 1   (2) 2   (3) 17   (4) 18

14 Ozone is an allotropic form of the element:
   (1) oxygen   (2) phosphorus (3) sulfur   (4) carbon

15 What is the total number of electrons found in the valence shell of a
    halogen in the ground state?
    (1) 1   (2) 2   (3) 7   (4) 8
16 The element found in Group 13 and in Period 2 is:
    (1) Be   (2) Mg   (3) B   (4) Al
17 An element that is liquid at STP is in Group
    (1) 1   (2) 2   (3) 11   (4) 12
18 Beryllium is classified as:
    (1) an alkaline earth metal   (3) a transition element
    (2) an alkali metal   (4) a noble gas
19 The reactivity of the metals in Groups 1 and 2 generally increases with:
    (1) increased ionization energy   (3) decreased nuclear charge
    (2) increased atomic radius   (4) decreased mass
20 Which group contains the atom with the highest first ionization energy?
    (1) 1   (2) 2   (3) 17   (4) 18
21 The pair of elements with the most similar chemical properties are:
    (1) Mg and S   (3) Mg and Ca
    (2) Ca and Br   (4) S and Ar
22 Which element is a member of the halogen family?
    (1) K   (2) B   (3) I   (4) S
23 Which group contains elements with a total of four electrons in the
    outermost principal energy level?
    (1) 1   (2) 18   (3) 16   (4) 14
24 Which element exhibits a crystalline structure at STP?
    (1) fluorine   (2) chlorine   (3) bromine   (4) iodine
25 Which is an alkaline earth metal?
    (1) Mg   (2) Zn   (3) Li   (4) Pb
26 The S^{2-} ion differs from the S^o atom in that the S^{2-} ion has a:
    (1) smaller radius, fewer electrons
    (2) smaller radius, more electrons
    (3) larger radius, fewer electrons
    (4) larger radius, more electrons
27 A reason why fluorine has a higher ionization energy than oxygen is that
    fluorine has a:
    (1) smaller nuclear charge   (3) smaller number of electrons
    (2) larger nuclear charge   (4) larger number of neutrons

C. Transition Elements

A transition element is an element whose atom has an incomplete *d* **sub-shell** or which gives rise to a cation or cations with an incomplete *d* subshell. Properties of the elements in this group include:

a. Transition elements are those elements in which electrons from the **two outermost sublevels** may be involved in a chemical reaction. Because of this, these elements generally exhibit **multiple positive oxidation states**.

b. The transition elements are found in *d* - **Block**, **Groups 3** through **11** (formerly known as the "B" groups and Group VIII) of the periodic table.

c. The ions of transition elements usually appear colored, both in solid compounds and in solution.

d-block

Transition Elements

GROUP									
3 IIIB	4 IVB	5 VB	6 VIB	7 VIIB	8	9 (VIII)	10	11 IB	12 IIB
Sc 21	Ti 22	V 23	Cr 24	Mn 25	Fe 26	Co 27	Ni 28	Cu 29	Zn 30
Y 39	Zr 40	Nb 41	Mo 42	Tc 43	Ru 44	Rh 45	Pd 46	Ag 47	Cd 48
La-Lu 57 71	Hf 72	Ta 73	W 74	Re 75	Os 76	Ir 77	Pt 78	Au 79	Hg 80
Ac-Lr 89 103	Unq* 104	Unp 105	Unh 106	Uns 107	Uno 108	Une 109			

IV. Chemistry Of A Period

A study of the **Blocks *s*** and ***p*** elements in a period, from left to right, leads to certain generalizations.

In each period, **as the atomic number increases**:

a. The radius of the atom generally decreases.
b. The ionization energy of the element generally increases.
c. The electronegativity of the element generally increases.
d. The elements generally change from very active metals, to less active nonmetals, and, finally, to an "inert" monoatomic gas molecules.
e. There is a transition from positive to negative oxidation states. Elements near the center of the period may exhibit both positive and negative oxidation states.
f. The metallic characteristics of the elements decrease.

The "rare earth" elements are found in the Lanthanoid Series.

V. Naming Elements
With Atomic Numbers Greater Than 100

The **name** is derived directly from the atomic number of the element using the following numerical roots:

0 = nil	2 = bi	4 = quad	6 = hex	8 = oct
1 = un	3 = tri	5 = pent	7 = sept	9 = enn

The roots are put together in order of the digits which make up the atomic number and are terminated by **"ium"** to spell out the complete name. The final "n" of "enn" is dropped when it occurs before "nil." Also, the final "i" of "bi" and of "tri" are dropped when they occur before "ium".

The **symbol** of the element is composed of the initial letters of the numerical roots which make up the name. The root "un" is **pronounced** with a long "u" to rhyme with "moon." In the element names, each root is pronounced separately. For example:

104*=	Un nil quad ium (Unq)		107 =	Un nil sept ium (Uns)
105 =	Un nil pent ium (Unp)		108 =	Un nil oct ium (Uno)
106 =	Un nil hex ium (Unh)		109 =	Un nil enn ium (Une)

As of August 30, 1997, IUPAC (International Union of Pure and Applied Chemistry) adopted nomenclature for the short-lived Transfermium elements 101 through 109:

Element	Name	Symbol	Element	Name	Symbol
101	Mendelevium	Md	106	Seaborgium	Sg
102	Nobelium	No	107	Bohrium	Bh
103	Lawrencium	Lr	108	Hassium	Hs
104	Rutherfordium	Rf	109	Meitnerium	Mt
105	Dubnium	Db			

Questions IV-3

1 Element M has an electronegativity of less than 1.2 and reacts with bromine to form the compound MBr_2. Element M could be:
   (1) Al   (2) Na   (3) Ca   (4) K

2 All elements in Period 3 are similar in that they all have the same number of
   (1) occupied principal energy levels   (3) $3p$ electrons
   (2) protons in the nucleus   (4) valence electrons

3 Which element in Period 2 is the most active metal?
   (1) neon   (2) beryllium   (3) fluorine   (4) lithium

4 Which element is a halogen?
   (1) iron   (2) nitrogen   (3) iodine   (4) neon

5 The elements in Period 2 are similar to each other in that each element has the same:
   (1) number of valence electrons
   (2) number of principal energy levels
   (3) atomic radius
   (4) atomic mass

6 Which is the atomic number of an alkali metal?
   (1) 10   (2) 11   (3) 12   (4) 13

7 In Period 2, the number of elements that are gases at STP is:
   (1) 1   (2) 2   (3) 3   (4) 4

8 In Period 2, as the elements are considered from left to right, there is a decrease in:
   (1) ionization energy   (3) metallic character
   (2) atomic mass   (4) nonmetallic character

9 All of the elements in Period 3 have a total of 2 electrons in the:
   (1) $2s$ sublevel   (2) $3s$ sublevel (3) $2p$ sublevel (4) $3p$ sublevel

10 At STP, which element is a solid?
   (1) hydrogen   (2) carbon   (3) nitrogen   (4) argon

11 The element in Period 3 with the most metallic character is:
   (1) sodium   (2) aluminum   (3) silicon   (4) phosphorus

12 Which of the following atoms will lose an electron most readily?
   (1) potassium   (2) calcium   (3) rubidium   (4) strontium

13 Which element exists as monatomic molecules at STP?
    (1) hydrogen   (3) argon
    (2) nitrogen   (4) chlorine
14 Which period contains three elements that commonly exist as diatomic molecules?
    (1) Period 1   (2) Period 2   (3) Period 3   (4) Period 4
15 A chloride dissolves in water to form a colored solution. The chloride could be:
    (1) HCl   (2) KCl   (3) $CaCl_2$ (4) $CuCl_2$
16 Which element forms a colored ion in solution?
    (1) Ni   (2) Li   (3) K   (4) Mg
17 The water solution of a compound is green. The compound could be:
    (1) Na_2SO_4 (2) K_2SO_4 (3) $FeSO_4$ (4) $MgSO_4$
18 An aqueous solution of XCl_2 contains colored ions. Element X is likely:
    (1) an alkaline earth   (3) a transition metal
    (2) a halogen   (4) a alkali metal
19 The elements of Period 2 have the same
    (1) atomic mass
    (2) atomic number
    (3) number of occupied principal energy levels
    (4) number of occupied sublevels
20 From which sublevel or sublevels can an atom of Fe lose electrons when forming the Fe^{3+} ion?
    (1) the 4*d, only*   (3) *both* the 3*d* and 4*s*
    (2) the 3*p, only*   (4) *both* the 3*s* and 4*d*
21 Compared to the covalent atomic radius of a sodium atom, the covalent atomic radius of a magnesium atom is smaller. The smaller radius is primarily a result of the magnesium atom having
    (1) a larger nuclear charge   (3) more principal energy levels
    (2) a smaller nuclear charge   (4) fewer principal energy levels

Self-Help Questions IV

Unit IV - Periodic Table

I. Properties of elements of the periodic table; including atomic radius, ionic radius in both metal and nonmetal atoms, metals, nonmetals, and metaloids.

1   Which element in Period 3 has the *least* tendency to lose an electron?
    (1) argon   (2) sodium   (3) phosphorus   (4) aluminum

2   At STP, which of the following substances is the best conductor of electricity?
    (1) hydrogen   (2) mercury   (3) oxygen   (4) helium

3   Which of the following periods contains the greatest number of metals?
    (1) 1   (2) 2   (3) 3   (4) 4

4   Which element will form an ion whose ionic radius is larger than its atomic radius?
    (1) K   (2) F   (3) Li   (4) Mg

5 Compared to the covalent atomic radius of a sodium atom, the covalent atomic radius of a magnesium atom is smaller. The smaller radius is primarily a result of the magnesium atom having
(1) a larger nuclear charge (3) more principal energy levels
(2) a smaller nuclear charge (4) fewer principal energy levels

6 Proceeding from left to right in Period 2 of the Periodic Table, the covalent radius of the elements generally
(1) decreases
(2) increases
(3) remains the same

7 Which element in Period 3 has the highest first ionization energy?
(1) Na (2) Ar (3) Cl (4) Mg

8 The arrangement of the elements in the present Periodic Table is based on atomic
(1) mass (2) number (3) radius (4) density

9 Which is the electron configuration of a metalloid in the ground state?
(1) $1s^2$ (2) $1s^22s^1$ (3) $1s^22s^22p^1$ (4) $1s^22s^12p^2$

10 Which element will form an ion whose radius is larger than its atomic radius?
(1) F (2) Fr (3) Ca (4) Cs

11 The elements that have the most pronounced nonmetallic properties are located toward which corner of the Periodic Table?
(1) upper right (2) upper left (3) lower right (4) lower left

12 Which element is a liquid at room temperature?
(1) K (2) I_2 (3) Hg (4) Mg

13 When a sodium atom becomes an ion, the size of the atom
(1) decreases by gaining an electron
(2) decreases by losing an electron
(3) increases by gaining an electron
(4) increases by losing an electron

II Chemistry of a group

14 On the Periodic Table of the Elements, all the elements within Group16 (VIA) have the same number of
(1) valence electrons (3) protons
(2) energy levels (4) neutrons

15 Which compound contains an alkali metal and a halogen?
(1) $CaCl_2$ (2) CaS (3) RbCl (4) Rb_2S

16 All atoms of Group 2 (IIA) elements in the ground state have the same number of electrons in which principal energy level?
(1) 1 (2) 2 (3) 3 (4) 4

17 Which represents the correct electron configuration of the outermost principal energy level of a Group 18 (O) element in the ground state?
(1) s^2p^2 (2) s^2p^4 (3) s^2p^6 (4) s^2p^8

18 The electron dot symbol X**:** represents
    (1) an alkali metal   (3) a halogen
    (2) an alkaline earth metal   (4) a transition element

19 Based on the Periodic Table of Elements, which Group 2 (IIA) element is most active?
    (1) Sr   (2) Mg   (3) Ca   (4) Ba

20 An element that has a high ionization energy and tends to be chemically inactive would most likely be
    (1) an alkali metal   (3) a noble gas
    (2) a transition element   (4) a halogen

21 An atom of the element in Period 2 Group 14 (IVA) is in the ground state. What total number of valence electrons does the atom have?
    (1) 1   (2) 2   (3) 3   (4) 4

22 Potassium forms an ion with a charge of
    (1) 1 + by losing one electron   (3) 1 + by gaining one electron
    (2) 1 - by losing one electron   (4) 1 - by gaining one electron

23 When oxygen combines with any alkali metal, M, the formula of the compound produced usually is
    (1) M_2O_3 (2) MO_2 (3) M_2O (4) M_3O_2

24 Which group contains elements in the solid, liquid, and gas phases at 25°C and 1 atmosphere?
    (1) 16 (VIA)   (2) 2 (IIA)   (3) 17 (VIIA)   (4) 18 (O)

25 A characteristic of the halogens is that they have relatively
    (1) low ionization energies   (3) high oxidation potentials
    (2) low reduction potentials   (4) high electronegativities

26 In the ground state, all atoms of Group 13 (IIIA) of the Periodic Table have the same number of
    (1) nuclear particles
    (2) occupied principal energy levels
    (3) electrons
    (4) valence electrons

27 Which Group 15 (VA) element has the most metallic character?
    (1) N   (2) P   (3) As   (4) Bi

28 The most active metal in Group 2 (IIA) is
    (1) Mg   (2) Sr   (3) Ba   (4) Ca

29 The atoms of which element require the greatest amount of energy to remove an electron?
    (1) helium   (2) neon   (3) argon   (4) krypton

30 Which halogen is a liquid at STP?
    (1) F_2 (2) Cl_2 (3) Br_2 (4) I_2

31 Which two elements have the most similar chemical properties?
    (1) aluminum and barium   (3) chlorine and sulfur
    (2) nickel and phosphorus   (4) sodium and potassium

32 Which element at STP exists as monatomic molecules?
    (1) N   (2) O   (3) Cl   (4) Ne

33 Element X forms the compounds XCl_3 and X_2O_3. In the Periodic Table, element X would most likely be found in Group
(1) 1 (IA)   (2) 2 (IIA)   (3) 13 (IIIA)   (4) 14 (IVA)

34 As the elements in Group 2 (IIA) of the Periodic Table are considered from top to bottom, the chemical reactivity of each succeeding element generally
(1) decreases   (2) increases   (3) remains the same

35 The highest ionization energies in any period are found in Group
(1) 1 (IA)   (2) 2 (IIA)   (3) 17 (VIIA)   (4) 18 (O)

36 The alkaline earth metals are found in Group
(1) 1 (IA)   (2) 2 (IIA)   (3) 11 (IB)   (4) 12 (IIB)

37 Which sequence of atomic numbers represents elements which have similar chemical properties?
(1) 19, 23, 30, 36   (3) 3, 12, 21, 40
(2) 9, 16, 33, 50   (4) 4, 20, 38, 88

38 Which of the following Group 17 (VIIA) elements has the highest melting point?
(1) fluorine   (2) chlorine   (3) bromine   (4) iodine

III Transition elements, chemistry of a period, and naming elements with atomic numbers greater than 100

39 The elements of Period 2 have the same
(1) atomic mass
(2) atomic number
(3) number of occupied principal energy levels
(4) number of occupied sublevels

40 Which element exists as a monatomic gas molecule at STP?
(1) nitrogen   (2) barium   (3) bromine   (4) neon

41 From which sublevel or sublevels can an atom of Fe lose electrons when forming the Fe^{3+} ion?
(1) the $4d$, only   (3) both the $3d$ and $4s$
(2) the $3p$, only   (4) both the $3s$ and $4d$

42 An element whose atoms have the electron configuration 2-8-18-1 is
(1) a transition element   (3) an alkali metal
(2) a noble gas   (4) an alkaline earth

43 Which element forms more than one binary compound with chlorine?
(1) Ca   (2) Fe   (3) Li   (4) Zn

44 As the elements in Period 4 of the Periodic Table are considered from left to right, the number of electrons in the $3p$ sublevel of each atom
(1) decreases   (2) increases   (3) remains the same

45 In Period 4 of the Periodic Table, the atom with the largest covalent radius is located in Group
(1) 1 (IA)   (2) 13 (IIIA)   (3) 3 (IIIB)   (4) 18 (O)

Important Terms and Concepts

Gram Atomic Mass (Gram–atom)
Gram Molecular Mass
Molar Volume
Stoichiometry
Empirical and Molecular Formula
Solutions – Solvent and Solute
Concentrations – Molarity
Miscible and Immiscible
Dilute and Concentrated

Saturated to Supersaturated
Percent Composition
Boiling Point Elevation
Freezing Point Depression
Electrolyte Abnormal Behavior
Calorimetry
Heat of Vaporization and Fusion
Combined Gas Laws
Graham's Law

I. Mole Interpretation

A mole contains Avogrado's number (6.02×10^{23}) of particles and may be used in calculations involving the number of particles (atoms, molecules, ions, electrons, or other particles) involved in chemical reactions, the mass of elements or compounds, or the volume relationships in gases.

A. Gram Atomic Mass (Gram-atom)

The gram atomic mass (gram-atom) of an element represents the mass in grams of Avogadro's number (6.02×10^{23}) of atoms of the element. The gram-atomic mass is numerically equal to the atomic mass as shown in the periodic table. (For example, the atomic mass of the element sulfur is 32.06 amu. A mole of this element weighs 32.06 grams.)

B. Gram Molecular Mass

The gram molecular mass (mass of 1 mole) is the sum of the gram atomic masses of the atoms that make up a particular molecule. The gram formula mass is the sum of the gram atomic masses of the atoms that make up a particular formula. The gram formula mass calculated from the empirical formula is used for the ionic substance and network solids, since they are not molecular substances.

Sample Problem: What is the gram molecular mass of water (H_2O)?

Solution:

The gram atomic mass of hydrogen is   1g mole x 2   =   2g

The gram atomic mass of oxygen is   16g mole x 1   =   16g

The gram molecular weight of water is   **18g per mole**

Question: What is the ratio by mass of hydrogen to oxygen in water?

Answer: _____

C. Molar Volume Of A Gas

A mole (6.02×10^{23}) of molecules of any gas occupies 22.4 liters at STP. It has a mass equal to the molecular mass expressed in grams.

Question: What does 22.4 liters of hydrogen gas weigh? (a mole of hydrogen gas molecules occupies a volume of 22.4 liters)

Answer: _____

Question: What is the volume of 10 grams of carbon monoxide? (a mole of carbon monoxide molecules occupies 22.4 liters)

Answer: _____

Question: What is the volume of 2.0 moles of carbon dioxide? (a mole of carbon dioxide gas occupies a volume of 22.4 liters)

Answer: _____

Questions V-1

1   What is the mass in grams of 1.0 mole of $(NH_4)_2S$?
     (1) 50   (2) 54   (3) 64   (4) 68

2   Which sample of O_2 contains a total of 3.01×10^{23} molecules at STP?
     (1) 1.00 mole   (3) 16.0 grams
     (2) 2.00 moles   (4) 32.0 grams

3   What is the gram atomic mass of the element chlorine?
     (1) 17 g   (3) 52 g
     (2) 35.5 g   (4) 70. g

4   The total number of molecules in 34.0 grams of NH_3 is equal to
     (1) 1.00×22.4 L   (3) $1.00 \times 6.02 \times 10^{23}$
     (2) 2.00×22.4 L   (4) $2.00 \times 6.02 \times 10^{23}$

5   A sample of helium gas at STP contains 3.01×10^{23} molecules. The amount of the sample is:
     (1) 0.500 gram   (3) 0.500 milliliter
     (2) 0.500 liter   (4) 0.500 mole

6   What is the volume of 4.00 moles of N_2 gas at STP?
     (1) 11.2 liter   (3) 44.8 liter
     (2) 22.4 liter   (4) 89.6 liter

7   The gram molecular mass of CO_2 is the same as the gram molecular mass of:
     (1) CO   (2) SO_2 (3) C_2H_6 (4) C_3H_8

8   Which sample of neon contains a total of 3.01×10^{23} molecules at STP?
     (1) 11.2 liters   (3) 20.0 grams
     (2) 22.4 liters   (4) 40.0 grams

9   Which quantity is equivalent to 39 grams of LiF?
     (1) 1.0 mole   (3) 0.50 mole
     (2) 2.0 moles   (4) 1.5 moles

10   What is the gram molecular mass of fluorine?
     (1) 38   (2) 9   (3) 18   (4) 14

II. Stoichiometry

Stoichiometry is the study of the quantitative relationships implied by chemical formulas and by chemical equations. In stoichiometry it is frequently convenient to use the mole interpretation and mole relationships in the solving of problems. In short, stoichiometry is the study of the molar proportions of:

a. atoms in a chemical formula
b. particles (molecules, ions, or compounds) in a chemical equation

A. Problems Involving Formulas

1. Percent composition. The percent composition by mass of an element in a compound can be calculated by dividing the total mass of that element in the formula by the total formula mass of the compound.

Sample Problems:
1) Calculate the percent composition by mass of water (H_2O). First, calculate the formula mass by the following method.

Each hydrogen atom has a mass of ... **1 amu** = **1 amu x 2** = **2 amu**

An oxygen atom has a mass of ... **16 amu** = **16 amu x 1** = **16 amu**

Total formula mass ... **= 18 amu**

Second, divide the total mass of each element by the formula mass, and multiply the resulting decimal by 100%.

$$hydrogen = \frac{2\ amu}{18\ amu} = 0.11 \times 100\% = 11\% \text{ (answer)}$$

$$oxygen = \frac{16\ amu}{18\ amu} = 0.89 \times 100\% = 89\% \text{ (answer)}$$

2) Calculate the percent, by mass, of water of hydration in gypsum ($CaSO_4 \cdot 2H_2O$).

First, calculate the formula mass:

Ca	=	40 x	1	=	40. amu
S	=	32 x	1	=	32. amu
O	=	16 x	6	=	96. amu
H	=	1 x	4	=	04. amu
					172. amu

Second, calculate the mass of water in the compound As stated above, each water molecule has a mass of 18 amu. Since there are two (2) molecules in the formula, the mass of the water molecules is 36 amu. Divide the mass of the water molecules by the formula mass of the compound and multiple the decimal by 100%.

$$\frac{36\ amu}{172\ amu} = 0.21 \times 100\% = 21\% \text{ (answer)}$$

2. Empirical formula. An empirical formula represents the simplest ratio in which the atoms combine to form a compound. For example, the empirical formula of H_2O_2 is HO.

The empirical formula of a compound that has a ratio composed of the number of atoms in 1.0 gram atomic mass (also called a gram-atom) of carbon atoms, to the number of atoms found in 2.0 gram atomic masses (2 gram-atoms) of hydrogen atoms, is CH_2.

If the empirical formula of a compound is CH, the **molecular formula** (the total number of atoms of each element in a compound) could be any simple multiple of CH such as C_2H_2 and C_3H_3.

Given the empirical formula and the molecular mass it is possible to derive the molecular formula:

1) Determine the formula mass of the empirical formula.
2) Divide this formula mass into the molecular mass.
3) Multiply this result by the subscript of each atom in the compound.

Questions V-2

1   Which is an empirical formula?

    (1) CH   (2) C_2H_2 (3) C_2H_4 (4) C_4H_8

2   A compound is found to contain 2 grams of hydrogen atoms to every 16 grams of oxygen atoms. The empirical formula of the compound is:

    (1) HO   (2) H_2O (3) H_2O_2 (4) HO_2

3   The percent by mass of nitrogen in $Mg(CN)_2$ is equal to:

    (1) $14/76$ x 100   (2) $14/50$ x 100   (3) $28/76$ x 100   (4) $28/50$ x 100

4   A compound whose empirical formula is CH_2O_2 could be:

    (1) HCOOH   (2) CH_3OH (3) CH_3COOH (4) CH_3CH_2OH

5   What is the percent by mass of oxygen in NaOH (formula mass = 40.)?

    (1) 80 %   (2) 40 %   (3) 32 %   (4) 16 %

6   A compound has an empirical formula of CH_2 and a molecular mass of 56. Its molecular formula is :

    (1) C_2H_4 (2) C_3H_6 (3) C_4H_8 (4) C_5H_{10}

7   The percent by mass of O_2 in $Ca(OH)_2$ (formula mass = 74) is closest to:

    (1) 16   (2) 22   (3) 43   (4) 74

8   The empirical formula of a compound is CH. Its molecular mass could be:

    (1) 21   (2) 40   (3) 51   (4) 78

9 .   The empirical formula of a compound is CH_2 and its molecular mass is 70. What is the molecular formula of the compound?

    (1) C_2H_2 (2) C_2H_4 .   (3) C_4H_{10} (4) C_5H_{10}

10   What is the percent by mass of oxygen in Fe_2O_3 (formula mass = 160)?

    (1) 16 %   (2) 30 %   (3) 56 %   (4) 70 %

11   What is the molecular formula of a compound whose empirical formula is CH_4 and molecular mass is 16?

    (1) CH_4 (2) C_2H_4 (3) C_4H_8 (4) C_8H_{18}

12   What is the percent by mass of hydrogen in NH_3 (formula mass = 17.0)?

    (1) 5.9%   (2) 17.6%   (3) 21.4%   (4) 82.4%

B. Problems Involving Equations

The following three step procedure should be applied to equation problems.

Step One. *Find the number of moles of the substance given.*
If grams are given, divide the grams by the molecular mass of the substance. If liters are given, divide the number of liters by 22.4 liters/mole. If molecules are given, divide the number by 6.02×10^{23} molecules/mole. If moles are given, go directly to **Step Two**.

Step Two. *Find the number of moles of the substance asked for.*
Put the number of moles calculated in **Step One** above the substance calculated in the equation. Put an "X" above the substance asked for. Set up a proportion or use the mole-ratio method. Solve for X.

Step Three. *Final calculation.*
If asked for grams, multiply answer in **Step Two** by the molecular mass of X. If asked for liters, multiply answer in **Step Two** by 22.4 liters/mole. If asked for molecules, multiply answer in **Step Two** by 6.02×10^{23} molecules/mole. If asked for moles, skip **Step Three**.

Sample Problem: *Mass – Mass*

A balanced equation shows the mole proportions of reactants and products:

$$Mg(s) + 2\,HCl(aq) \rightarrow MgCl_2(aq) + H_2(g) \uparrow$$

How many grams of $MgCl_2$ are produced when 48 grams of Mg(s) are used?

Step One. Find the number of moles given by dividing the mass of one mole into the amount given: mole of Mg = 24 g/mole

$$\frac{48\ g\ of\ Mg(s)}{24\ g/mole\ of\ Mg(s)} = 2\ moles\ Mg(s)$$

Step Two.

$$\overset{\text{6 moles}}{Mg(s)} + 2\,HCl(aq) \rightarrow \overset{\text{X moles}}{MgCl_2(aq)} + H_2(g) \uparrow$$
1 mole   1 mole

Proportion Method: $\dfrac{2\ moles}{1\ moles} = \dfrac{X\ moles}{1\ moles}$ = 2 moles $MgCl_2$ (answer)

Step Three.

$$2\ moles\ MgCl_2 \times 96\ g/mole = 192\ g\ \text{(answer)}$$

Sample Problem: *Mass – Volume*

How many liters of carbon dioxide at STP are produced by the combustion of 684 grams of octane (C_8H_{18}) according to the equation:

$$2C_8H_{18} + 25O_2 \rightarrow 16CO_2 + 18H_2O$$

Step One. (Division)

$$mole\ of\ C_8H_{18} = (8 \times 12) + (18 \times 1) = 114\ g/mole\ C_8H_{18}$$

$$\frac{684\ g\ C_8H_{18}}{114\ g/mole\ C_8H_{18}} = 6\ moles\ C_8H_{18}$$

Step Two.

$$\underset{2 \text{ moles}}{\overset{6 \text{ moles}}{2C_8H_{18}}} + 25O_2 \longrightarrow \underset{16 \text{ moles}}{\overset{X \text{ moles}}{16CO_2}} + 18H_2O$$

Proportion Method: $\dfrac{6 \text{ moles}}{2 \text{ moles}} = \dfrac{X \text{ moles}}{16 \text{ moles}}$ = **48 moles** (answer)

Mole-Ratio Method:
(Alternate Method) $\overset{3}{\cancel{6C_8H_{18}}}$ x $\dfrac{16CO_2}{\cancel{2C_8H_{18}}}$ = **48 moles CO_2** (answer)

Step Three. (Multiplication)

48 moles CO_2 x **22.4 liters/mole = 1075.2 liters**

Sample Problem: *Volume – Volume*

Since one mole of any gas occupies the same volume as one mole of any other gas at the same temperature, the volume of the gases involved in a reaction is proportional to the number of moles indicated by the numerical coefficients in a balanced equation. When reacting gases are at the same temperature and pressure, the volume ratio is equal to the mole ratio.

In the reaction: $N_2 + 3H_2 \rightarrow 2NH_3$ at STP, calculate the volume of hydrogen gas required to form 100 liters of NH_3 at STP.

Solution:
Use same **Three Step** process. $\underset{X}{\overset{3 \text{ moles}}{N_2 + 3H_2}} \longrightarrow \underset{100 \text{ liters}}{\overset{2 \text{ moles}}{2NH_3}}$

X = 150 liters (answer)

Questions V-3

1 Given the reaction: $N_2(g) + O_2(g) \rightarrow 2NO(g)$, what is the total number of liters, of NO(g) produced when 2.0 liters of N_2 reacts completely with oxygen?
   (1) 1.0 L   (2) 2.0 L   (3) 0.50 L   (4) 4.0 L

2 Given the reaction: $N_2 + 3H_2 \rightarrow 2NH_3$, what is the total number of grams of H_2 that reacts when 14 grams of N_2 are completely consumed?
   (1) 6.0 g   (2) 2.0 g   (3) 3.0 g   (4) 4.0 g

3 Given the reaction: $Cu + 4HNO_3 \rightarrow Cu(NO_3)_2 + 2H_2O + 2NO_2$, what is the total mass of H_2O produced when 32 g of Cu is completely consumed?
   (1) 9.0 g   (2) 18 g   (3) 36 g   (4) 72 g

4 Given the equation: $2C_2H_6 + 7O_2 \rightarrow 4CO_2 + 6H_2O$, when 30. grams of C_2H_6 (molecular mass = 30) are completely burned, the total number of moles of CO_2 produced is:
   (1) 1.0   (2) 2.0   (3) 8.0   (4) 4.0

5 Given the reaction: $2Na + 2H_2O \rightarrow 2NaOH + H_2$, what is the total number of moles of hydrogen produced when 4 moles of sodium react completely'
   (1) 1   (2) 2   (3) 3   (4) 4

6 Given the reaction: $2C_2H_6 + 7O_2 \rightarrow 4CO_2 + 6H_2O$, what is the total number of CO_2 molecules produced when one mole of C_2H_6 is consumed?
(1) $1 \times 6.02 \times 10^{23}$ (3) $3 \times 6.02 \times 10^{23}$
(2) $2 \times 6.02 \times 10^{23}$ (4) $4 \times 6.02 \times 10^{23}$

7 Given the reaction: $4NH_3(g) + 5O_2(g) \rightarrow 4NO(g) + 6H_2O(g)$ at constant pressure, how many liters of $O_2(g)$ would be required to produce 40. liters of $NO(g)$?
(1) 5.0   (2) 9.0   (3) 32   (4) 50

8 In the reaction $Zn + 2HCl \rightarrow ZnCl_2 + H_2$, how many moles of hydrogen will be formed when 4 moles of HCl are consumed?
(1) 6   (2) 2   (3) 8   (4) 4

III. Solutions

A solution is a homogeneous mixture of two or more substances, the composition of which may vary within limits. The component of a solution which is usually a liquid and is present in excess is called the **solvent**, while the other component is called the **solute**.

The dissociation of solute particle by a solvent is called **solvation**.

Most solutions dealt with in beginning courses in chemistry are aqueous solutions. When water is the solvent, the dissociation of solute particles is called **hydration**.

A. Methods Of Indicating Concentrations

Concentration of solutions may be indicated in a variety of ways:

1. Molarity. The molarity (M) of a solution is the number of moles of solute contained in a liter (1000 mL) of solution. As indicated by the formula:

$$\textbf{Molarity (M)} \; = \; \frac{\textbf{number of moles of solute}}{\textbf{1 liter of solution}}$$

Therefore, a two molar (2M) solution contains 2 moles of solute per liter of solution and 0.1 molar solution, (0.1M) contains 0.1 mole of solute per liter of solution. Rearranging the above equation we can also state that the concentration in moles per liter multiplied by the volume in liters equals the number of moles of solute in the solution, or:

Moles of solute = molarity x volume in liters

The mass in grams of solute can be determined by multiplying the number of moles of solute by the mass of 1 mole.

Grams of solute = number of moles x mass of 1 mole

Sample Problem 1:
How many moles of NaOH are contained in 200 mL of 0.1M solution of NaOH?

Solution:
Substituting in the above equation:

Moles of solute = molarity x volume in liters
Moles of NaOH = 0.1 M x 0.2 liters
= 0.02 moles

Sample Problem 2:
How many grams of NaOH are contained in 500 mL of 0.5M solution of NaOH?

Solution:
In order to find the mass of NaOH we must first find the moles of NaOH used by substituting in the same equation.

Moles of solute = molarity x volume of solution
Moles of NaOH = 0.5 M x 0.5 liters
= 0.25 moles

We now need to find the grams of NaOH used by using the formula:

Grams of solute = no. of moles x mass of one mole
= 0.25 x 40g
= 10g

Sample Problem 3:
What is the molarity of a solution of KOH if 500 mL of the solution contains 5.6 grams of KOH?

Solution:
First - find the moles of KOH by substituting in the equation:

Grams of solute = no. of moles x mass of one mole
5.6 g = X x 56 g/mole

$$\frac{5.6\ g}{56\ g/mole} = X$$

0.1 mole = X

Second - Find the molarity by substituting in the formula:

$$\text{Molarity (M)} = \frac{\text{moles of solute}}{\text{liter of solution}}$$

$$X = \frac{0.1\ mole}{0.5\ liter} = 0.2\ M$$

B. Additional Terms: Concentration
You should also be familiar with the following terms:

1. Miscible Solution is a solution of liquid solutes that are soluble in liquid solvents, such as alcohol in water.

2. Immiscible Solution is a solution of liquid solutes that are insoluble in liquid solvents, such as oil in water.

3. Dilute Solution is a solution in which a large amount of solvent is required to dissolve a small amount of solute. For example, at 30°C, only about 12 grams of Cesium Sulfate will dissolve in 100 grams of water.

4. Concentrated Solution is a solution in which a large amount of solute can be dissolved in a small amount of solvent. For example, at 30°C, about 97 grams of Sodium Nitrate will dissolve in 100 grams of water.

5. Saturated Solution is a solution, which under specific conditions, holds all of the solute that it is capable of holding in a dissolved state. At this point, the liquid and solid phase are in a state of equilibrium. For example, according to Table *D* of the Reference Tables, any points on the line graphs represent saturated solutions at that temperature.

6. Unsaturated Solution is a solution in which less solute is dissolved than is capable of being dissolved under specific conditions. For example, at a specific temperature, a point below any line in Table *D* of the Reference Tables indicates that the solution is not in equilibrium and, therefore, unsaturated at that temperature.

7. Supersaturated Solution is a solution in which more solute is dissolved than can be dissolved under specific conditions. For example, at a specific temperature, a point above any line graph on Table *D* of the Reference Tables indicates that the amount of solute dissolved in solution is greater than is normally dissolved and is called supersaturated.

Questions V-4
1 How many grams of KNO_3 are needed to saturate 50.0 grams of water at 70°C?
   (1) 30. g   (2) 65 g   (3) 130. g   (4) 160. g
2 How many grams of $NaNO_3$ per 100 grams of H_2O would produce a supersaturated solution?
   (1) 110. g at 40°C   (3) 80. g at 20°C
   (2) 90. g at 30°C   (4) 60. g at 10°C
3 As additional $KNO_3(s)$ is added to a saturated solution of KNO_3 at constant temperature, the concentration of the solution:
   (1) decreases   (2) increases   (3) remains the same
4 As additional solid KCl is added to a saturated solution of KCl, the conductivity of the solution:
   (1) decreases   (2) increases   (3) remains the same
5 Based on Reference Table *D*, which of the following substances is most soluble at 60°C?
   (1) NH_4Cl (2) KCl   (3) NaCl   (4) NH_3

Additional Materials
in the Mathematics of Chemistry

AM 1. The Mole - Additional Problems

1. The number of molecules contained in a specific volume of gas at STP?

Sample Problem:
What is the number of molecules contained in 5.6 liters of nitrogen gas at STP?

Answer: _____

2. The number of molecules in a given mass of substance.

Sample Problem:
What is the number of molecules contained in 14 grams of nitrogen gas?

Answer: _____

3. The number of atoms of a particular element in a given number of moles of the substance.

Sample Problem:
How many atoms of hydrogen are contained in 0.5 mole of NH_3 gas?

Answer: _____

4. The number of grams of a substance that can be reacted with or produced from a known number of moles of a reactant.

Sample Problem:
Given the reaction: $N_2 + O_2 \longrightarrow 2NO$

What is the total number of grams of oxygen needed to react completely with 2 moles of nitrogen?

Answer: _____

5. The volume of a gas at STP that can be reacted with or produced from a known number of moles of a reactant.

Sample Problem:
Given the reaction: $N_2 + 3H_2 \longrightarrow 2NH_3$

How many liters of NH_3 gas at STP can be produced from 1.0 mole of N_2?

Answer: _____

6. The number of molecules that can be reacted with or produced from a known number of moles of a reactant.

Sample Problem:

Given the reaction: $2H_2 + O_2 \longrightarrow 2H_2O$

How many molecules of hydrogen are needed to produce 6.0 moles of water?

Answer:_____

You should also be able to solve the following problems:

7. Given the reaction: $Zn + 2HCl \longrightarrow ZnCl_2 + H_2$

How many molecules of H_2 are produced by the reaction of 6.54 grams of Zn with excess dilute HCl (aq)?

Answer:_____

8. Given the reaction: $2C_2H_6 + 7O_2 \longrightarrow 4CO_2 + 6H_2O$

How many liters of CO_2 at STP are produced by the combustion of 22.4 liters of C_2H_6 at STP?

Answer:_____

AM 2. Formula from Percent Composition

The empirical formula of a compound can be determined from the percent composition of the compound, and the atomic masses of the elements. If the compound is molecular, the molecular formula can be determined if the gram molecular mass is known.

Sample Problem:

A compound was found by analysis to consist of 80% carbon and 20% hydrogen by mass. What is the empirical formula of the compound?

First, to find the number of moles, you must convert the percentages to mass per hundred grams, then divide by its gram atomic mass:

Carbon	-80%	= 80 g / 12g/mole	= 6.67 moles
Hydrogen	- 20%	= 20 g / 1g per mole	= 20 moles

Second, in order to obtain the smallest whole number ratio, divide by the smallest value:

Carbon	= 6.67 moles /6.67 moles		= 1
Hydrogen	= 20 moles /6.67 moles		= 3
(answer)	= the empirical formula is CH_3		

The gram molecular mass of the compound in the above problem is 30 grams. What is the molecular formula of the compound?

The empirical formula CH_3 has the gram molecular mass of 15 g. In order to obtain the number of formula units, divide the quantity given by the mass of the empirical formula:

$$\frac{30 \text{ g}}{15 \text{ g}} = 2 \text{ formula units}$$

Then, multiply each element by the number of formula units to get the molecular formula C_2H_6

AM 3. Gram Molecular Mass From Gas Density

The molecular mass of a gas can be determined from the gas density (mass per unit volume, usually expressed in grams/liter) by using the relationship between the molar volume and the mass of one mole of the gas.

Sample Problem:
The density of a gas is 1.35 grams/liter at STP. Calculate the gram molecular mass of the gas.

Solution:
Knowing the volume of one mole of any gas is 22.4 liters, we need only to multiply the fraction by 22.4 liters/mole.

$$\frac{1.35 \text{ grams}}{1 \text{ liter}} \quad x \quad \frac{22.4 \text{ liters}}{\text{mole}} = \frac{30.24 \text{ grams}}{\text{mole}}$$

AM 4. Effect of Solute on Solvent

The presence of dissolved particles affects some properties of the solvent. Properties which depend on the relative number of particles rather than on the nature of the particles are called **colligative properties**.

Colligative properties, as related to solutions, include changes in boiling point, freezing point, vapor pressure, and osmotic pressure.

The effect on the boiling point and the freezing point on a solvent by the addition of a solute is measured by knowing the **molality** of the solution. The **molality** of a solution is an expression of the solution's concentration and is defined as the number of moles of a solute dissolved in 1,000 grams (1kg) of solvent. It is arrived at by using the following formula:

$$\text{Molality (m)} = \frac{\text{moles of solute}}{\text{kg of solvent}}$$

A. Boiling Point Elevation

The presence of a nonvolatile solute raises the boiling point of the solvent. The amount of increase is proportional to the concentration of dissolved solute particles.

One mole of particles per 1000 grams of water raises the boiling point of water 0.52°C.

The relationship between moles of solute and grams of solvent is expressed in the concentration unit, **molality**.

B. Freezing Point Depression

The presence of a solute lowers the freezing point of the solvent by an amount that is proportional to the concentration of dissolved solute particles. One mole of particles per 1,000 grams of water lowers the freezing point of water 1.86°C.

C. Abnormal Behavior of Electrolytes

Non-electrolytes dissolve in solution to form molecules which are not charged and will not conduct an electric current.

Electrolytes dissolve in solution and dissociate into ions that carry an ionic charge and, therefore, do conduct an electric current.

A mole of sugar ($C_{12}H_{22}O_{11}$), which is a non-electrolyte, will dissolve in 1 kg of water to give a mole of sugar molecules. A mole of salt (NaCl), which is an electrolyte, will dissolve in the same amount of water and also dissociate to give two moles of ions (1 mole of Na+ plus 1 mole of Cl-). Therefore, a 1 mole solution of NaCl will increase the boiling point of the solvent by +1.04°C. Whereas, a 1 mole solution of sugar will increase the boiling point of the solvent by +0.52°C. The same solution of NaCl will lower the freezing point of the solvent by -3.72°C. Also, a 1 mole solution of sugar will lower the freezing point of the solvent by -1.86°C.

This behavior of electrolytes in solution gives evidence to the existence of ions.

Questions V-5 and AM 1—4

1 How many moles of hydrogen atoms are there in one mole of $C_6H_{12}O_6$ molecules?
   (1) 24 (6.0×10^{23})   (3) 24
   (2) 12 (6.0×10^{23})   (4) 12
2 What is the total number of molecules of hydrogen in 0.25 mole of hydrogen?
   (1) 6.0×10^{23} (3) 3.0×10^{23}
   (2) 4.5×10^{23} (4) 1.5×10^{23}
3 The heat of vaporization for water is 540.0 calories per gram. What is the minimum number of calories needed to change 40.0 grams of water at 100°C to steam at the same temperature and pressure?
   (1) 43,200   (2) 21,600   (3) 540   (4) 40.0
4 A 15 gram sample of a gas has a volume of 30.0 liters at STP. What is the density of the gas?
   (1) 30. g/L   (2) 2.0 g/L   (3) 15 g/L   (4) 0.50 g/L

5   If the density of a gas at STP is 2.450 grams per liter, what is the gram molecular mass of the gas?
    (1) 2.50   (2) 22.4   (3) 56.0   (4) 54.9

6   Which solution will freeze at the lowest temperature?
    (1) 1 mole of sugar in 500 g of water
    (2) 1 mole of sugar in 1,000 g of water
    (3) 2 moles of sugar in 500 g of water
    (4) 2 moles of sugar in 1,000 g of water

7   Which gas has a greater density at STP than air at STP?
    (1) H_2 (2) NH_3 (3) Cl_2 (4) CH_4

8   The density of a gas is 2.93 grams per liter at STP. The gram-molecular mass of this gas is:
    (1) 1.46 g   (2) 7.65 g   (3) 32.0 g   (4) 65.6 g

9   The density of a gas is 2.0 grams per liter at STP. Its molecular mass is approximately:
    (1) 67   (2) 45   (3) 22   (4) 8.0

10   A 2.00-gram sample of helium gas at STP will occupy:
    (1) 11.2 liters   (3) 33.6 liters
    (2) 22.4 liters   (4) 44.8 liters

11   A compound contains 40% calcium, 12% carbon, and 48% oxygen by mass. What is the empirical formula of this compound?
    (1) $CaCO_3$ (2) CaC_2O_4 (3) CaC_3O_6 (4) $CaCO_2$

12   A compound consists of 85% silver and 15% fluorine by mass. What is its empirical formula?
    (1) AgF (2) AgF_2 (3) Ag_2F (4) Ag_6F

13   Which 1 molal solution of the following compounds will have the highest boiling point?
    (1) $NaCl$ (2) Na_2SO_4 (3) $NaClO_3$ (4) $NaOH$

AM 5. Calorimetry

The energy changes involved in chemical reactions are measured in calorimeters. These reactions do not normally use external forms of energy, except in cases of combustion reactions, when electrical energy is used to spark the reaction forward.

The construction of a calorimeter is very simple. It consists of a reaction chamber constructed of a metal, which is a good conductor of heat energy. Surrounding the reaction chamber is a known mass of water, which is held in an insulated container. The reaction takes place in the reaction chamber and raises the temperature of the water. Knowing the mass of the water and the temperature difference, we can use the formula:

Calories gained or lost	=	mass of water in grams	x	change in temperature in grams	x	specific heat or water

In using a calorimeter, the heat gained or lost by the container is disregarded. Also the heat capacity (specific heat) of the water (in the liquid phase) is assumed to be one calorie per gram, per Celsius degree.

Although we continue to use calories for measuring heat energy, the basic SI unit of energy is the **joule**. Since one calorie = 4.18 joules, it is easy to convert the heat energy to joules by simply multiplying the number of calories by 4.18joules/1 calorie.

A. Heat Of Vaporization

As stated in Unit I, the energy required to vaporize a unit mass (one gram) of a liquid at constant temperature is called its heat of vaporization.

Therefore, we can calculate the number of calories of energy absorbed during vaporization at the boiling point by multiplying the mass of the liquid in grams times its heat of vaporization (the heat of vaporization of water is 540 cal./gram).

Question: If we wanted to vaporize one liter of water...

1. How many calories would we require? (Remember that 1 mL of water weighs 1 gram.)

        Answer: _____ (give your answer in scientific notation)

2. How many joules are needed?

        Answer: _____ (give your answer in scientific notation)

B. Heat Of Fusion

The amount of energy required to convert one gram of any solid substance to a liquid at its melting point is called its **heat of fusion**. This amount of heat energy required to make the change from a liquid to a solid is found by multiplying its mass times its heat of fusion. (*Note. The heat of fusion of water is 80 cal./gram.*)

Question: If we were to melt 70 g of $H_2O(s)$ to $H_2O(l)$,

1. How much heat energy would the ice need to absorb to change to water?

        Answer: _____ (give your answer in scientific notation)

2. How many joules are needed?

        Answer: _____ (give your answer in scientific notation)

AM 6. Combined Gas Laws

In studying the gases, we are concerned with three variables: volume, pressure, and temperature.

Boyle's Law is concerned with the relationship of volume and pressure, when temperature is a constant. Charles' Law is concerned with the relationship of volume and absolute temperature, when pressure is a constant.

Since changes in volume, pressure, and temperature often occur simultaneously, it is convenient to combine the two equations of Boyle and Charles into a single equation.

The combined gas law equation may be written:
$$\frac{P_1 V_1}{T_1} = \frac{P_2 V_2}{T_2}$$

P_1, V_1, and T_1 are the original conditions of pressure, volume and Kelvin temperature. P_2, V_2, and T_2 are the corresponding values of the final conditions.

Note. The definitive units given for both pressures must be the same, and both temperatures must be in Kelvin degrees.

AM 7. Graham's Law

Because gases possess translational movement, as well as vibrational and rotational movement, they will eventually spread out and fill their container. The "spreading out" of a substance is called diffusion. As we have stated before, temperature is a measure of the kinetic energy in a system. Therefore, all the molecules in a system at a specific temperature will have the same amount of kinetic energy, regardless of their size.

The equation that is concerned with kinetic energy (**K.E.**), mass (**m**), and velocity (**v**), is:

$$\textbf{K.E.} = \frac{1}{2}\, \textbf{mv}^2$$

If the kinetic energy for all the molecules in a system is equal, whether large or small, the value for $\frac{1}{2}\, \textbf{mv}^2$ is the same for all of them. Therefore, it follows that if a gas has less mass than another, the value of its velocity must be greater. Since different particles have different masses, each particle will move at a different rate.

This relationship between mass and velocity (which is called the rate of diffusion) was studied by Thomas Graham, a British chemist, who concluded:

"Under the same conditions of temperature and pressure, gases diffuse at a rate inversely proportional to the square roots of their molecular masses."

Therefore, $H_2 \left(\frac{1}{\sqrt{2}} \right)$ diffuses faster than $O_2 \left(\frac{1}{\sqrt{16}} \right)$.

Questions V-6 and AM 5—7

1 A solution contains 50 grams of solute per 100 grams of water at 80°C. This solution could be a saturated solution of:

    (1) NaCl   (2) $NaNO_3$ (3) KCl   (4) $KClO_3$

2 What is the total number of molecules in 1.0 mole of $Cl_2(g)$?

    (1) 35   (2) 70   (3) 6.0×10^{23} (4) 12×10^{23}

3 According to Reference Table *L*, which 0.1 molar solution will have the lowest freezing point?

    (1) HF   (2) HNO_2 (3) HNO_3 (4) CH_3COOH

4   A 1 kilogram sample of water will have the highest freezing point when it
     contains:
     (1) 1×10^{17} dissolved particles   (3) 1×10^{21} dissolved particles
     (2) 1×10^{19} dissolved particles   (4) 1×10^{23} dissolved particles
5   If 0.50 liter of a 12-molar solution is diluted to 1.0 liter, the molarity of
     the new solution is:
     (1) 2.4   (2) 6.0   (3) 12   (4) 24

Self-Help Questions V

Unit V - Mathematics of Chemistry

Mole interpretation; including gram atomic mass, gram molecular mass, molar volume of a gas.

1   What is the gram formula mass of $CuSO_4 \cdot 5H_2O$?
     (1) 160. g   (2) 178 g   (3) 186 g   (4) 250. g
2   What is the total number of molecules contained in 0.50 mole of O_2 at
     STP?
     (1) 6.0×10^{23} (2) 4.5×10^{23} (3) 3.0×10^{23} (4) 1.5×10^{23}
3   If 1.00 mole of $H_2(g)$ at STP is compared to 1.00 mole of He(g) at STP, the
     volumes of the gases would be
     (1) equal and their masses unequal
     (2) equal and their masses equal
     (3) unequal and their masses unequal
     (4) unequal and their masses equal
4   At STP, what mass of CH_4 has the same number of molecules as
     64 grams of SO_2?
     (1) 16 g   (2) 32 g   (3) 64 g   (4) 128 g
5   What is the total number of moles contained in 115 grams of C_2H_5OH?
     (1) 1.00   (2) 1.50   (3) 3.00   (4) 2.50
6   What is the mass of 2 moles of nitrogen gas?
     (1) 14   (2) 56   (3) 24   (4) 28

Stoichiometry; including percent composition, finding the empirical formula given the mass of each element in a compound along with its molecular formula.

7   The percent by mass of carbon in CO_2 is equal to

     (1) $44/12 \times 100$ (2) $12/44 \times 100$ (3) $28/12 \times 100$ (4) $12/28 \times 100$

8   The percentage by mass of hydrogen in NH_3 is equal to

     (1) $1/17 \times 100$ (2) $3/17 \times 100$ (3) $17/3 \times 100$ (4) $6/17 \times 100$

9   The empirical formula of a compound is CH_4. The molecular formula of the compound could be
    (1) CH_4 (2) C_2H_6 (3) C_3H_8 (4) C_4H_{10}

10 According to the equation $HCl + NaOH \rightarrow NaCl + H_2O$,   the total number of moles of HCl that can be neutralized by 80. grams of NaOH is
    (1) 1.0   (2) 2.0   (3) 36   (4) 72

11 Given the reaction: $N_2(g) + 3H_2(g) \leftrightarrow 2NH_3(g)$ What is the ratio of moles of $H_2(g)$ consumed to moles of $NH_3(g)$ produced?
    (1) 1:2   (2) 2:3   (3) 3:2   (4) 6:6

12 A compound with an empirical formula of CH_2 has a molecular mass of 70. What is the molecular formula?
    (1) CH_2 (2) C_2H_4 (3) C_4H_8 (4) C_5H_{10}

13 What is the percent by mass of oxygen in CH_3OH?
    (1) 50.0   (2) 44.4   (3) 32.0   (4) 16.0

14 Given the reaction: $(NH_4)_2CO_3 \rightarrow 2NH_3 + CO_2 + H_2O$
    What is the minimum amount of ammonium carbonate that reacts to produce 1.0 mole of ammonia?
    (1) 0.25 mole   (3) 17 moles
    (2) 0.50 mole   (4) 34 moles

15 The approximate percent by mass of potassium in $KHCO_3$ is
    (1) 19%   (2) 24%   (3) 39%   (4) 61%

16 How many moles of water are contained in 0.250 mole of $CuSO_4 \cdot 5H_2O$?
    (1) 1.25   (2) 4.50   (3) 40.0   (4) 62.5

17 What is the empirical formula of a compound whose composition by mass is 40.% sulfur and 60.% oxygen?
    (1) SO_2 (2) SO_3 (3) S_2O_3 (4) S_2O_7

Stoichiometry; involving equations
including mass-mass, mass-volume, volume-volume problems

18 Given the reaction: $2C_2H_6(g) + 7O_2(g) \rightarrow 4CO_2(g) + 6H_2O(g)$
    At STP, what is the total volume of $CO_2(g)$ formed when 6.0 liters (L) of $C_2H_6(g)$ are completely oxidized?
    (1) 24 L   (2) 12 L   (3) 6.0 L   (4) 4.0 L

19 Given the reaction at STP: $N_2(g) + 3H_2(g) \rightarrow 2NH_3(g)$
    What is the total number of liters of NH_3 formed when 20 liters of N_2 reacts completely?
    (1) 10 L   (2) 20 L   (3) 30 L   (4) 40 L

20 Given the reaction: $2H_2(g) + O_2(g) \rightarrow 2H_2O(g)$ How many liters of H_2 (g) are required to produce a total of 10. liters of $H_2O(g)$?
    (1) 1.0   (2) 2.0   (3) 10.   (4) 20.

21 Given the balanced equation: $NaOH + HCl \rightarrow NaCl + H_2O$
What is the total number of grams of H_2O produced when 116 grams of the product, NaCl, is formed?
(1) 9.0 g   (2) 18 g   (3) 36 g   (4) 54 g

Solutions; Including concentration expressions such as molarity, miscible,immiscible, dilute, concentrated, saturated, unsaturated, and supersaturated solutions.

22 How many grams of KOH are needed to prepare 250. milliliters of a 2.00 M solution of KOH (formula mass = 56.0)?
(1) 1.00   (2) 2.00   (3) 28.0   (4) 112

23 What is the maximum number of grams of NH_4Cl that will dissolve in 200 grams of water at 70°C?
(1) 60   (2) 70   (3) 100   (4) 120

24 A solution in which an equilibrium exists between dissolved and undissolved solute must be
(1) saturated   (2) unsaturated (3) dilute   (4) concentrated

25 Which solution contains the greatest number of moles of solute?
(1) 0.5 L of 0.5 M   (3) 2 L of 0.5 M
(2) 0.5 L of 2 M   (4) 2 L of 2 M

26 Which quantity of salt will form a saturated solution in 100 grams of water at 45°C?
(1) 30 g of KCl   (3) 60 g of KNO_3
(2) 35 g of NH_4Cl (4) 110 g of $NaNO_3$

27 How many grams of ammonium chloride (gram formula mass = 53.5 g) are contained in 0.500 L of a 2.00 M solution?
(1) 10.0 g   (2) 26.5 g   (3) 53.5 g   (4) 107 g

Extended materials on the mole, formula from percent composition, gram molecular mass from gas density, colligative effects of solute on the solvent; which includes molality, boiling point elevation, freezing point depression, and abnormal behavior of electrolytes.

28 Which of the following gases has the greatest density at STP?
(1) SO_2 (2) CO_2 (3) Cl_2 (4) N_2

29 What is the empirical formula of a compound whose composition by mass is 50.% sulfur and 50.% oxygen?
(1) SO   (2) SO_2 (3) SO_3 (4) S_2O_3

30 The density of a gas is 3.00 grams/Liter at STP. What is the gram molecular mass of the gas?
(1) 7.47 g   (2) 11.2 g   (3) 22.4 g   (4) 67.2 g

31 Which solution has the highest boiling point?
(1) 1 mole of $NaNO_3$ in 250 g of water
(2) 1 mole of $NaNO_3$ in 500 g of water
(3) 1 mole of $NaNO_3$ in 750 g of water
(4) 1 mole of $NaNO_3$ in 1000 g of water

32 At STP, the volume occupied by 32 grams of a gas is 11.2 liters. The gram molecular mass of this gas is closest to
(1) 8.0 g   (2) 16 g   (3) 32 g   (4) 64 g

33 Which ratio of solute-to-solvent could be used to prepare a solution with the highest boiling point?
(1) l g of NaCl dissolved per 100 g of water
(2) 1 g of NaCl dissolved per 1000 g of water
(3) 1 g of $C_{12}H_{22}O_{11}$ dissolved per 100 g of water
(4) 1 g of $C_{12}H_{22}O_{11}$ dissolved per 1000 g of water

Extended materials on calorimetry, heat of vaporization, heat of fusion, combined gas laws, and Graham's Law.

34 At STP, which of the following gases diffuses most rapidly?
(1) H_2 (2) F_2 (3) Ne   (4) Xe

35 What is the total number of kilocalories of heat needed to change 150grams of ice to water at 0°C? (Heat of fusion = 80. cal/g)
(1) 12   (2) 2.0   (3) 70.   (4) 230

36 A gas has a volume of 1400 milliliters at a temperature of 20. K and a pressure of 760 mm Hg. What will be the volume when the temperature is changed to 40. K and the pressure is changed to 380mm Hg?
(1) 350 mL   (2) 750 mL   (3) 1400 mL   (4) 5600 mL

37 How many calories of heat are absorbed when 50. grams of water at 100°C are completely vaporized? (Heat of vaporization = 540 cal/g)
(1) 590   (2) 5400   (3) 27000   (4) 54000

38 When 20. calories of heat is added to 2.0 grams of water at 15°C, the temperature of the water increases to
(1) 5.0°C   (2) 15°C   (3) 25°C   (4) 50.°C

39 At STP, which gas will diffuse most rapidly?
(1) NH_3 (2) CO   (3) He   (4) Ar

40 A 20.-milliliter sample of a gas is at 546 K and has a pressure of 6.0 atmospheres. If the temperature is changed to 273 K and the pressure to 2.0 atmospheres, the new volume of the gas will be
(1) 3.3 mL   (2) 13 mL   (3) 30. mL   (4) 120. mL

41 Which gas would diffuse most rapidly under the same conditions of temperature and pressure?
(1) gas *A*, molecular mass = 4   (3) gas *C*, molecular mass = 36
(2) gas *B*, molecular mass = 16   (4) gas *D*, molecular mass = 49

42 When 20 grams of water is cooled from 20°C to 10°C, the number of calories of heat released is
(1) 10   (2) 20   (3) 30   (4) 200

6

Unit

Kinetics and Equilibrium

Important Terms and Concepts

Kinetics	Saturated Solutions - Solubility
Activation Energy	Closed and Open Systems
Heat (Enthalpy) of Reaction	Le Chatelier's Principle
Exothermic and Endothermic	Law of Chemical Equilibrium
Potential Energy	Spontaneous Reactions
Effective Collisions	Energy and Entropy Changes
Temperature and Catalysts	Free Energy Change
Equilibrium	Solubility Product Constant
Phase, Dynamic, Motion	Common Ion Effect

I. Kinetics

Chemical kinetics is the branch of chemistry concerned with the rate of chemical reactions and the mechanisms by which chemical reactions occur.

The **rate** of a chemical reaction is measured in terms of the number of moles of reactant consumed (or moles of product formed) per unit volume in a unit of time.

The **mechanism** of a chemical reaction is a sequence of stepwise reactions by which the overall change occurs. Though many reactions take place because of a series of steps, only the net reaction is often observable. The net reaction represents a summation of all the changes that occur.

A. Role Of Energy In Reactions

To *initiate a chemical reaction, energy is required,* and once a chemical reaction begins, *energy may be released or absorbed.*

By graphing the potential energy of the reactants, the activation energy, and the potential energy of the products against a time sequence, one may describe the energies involved in a chemical reaction (see Figure VI-1 on page 101). They are:

1. Activation Energy. Activation energy is the minimum energy required to initiate a reaction by forming an activated complex.

2. Heat (Enthalpy) of Reaction. Heat (enthalpy) of reaction (ΔH) is the heat energy released or absorbed in the formation of the products. It represents the difference in heat content between the products and reactants.

$$\Delta H = H \text{ products } - H \text{ reactants}$$

In an **exothermic** reaction, energy is released. The products have a lower potential energy than the reactants, and the sign of ΔH **is negative.**

The sign that may be used when energy is included in a chemical equation should not be confused with the sign for ΔH. For example, the equation for the reaction of hydrogen and oxygen to form water may be written:

$$H_2(g) + \tfrac{1}{2}O_2(g) \longrightarrow H_2O(l) + 68.3 \text{ kcal}$$

Since this is an exothermic reaction (because energy is being produced and is written with the products), the sign of ΔH **is negative** (value: −68.3 kcal.) In equations that include heat, the phase of each species should be specified, such as **(g)** for gas, **(l)** for liquid, and **(s)** for solid.

When water is formed from gaseous hydrogen and oxygen, the first product that is formed is water vapor $[H_2O(g)]$, which condenses into liquid water $[H_2O(l)]$. One possible mechanism for the reaction is:

$$H_2(g) + O_2(g) \longrightarrow HOH(g) + O(g) + 57.8 \text{ kcal}$$

$$O(g) + H_2(g) \longrightarrow HOH(g) + 57.8 \text{ kcal}$$

$$2H_2O(g) \longrightarrow 2H_2O(l) + 21.0 \text{ kcal}$$

Summarizing:

$$2\,H_2(g) + O_2(g) \longrightarrow 2H_2O(l) + 136.6 \text{ kcal}$$

The above reaction shows that when two moles of $H_2O(l)$ are formed, the heat energy produced is 136.6 kcal. Table *G* of the Reference Tables for Chemistry lists the standard energy of formation of one mole of various compounds under "standard condition" (1 atm pressure and 298 K) as:

- energy of formation for one mole of $H_2O(g)$ is 57.8 kcal

- energy of formation for one mole of $H_2O(l)$ is 68.3 kcal

In an **endothermic** reaction, energy is absorbed, the products have a higher potential energy than the reactants, and the sign of ΔH **is positive**.

A good example of this type of reaction is the reverse of the above reaction. If 68.3 kcal is released when one mole of water molecules is formed, when a mole of water molecules is broken down so as to form 1 mole of hydrogen gas and $\tfrac{1}{2}$ mole of oxygen gas, the reaction requires 68.3 kcal of energy.

For an endothermic reaction, the energy is written with the reactant. For example:

$$H_2O(l) + 68.3 \text{ kcal} \longrightarrow H_2(g) + \tfrac{1}{2}O_2(g)$$

3. Potential Energy Diagram. For a given reaction, the activation energy and heat of reaction can be shown graphically in a potential energy diagram by plotting potential energy against a reaction coordinate representing the process of the reaction.

If the potential energy of the products is **higher** than the potential energy of the reactants, energy has been **absorbed** (**endothermic reaction**).

If the potential energy of the products is **lower** than the potential energy of the reactants, energy has been **liberated** (**exothermic reaction**).

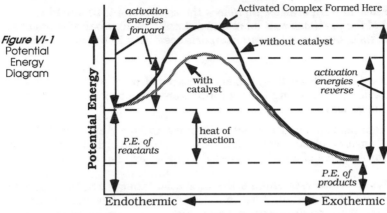

Figure VI-1
Potential
Energy
Diagram

The graph shown above, when read from left to right, represents an exothermic reaction with the products at a lower potential energy than the reactants. If the graph is reversed, or read from right to left, it would represent an endothermic reaction.

The highest point in the curve represents the potential energy of the **activated complex**. The difference between this point and the initial potential energy of the reactants represents the **activation energy** of the reaction.

B. Factors Affecting Rate Of Reaction

Chemical reactions depend on collisions between the reacting particles, atoms, molecules, or ions. It is these collisions that will produce an activated complex so the reaction will take place. The rate of reaction is affected by the number of collisions occurring and the fraction of these collisions that are effective. These **effective collisions** depend on:

1. Nature of the Reactants. Since bonds may be broken or formed in a reaction, the nature of the bond is an important fact for affecting reaction rates.

Reactions that involve **negligible bond rearrangements** are usually **rapid** at room temperature, such as the reactions of ionic substances in aqueous solutions. Reactions that involve the **breaking of bonds** tend to be **slow** at room temperature, such as the reaction between hydrogen and oxygen.

2. Concentration Affects the Rates of Reaction as Follows:

- An increase in the concentration of one or more reactants generally increases the rate of reaction.

- An increase in the concentration of a reactant increases the frequency of collisions by increasing the number of collisions per unit time.

- In a gaseous system, an increase in pressure will result in an increase in concentration and thus an increase in the rate of reaction.

3. Temperature Affects the Rates of Reaction as Follows:

• An increase in temperature increases the rate of all chemical reactions.

• An increase in temperature increases the speed (and thus the kinetic energy) of the particles and increases not only the number of collisions per unit time, but also of greater importance, the effectiveness of the collisions.

4. Surface Area. Increasing the surface area of reactants increases their rate of reaction. By increasing the concentration of the reactants, the number of collisions increases.

In heterogeneous reactions, surface area plays an important role. For example, a given amount of zinc will react more readily with dilute hydrochloric acid if the surface area of zinc is increased by using smaller pieces.

5. Catalysts. Catalysts change the activation energy required and thus change the rate of reaction. Also, a catalyst changes the mechanism of a reaction to one involving less activation energy, but does not change the overall process. **A catalyst does not initiate a chemical reaction.**

Questions VI-1

Base the answers to the following questions, 1 and 2, on the potential diagram and reaction shown.

1 Compared to the activation energy of the forward reaction, the activation energy of the reverse reaction is:
(1) less
(2) greater
(3) the same

2 Compared to the potential energy of the activated complex of the forward reaction, the potential energy of the activated complex of the reverse reaction is
(1) less   (2) greater   (3) the same

3 As the temperature of a chemical reaction increases, the rate of reaction:
(1) decreases   (2) increases   (3) remains the same

4 Which diagram shows the potential energy of an exothermic reaction?

(1)   (2)   (3)   (4)

5 Which will occur if a catalyst is added to a reaction mixture?
(1) the activation energy will be changed
(2) only the rate of the forward reaction will be increased
(3) only the rate of the reverse reaction will be increased
(4) the energy charge (ΔH) of the reaction will be decreased

6   In a reversible reaction, the difference between the activation energy of
    the forward reaction and the activation energy of the reverse action is
    equal to the
    (1) activation complex   (3) potential energy of reactants
    (2) heat of reaction   (4) potential energy of products

7   If the concentration of one of the reactants in a chemical reaction is
    increased, the rate of the reaction usually:
    (1) decreases   (2) increases   (3) remains the same

8   In a chemical reaction, the use of a catalyst usually results in a decrease
    in the:
    (1) activation energy
    (2) potential energy of the reactants
    (3) heat of reaction
    (4) amount of products

9   In a gaseous system, temperature remaining constant, an increase in
    pressure will:
    (1) increase activation energy   (3) increase reaction rate
    (2) decrease activation energy   (4) decrease reaction rate

10 The graph on the right represents the
    potential energy changes that occur in
    a chemical reaction. Which letter
    represents the activated complex?
    (1) A
    (2) B
    (3) C
    (4) D

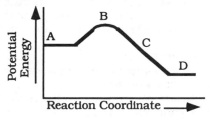

*Base your answer to questions 11 and 12
on the potential energy diagram at the
right.*

11 The reaction
    $A(g) + B(g) \rightarrow C(g) + D(g) + 30$ kcal
    has a forward activation energy of 20
    kcal. What is the activation energy
    for the reverse reaction?
    (1) 10 kcal   (3) 30 kcal
    (2) 20 kcal   (4) 50 kcal

12 The potential energy of the activated complex is equal to the sum of:
    (1) X + Y   (2) X + W   (3) X + Y + W   (4) X + W + Z

*Base your answers to the following
questions, 13 (below) and 14 (next page), on
the diagram at the right.*

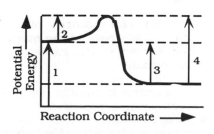

13 The activation energy for the reverse
    reaction is represented by
    (1) 1
    (2) 2
    (3) 3
    (4) 4

14 The heat of reaction (ΔH) is represented by:
  (1) 1   (2) 2   (3) 3   (4) 4

15 The effect of a catalyst on a chemical reaction is to change the:
  (1) activation energy   (3) potential energy of the product
  (2) heat of reaction   (4) potential energy of the reactant

16 Consider the reaction: $H_2O(l)$ + energy $\rightarrow$ $H_2(g)$ + $\frac{1}{2}O_2(g)$.

  Which phrase best describes this reaction?
  (1) exothermic, releasing energy
  (2) exothermic, absorbing energy
  (3) endothermic, releasing energy
  (4) endothermic, absorbing energy

II. Equilibrium

When a quantity of water is placed in a closed container, it is not long before evaporation begins. The result is small beads of water collecting on the bottom of the cap of the container, and eventually dropping back down to the bottom of the container. This indicates that some of the water vaporized, and because of its added translational movement, hit the bottom of the cap, lost some of its kinetic energy, and condensed.

When the condensed water droplet becomes heavy enough, it drops back into the bottom of the container, where it will start once again on the odyssey of evaporation and condensation. This change of phase *from liquid to vapor to liquid* is called **phase equilibrium.** If the temperature remains constant, it will continue indefinitely as long as the container is closed.

Equilibrium is a state of balance between two opposing reactions (physical or chemical) occurring at the same rate. Most reactions in closed systems are reversible. **Equilibrium is dynamic** and only describes the overall appearance of the system. It does not describe the activity of individual particles.

The word **dynamic implies motion,** and dynamic equilibrium is that condition where the interaction of the particles of the reactants in one direction, is balanced by the interaction of the particles of the products in the opposite direction. Although the reaction rates for the opposing reactions are equal, a state of equilibrium may exist where the quantities of reactants and products are not equal. Thus equilibrium may be reached when only a small quantity of the products has been formed or when only a small quantity of reactants remains.

> **Note: The quantity of water on the bottom of the cap may not be as great as the quantity of water in the bottom container. However, the rates of evaporation and condensation will be equal.**

For a system in equilibrium, a change in conditions (such as temperature, concentration, or pressure) may result in a change in the equilibrium point. Because the reactions in an equilibrium are reversible, it follows that equilibrium may be attained either from the forward or the reverse reaction.

A. Phase Equilibrium

In general, phase changes (solid to liquid or liquid to gas) are reversible, and, in a closed system, equilibrium may be attained.

Normally, if a solid or a liquid is confined in a closed container, eventually there will be enough particles in the vapor phase so that rate of return is equal to the rate of escape. Therefore, a dynamic equilibrium results where there is an equilibrium vapor pressure, characteristic of the solid or the liquid. The word dynamic refers to the fact that every molecule in the system is participating in a phase change.

The basic example is the one stated above, as water changes to a vapor and then returns to its liquid form.

B. Solution Equilibrium

1. Gases in Liquids. In a closed system, equilibrium may exist between a gas dissolved in a liquid and the undissolved gas above the liquid. The equilibrium between dissolved and undissolved gas is affected by temperature and pressure in the following ways:

- Increased temperature *decreases* the solubility of gases in liquids.
- Increased pressure *increases* the solubility of gases in liquids.

An example that points out this type of equilibrium is a bottle of soda. As you decrease the pressure on the top of the soda, by removing the cap, the gas becomes less soluble. You hear the resulting "swish..." as the gas escapes.

2. Solids in Liquids. A solution equilibrium exists when the opposing processes of dissolving and crystallizing a solute occur at equal rates. Such a solution exhibiting equilibrium between the dissolved and recrystallized solute must be a **saturated solution**.

For example, if we added salt to water in a beaker, the salt would dissolve. If we continued to add more salt, the solution will become so saturated with the compound that for every sodium chloride particle going into solution, one would recrystallize out of solution. At this point, solution equilibrium exists, and we have a saturated solution.

3. Solubility. Solubility is an expression of concentration of a solute in a solvent and can be given under **two** conditions. They are:

- The solubility of a solute is defined as the maximum mass of the solute dissolved in a given volume of solvent under specified conditions, not necessarily a saturated solution.
- Solubility may also be defined as the concentration of solute in a saturated solution.

Reactions Which Go To Completion. Continuous removal of the product in a **closed system** may destroy the equilibrium system by removing all of that substance necessary for the reverse reaction. Removal of a product usually causes the reactants to exhaust themselves and the forward reaction to go nearer to "**completion** or to an **end**."

In an **open system**, products may be removed from a reaction, wholly or in part by:

- The formation of a gas.
- The formation of an insoluble product (precipitate).
- The formation of an essentially un-ionized product, such as water in an ionic reaction.

Such reactions are said to go to "**completion**."

C. Chemical Equilibrium

Chemical equilibrium is attained when the **concentration of the reactants and products remains constant**. When **observable changes** (such as color, pressure, and temperature) no longer occur in a reacting chemical system, the system has reached a state of equilibrium. At this point the forward reaction and the reverse reaction are occurring at equal rates.

Le Chatelier's Principle. If a stress, such as a change in concentration, pressure, or temperature, is applied to a system at equilibrium, the equilibrium is shifted in a way that usually relieves the effects of the stress.

When a chemical system at equilibrium is disturbed, chemical reaction occurs and equilibrium is re-established at a different point (such as with new concentrations of reactants and products).

Effect of Concentration

Increasing the concentration of one substance in a reaction at equilibrium will cause the reaction to go in such a direction as to consume the increase. Eventually, a new equilibrium will be established.

For example, in the equilibrium reaction, $N_2 + 3H_2 \leftrightarrow 2NH_3$, increasing the concentration of either nitrogen or hydrogen will increase the rate of ammonia formation. If the system remains closed, the increased concentration of ammonia that results will increase the rate of decomposition of ammonia, and a new equilibrium point will be established.

Removal of one product of a reaction results in a decrease in its concentration. This will cause the reaction to go in such a direction as to increase the concentration of the products.

Effect of Pressure

A change in pressure affects chemical equilibria in which gases are involved. An increase in pressure will displace the point of equilibrium in the direction that favors the formation of a lesser number of moles (lower volume). If no change in the number of moles is involved (equal volumes of reactants and products), a change in pressure has no effect on the equilibrium.

In the Haber process, $N_2(g) + 3H_2(g) \leftrightarrow 2\,NH_2(g)$, four moles (volume) of reactants form two moles (volume) of product. The forward reaction thus results in a decrease in the total number of moles (lower volume). Therefore, an increase in pressure will result in an increased production of ammonia, which relieves the effect of increased pressure.

However, in the reaction: $H_2(g) + Cl_2(g) \leftrightarrow 2HCl(g)$ a change in pressure does not affect the equilibrium, since there is an equal volume of gases on either side.

Note: Solids and liquids are not affected by a pressure change to any appreciable amount.

Effect of Temperature
When the temperature of a system in an equilibrium is raised, the equilibrium is displaced in such a way that heat is absorbed.

Chemical changes involve either the evolution or the absorption of energy. In every system in equilibrium, an endothermic and exothermic reaction are taking place simultaneously. The endothermic reaction is favored by an increase in temperature, the exothermic reaction by a decrease in temperature. Keep in mind that the rates of all reactions, both endothermic and exothermic, are increased by a rise in temperature.

However, the opposing reactions are increased unequally, resulting in a displacement of the equilibrium.

For example, in the Haber process, $N_2(g) + 3H_2(g) \leftrightarrow 2NH_3(g) + 22\ kcal$, raising the temperature favors the decomposition of ammonia.

Effect of Catalyst
In a system in equilibrium, a catalyst increases the rate of both the forward and reverse reactions equally, and produces no net change in the equilibrium concentrations. A catalyst may cause equilibrium to be reached more quickly, but does not affect the point of equilibrium.

Questions VI-2
1   Given the reaction at equilibrium: $X(g) + Y(g) \leftrightarrow 2Z(g)$ as the pressure on the system increases, the temperature remaining constant, the amount of $Z(g)$:
    (1) decreases   (2) increases   (3) remains the same
2   Which is true in all chemical reactions in which equilibrium has been reached?
    (1) forward reaction stops
    (2) reverse reaction stops
    (3) concentrations of reactants and products are equal
    (4) rates of the forward and reverse reactions are equal
3   When a catalyst is added to a chemical reaction, what will remain constant?
    (1) heat of the reaction
    (2) rate of the reaction
    (3) potential energy of the activated complex
    (4) activation energy of the forward reaction
4   As a catalyst is added to a system at equilibrium, the value of the equilibrium constant:
    (1) decreases   (2) increases   (3) remains the same

5   A sample of $H_2O(l)$ at 20°C is in equilibrium with its vapor in a sealed container. When the temperature decreases to 14°C, the amount of liquid within the system will:
    (1) decrease   (2) increase   (3) remain the same
6   Given the reaction at equilibrium: $BaSO_4(s) \leftrightarrow Ba^{2+}(aq) + SO_4^{2-}(aq)$, as the concentration of the SO_4^{2-} ions is increased, at constant temperature, the concentration of Ba^{2+} ions:
    (1) decreases   (2) increases   (3) remains the same

D. Law of Chemical Equilibrium

When a reversible reaction has attained equilibrium at a given temperature, the following mathematical relationship occurs. The product of the molar concentrations of the substances on the right side of the equation (products), divided by the product of the molar concentrations of the substances on the left side of the equation (reactants), is a **CONSTANT**. Note: Each concentration is raised to the power equal to the coefficients in the balanced equation.

For the reaction: **aM(g) + bN(g)** ⇌ **cP(g) + dQ(g)**

the equilibrium expression is written:

$$\frac{[P]^c \ [Q]^d}{[M]^a \ [N]^b} = K = \text{a constant at constant temperature}$$

This constant is called the equilibrium constant.

In the mathematical expression of this law:
- The equation must be balanced.

- Square brackets " [ ] " are used to indicate "*concentrations measured in moles per liter*."

- The concentrations of the products on the right of the chemical equation form the numerator, and the concentrations of reactants on the left form the denominator.

- The power of each concentration is derived from its coefficient.

- Finally, the mathematical proportion obtained in the first four steps, at a specific temperature, is equal to a constant which we designate as follows:

     K_{eq} for all forms of chemical reactions, and
     K_a for acids,
     K_b for bases,
     K_{sp} for solubility of solids, and
     K_w for the ionization of water.

Also, note the following:

1) The concentration of a solid or a liquid is essentially constant. In the expression of the equilibrium constant for any reaction involving a solid or a liquid, the concentration of the solid or liquid can be included in the constant. Therefore, it does not appear in the equation.

2) The magnitude of K is used by chemists to predict the extent of chemical reactions. A large value for **K** indicates the products are favored; that is, the equilibrium mixture consists largely of products. A small value for **K** indicates that reactants are favored.

3) The equilibrium constant has a numerical value for any given chemical reaction at a particular temperature. This value remains constant even though the concentrations of the substances involved may increase or decrease. In general, in the reaction given, an increase in the concentration of P would cause the reaction to go to the left. This decreases the concentration of Q and increases the concentrations of M and N. The value of K would remain constant.

4) The equilibrium constant changes with a change of temperature. In a reversible reaction, the reaction rates of the forward and reverse reactions are not affected equally by change in temperature.

Questions VI-3

1 What is the equilibrium expression for the reaction:
$$4Al(s) + 3O_2(g) \leftrightarrow 2Al_2O_3(s) ?$$

   (1) $K_{eq} = [O_2]^3$ (3) $K_{eq} = 1$ over $[O_2]^3$

   (2) $K_{eq} = [3O_2]$ (4) $K_{eq} = 1$ over $[3O_2]$

2 Which equilibrium constant indicates an equilibrium mixture that favors the formation of products?

   (1) $K_{eq} = 1 \times 10^{-5}$ (3) $K_{eq} = 1 \times 10^0$

   (2) $K_{eq} = 1 \times 10^{-1}$ (4) $K_{eq} = 1 \times 10^5$

3 Which equilibrium constant indicates an equilibrium mixture that most favors the formation of products?

   (1) A + B = C; (K = 1.8×10^{-5})   (3) R + S = T; (K = 1.2×10^2)

   (2) D + E = F; (K = 6.2×10^{-8})   (4) X + Y = Z; (K = 3.5×10^3)

4 Given the reaction at equilibrium: $2CO(g) + O_2(g) \leftrightarrow 2CO_2(g)$, the correct equilibrium expression for this reaction is:

   (1) $K_{eq} = \dfrac{[2CO][O_2]}{[2CO_2]}$ (3) $K_{eq} = \dfrac{[2CO_2]}{[2CO][O_2]}$

   (2) $K_{eq} = \dfrac{[CO]^2[O_2]}{[CO_2]^2}$ (4) $K_{eq} = \dfrac{[CO_2]^2}{[CO]^2[O_2]}$

5 Given the system at chemical equilibrium:
$$2O_3(g) \leftrightarrow 3O_2(g) (K_{eq} = 2.5 \times 10^{12}),$$
   the concentration of O_3 and O_2 must be:

   (1) constant   (2) equal   (3) increasing   (4) decreasing

6 Given the equilibrium: $AgCl(s) \leftrightarrow Ag^+(aq) + Cl^-(aq)$, the equilibrium constant will change if there is an increase in the:

   (1) concentration of Ag^+ ions   (3) pressure

   (2) concentration of Cl^- ions   (4) temperature

III. Spontaneous Reactions

Under specific conditions we see ice melt to water; other conditions will turn water back to ice. The initiating ingredient here is temperature; the rest occurs spontaneously. Once the activation energy is applied, in the form of a match, paper will combust rapidly under specific conditions. These are forms of physical and chemical spontaneous reactions.

After studying these phenomenas for many years, it has been discovered that spontaneous reactions depend on the balance between the two fundamental tendencies in nature:

- toward a lower energy state $\Delta H = (-)$
- toward randomness $\Delta S = (+)$(greater entropy).

A. Energy Changes

As you have learned, the activation energy required for an exothermic reaction is less than that required for an endothermic reaction. That difference is the energy of the reaction (ΔH). Therefore, when particles collide, they require less energy to go in the exothermic direction than in the endothermic direction.

This tendency in nature favors the exothermic reaction, in which ΔH is negative. So, we can generalize by stating, "at constant temperature and pressure, a system tends to change from one of high energy to one of low energy."

B. Entropy Changes

Entropy is a measure of the disorder, randomness, or lack of organization of a system. The solid phase, in regular crystalline arrangement, is more organized than the liquid phase; the liquid phase is more organized than the gaseous phase. Entropy is so defined that the more random a system is, the higher the entropy.

An increase in entropy during a change in the state of a system means that in its final state the system is more disordered (random) than in its initial state. High entropy (randomness) is favored by high temperatures. High temperatures increase the rate of motion of the particles and, therefore, increases randomness.

At constant temperature, a system tends to undergo a reaction so that in its final state it has higher entropy (greater randomness) than in its initial state. Therefore, we state:

> **Note: A system tends to change from a state of great order to a state of less order.**

For chemical systems this change in entropy is represented mathematically as ΔS.

Additional Materials In Kinetics And Equilibrium
AM 1. Free Energy Change

The difference between energy change and entropy change is the free energy change (ΔG). It is expressed by the following equation:

$$\Delta G = \Delta H - T\Delta S$$

Where,

ΔG is the free energy change and represents the net energy of the reaction that can perform work.

ΔH represents the energy factor, or the total heat energy, given off or taken on during the reaction.

T   is the temperature in degrees Kelvin.

ΔS is the measurement of entropy and represents the change in randomness of the reactants to the products.

Since ΔS is multiplied by the temperature, the whole expression $T\Delta S$ is considered the entropy factor.

For a **spontaneous** change to occur in a system, the free energy change (ΔG) **must be negative.**

Mathematically, in order for ΔG to be negative, the ideal condition occurs when ΔH is negative and ΔS is positive (as in combustion reactions). However, other conditions can occur and also result in a negative ΔG. These cases depend on the value of the temperature. If the temperature is high enough, the significance of ΔH diminishes, and the $T\Delta S$ term dominates. If instead, the temperature is very low, the $T\Delta S$ term is small, and the influence it has on the value of ΔG is diminished. It is overbalanced by a large negative value for ΔH.

AM 2. Predicting Spontaneous Reactions

As stated above, for a spontaneous change to occur in a system, the free energy change (ΔG) must be negative. In a system at **equilibrium**, the free energy change is **zero.**

When the two factors (tendency toward lower energy content, and tendency toward higher entropy in a system) cannot be satisfied simultaneously, the spontaneous change that may take place will be determined by the factor that is dominant at the temperature of the system. From energy changes alone, **exothermic reactions** would always be expected to occur spontaneously, and **endothermic reactions** would never be expected to occur spontaneously. Exceptions to both of these predictions may occur when a change in entropy opposes an exothermic reaction or favors an endothermic reaction. For example:

1. The change in phase from water to ice is an exothermic reaction and, from the consideration of energy only, water might be expected to freeze spontaneously at any temperature. However, the tendency toward higher entropy favors the reaction from ice to water. At temperatures below the freezing

point the energy change is dominant and water will freeze spontaneously. At temperatures above the freezing point the entropy change becomes the dominant factor, and ice melts.

2. The reaction, $2KClO_3(s) \longrightarrow 2KCl(s) + 3O_2(g)$, is an endothermic reaction, and the energy change would oppose a spontaneous reaction. However, the reaction results in an increase in entropy due to the formation of a gas and a solid. At high temperatures the effect of the entropy changes becomes sufficient to overcome the effect of the energy change, and the reaction takes place. If the temperature is not high enough, the reaction will not take place spontaneously.

AM 3. Solubility Product Constant (K_{sp})

The solubility product constant, K_{sp}, is a measure of the concentration of slightly soluble salts in water. In a saturated solution of an ionic solid, an equilibrium is established between the ions in the saturated solution and the excess solid phase. For the reaction:

$$AB(s) \rightleftharpoons A^+(aq) + B^-(aq)$$
$$K_{sp} = [A^+][B^-]$$

The value of the K_{sp} changes with a change of temperature.

The concentration (mass/unit volume) of a solid is essentially constant. In the expression of the equilibrium constant for any reaction involving a solid, the concentration of the solid can be included in the constant. It does not appear in the equation.

Therefore, in the case of the solubility equilibrium (below), application of the law of chemical equilibrium would give:

$$K = \frac{[A^+][B^-]}{[AB]}$$

Since (AB) itself is constant, we may say

$$[A^+][B^-] = K[AB] = K_{sp}.$$

The magnitude of K_{sp} is used in comparing the solubilities of slightly soluble salts. For example, at room temperature

$$K_{sp}\ CaSO_4 = 2.4 \times 10^{-5}$$
$$K_{sp}\ BaSO_4 = 1.6 \times 10^{-9}$$

Since the K_{sp} of $BaSO_4$ is less than that of $CaSO_4$, $BaSO_4$ must be less soluble than $CaSO_4$ and would be precipitated at a lower concentration.

Common Ion Effect

The addition of a common ion to the solution of a slightly soluble salt results in a decrease in the solubility of the salt. Consider the equilibrium:

$$AgCl(s) \rightleftharpoons Ag^+(aq) + Cl^-(aq)$$

Addition of NaCl, or any other soluble chloride salt, to this equilibrium system increases the $Cl^-(aq)$ concentration and, according to Le Chatelier's principle shifts the equilibrium to the left. This results in a decrease in the solubility of AgCl.

Questions VI-4 and AM 1—3

1 According to Reference Table M, which compound is more soluble than $BaSO_4$ at 1 atmosphere and 298 K?

(1) AgBr   (2) $PbCl_2$ (3) AgI   (4) $ZnCO_3$

2 Which is the correct solubility product constant expression for the reaction: $PbI_2(s) \leftrightarrow Pb^{2+}(aq) + 2I^-(aq)$?

(1) $K_{sp} = \dfrac{[Pb^{2+}]}{[I^-]^2}$ (3) $K_{sp} = [Pb^{2+}] - [I^-]^2$

(2) $K_{sp} = \dfrac{[I^-]^2}{[Pb^{2+}]}$ (4) $K_{sp} = [Pb^{2+}][I^-]^2$

3 Given the reaction at equilibrium: $Mg(OH)_2(s) \leftrightarrow Mg^{2+}(aq) + 2OH^-(aq)$ What is the correct expression for the solubility product constant for this reaction?

(1) $K_{sp} = [Mg^{2+}][2OH^-]$ (3) $K_{sp} = [Mg^{2+}][OH^-]^2$

(2) $K_{sp} = [Mg^{2+}] + [2OH^-]$ (4) $K_{sp} = [Mg^{2+}] + [OH^-]^2$

4 Based on Reference Table M, a saturated solution of which salt would be most dilute?

(1) AgCl   (2) $BaSO_4$ (3) $ZnCO_3$ (4) $PbCrO_4$

5 Based on Reference Table M, in a saturated solution of $BaSO_4$, at 1 atmosphere and 298 K, the product of $[Ba^{2+}] \times [SO_4^{2-}]$ is equal to:

(1) 1.1×1^{10} (2) 1.1×10^{-5} (3) 1.1×10^5 (4) 1.1×10^{-10}

6 Given the reaction at equilibrium: $NaCl(s) \leftrightarrow Na^+(aq) + Cl^-(aq)$, the addition of KCl to this system will cause a shift in the equilibrium to the:

(1) left, and the concentration of $Na^+(aq)$ ions will increase

(2) right, and the concentration of $Na^+(aq)$ ions will increase

(3) left, and the concentration of $Na^+(aq)$ ions will decrease

(4) right. and the concentration of $Na^+(aq)$ ions will decrease

7 Which change of phase results in a decrease in entropy?

(1) $H_2O(l) \rightarrow H_2O(g)$ (3) $H_2O(s) \rightarrow H_2O(g)$

(2) $H_2O(l) \rightarrow H_2O(s)$ (4) $H_2O(s) \rightarrow H_2O(l)$

8 The free energy change, ΔG, must be negative when:
(1) ΔH is positive and ΔS is positive
(2) ΔH is positive and ΔS is negative
(3) ΔH is negative and ΔS is positive
(4) ΔH is negative and ΔS is negative

9 A chemical reaction will always occur spontaneously if the reaction has a:
(1) $+ \Delta G$ (2) $- \Delta G$ (3) $- \Delta H$ (4) $+ \Delta H$

10 Which statement is true if the free energy (ΔG) of a reaction is zero?
(1) the rate of the forward reaction is zero
(2) the rate of the reverse reaction is zero
(3) the reaction is approaching equilibrium
(4) the reaction is at equilibrium

11 According to Reference Table G, which compound is formed spontaneously from its elements?
(1) NO   (2) ICl   (3) NO_2 (4) HI

12 Why does the reaction: $K(s) + \frac{1}{2} Cl_2(g) \rightarrow KCl(s)$ occur spontaneously? (Refer to Reference Table G).
(1) ΔS is +   (2) ΔS is −   (3) ΔG is +   (4) ΔG is −

13 For a chemical reaction, the free energy change, ΔG, is equal to:
(1) $\Delta H + T\Delta S$ (2) $\Delta H - T\Delta S$ (3) $T\Delta H + \Delta S$ (4) $T\Delta S - \Delta H$

14 In a chemical reaction, the difference in potential energy between the products and the reactants is equal to:
(1) ΔS (2) ΔG (3) ΔH (4) ΔT

15 Based on Reference Table M, which of the following compounds is least soluble in water?
(1) AgI   (2) AgCl   (3) $PbCl_2$ (4) PbI_2

16 According to Reference Table G, which substance will form spontaneously from its elements in their standard states at 1 atmosphere and 298 K?
(1) ethene   (3) hydrogen iodide
(2) ethyne   (4) hydrogen fluoride

17 Based on Reference Table E, which of the following compounds would most likely have the *smallest* K_{sp}?
(1) barium carbonate   (3) magnesium nitrate
(2) calcium sulfate   (4) silver acetate

18 Given the reaction at equilibrium and 298 K:

$$CH_3COOH(aq) \leftrightarrow H^+(aq) + CH_3COO^-(aq)$$

The equilibrium constant, K_a, for the reaction will change if there is an increase in the
(1) pressure   (3) concentration of H^+ ions
(2) temperature   (4) concentration of CH_3COO^- ions

19 As products are formed in the reaction

$$NH_4Cl(s) + 3.5\ kcal \xrightarrow{H_2O} NH_4^+(aq) + Cl^-(aq),$$

the entropy of the system
(1) decreases and heat is absorbed
(2) decreases and heat is released
(3) increases and heat is absorbed
(4) increases and heat is released

Self-Help Questions VI
Unit VI - Kinetics And Equilibrium
I. Kinetics - energy; including activation energy, heat of reaction, and potential energy diagram. factors that affect the rate of reaction; including nature of reactants, concentration, temperature, surface area, and catalyst.

1 In a chemical reaction, the products have a lower potential energy than the reactants. This reaction must have a negative
   (1) ΔG (2) ΔS (3) ΔH (4) ΔX

2 The heat of formation of $H_2O(g)$ is
   (1) -57.8 kcal/mole   (3) -54.6 kcal/mole
   (2) +68.3 kcal/mole   (4) +56.7 kcal/mole

3 As the surface area of the Zn(s) used in the reaction
         $Zn(s) + 2HCl(aq) \rightarrow ZnCl_2(aq) + H_2(g)$
   is increased, the rate of the reaction will
   (1) decrease   (2) increase   (3) remain the same

4 Given the reaction: $Zn(s) + 2HCl(aq) \rightarrow ZnCl_2(aq) + H_2(g)$
   The reaction occurs more slowly when a single piece of zinc is used than when the same mass of powdered zinc is used. Why does this occur?
   (1) The powdered zinc is more concentrated.
   (2) The powdered zinc has a greater surface area.
   (3) The powdered zinc required less activation energy.
   (4) The powdered zinc generates more heat energy.

5 As the number of effective collisions between reacting particles increases, the rate of the reaction
   (1) decreases   (2) increases   (3) remains the same

6 According to the potential energy diagram for the reaction A + B $\rightarrow$ C + D, the activation energy is highest for the

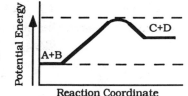

   (1) forward reaction, which is endothermic
   (2) forward reaction, which is exothermic
   (3) reverse reaction, which is endothermic
   (4) reverse reaction, which is exothermic

7 As a solid is dissolving in water, the temperature of the water decreases. The dissolving of this solid is
   (1) exothermic and spontaneous   (3) exothermic and *not* spontaneous
   (2) endothermic and spontaneous (4) endothermic and *not* spontaneous

8 In order for a chemical reaction to occur, there must *always* be
   (1) an effective collision between reacting particles
   (2) a bond that breaks in a reactant particle
   (3) reacting particles with a high charge
   (4) reacting particles with a high kinetic energy

9 What is the heat of formation of $H_2O(l)$, in kilocalories per mole, at 1atmosphere and 298 K?
   (1) -79.7   (2) -68.3   (3) -56.7   (4) -54.6

10 A student adds two 50-milligram pieces of Ca(s) to water. A reaction takes place according to the following equation:

$$Ca(s) + 2H_2O(l) \rightarrow Ca(OH)_2(aq) + H_2(g)$$

Which change could the student have made that would most likely have increased the rate of the reaction?
(1) used ten 10-mg pieces of Ca(s)
(2) used one 100-mg piece of Ca(s)
(3) decreased the amount of the water
(4) decreased the temperature of the water

11 In a chemical reaction, the difference between the potential energy of the products and the potential energy of the reactants is the
(1) heat of reaction   (3) free energy
(2) heat of fusion   (4) activation energy

II. Equilibrium - Phase, solution, (including reaction that run to an "end") and chemical equilibrium, LeChatelier's principle. Effect of concentration, pressure, temperature, and catalyst, The law of chemical equilibrium

12 Which change takes place when a catalyst is added to a reaction at equilibrium?
(1) The point of equilibrium is shifted to the right.
(2) The point of equilibrium is shifted to the left.
(3) The rates of the forward and reverse reactions are increased unequally.
(4) The rates of the forward and reverse reactions are increased equally.

13 Given the equilibrium: A + B ↔ C
The greatest concentration of C would be produced if the equilibrium constant of the reaction is equal to
(1) 1×10^3 (2) 1×10^9 (3) 1×10^{-3} (4) 1×10^{-9}

14 Which is the equilibrium expression for the reaction
$$3A(g) + B(g) \leftrightarrow 2C(g)?$$

$K = \dfrac{[3A][B]}{[2C]}$ $K = \dfrac{[A]^3[B]}{[C]^2}$ $K = \dfrac{[2C]}{[3A][B]}$ $K = \dfrac{[C]^2}{[A]^3[B]}$
(1)   (2)   (3)   (4)

15 Given the equilibrium reaction:
$$2SO_2(g) + O_2(g) \leftrightarrow 2SO_3(g) + heat$$
When the pressure on the system is increased, the concentration of the SO_3 will
(1) decrease   (2) increase   (3) remain the same

16 Given the reaction at equilibrium and 298 K:
$$CH_3COOH(aq) \leftrightarrow H^+(aq) + CH_3COO^-(aq)$$
The equilibrium constant, K_a, for the reaction will change if there is an increase in the
(1) pressure   (3) concentration of H^+ ions
(2) temperature   (4) concentration of CH_3COO^- ions

17 Given the reaction at equilibrium:
   $A(g) + B(g) \leftrightarrow C(g) + D(g) + heat$ As additional $A(g)$ is added to the system at constant temperature, the concentration of $B(g)$
   (1) decreases   (2) increases   (3) remains the same

18 Given the reaction at equilibrium: $AgCl(s) \leftrightarrow Ag^+(aq) + Cl^-(aq)$

   The addition of Cl^- ions will shift the equilibrium to the
   (1) right, decreasing the solubility of $AgCl(s)$
   (2) right, increasing the solubility of $AgCl(s)$
   (3) left, decreasing the solubility of $AgCl(s)$
   (4) left, increasing the solubility of $AgCl(s)$

19 Given the reaction at equilibrium:
   $$CH_4(g) + 2O_2(g) \leftrightarrow CO_2(g) + 2H_2O(g)$$
   An increase in the concentration of $O_2(g)$ at constant temperature and pressure will result in
   (1) an increase in the concentration of $CH_4(g)$
   (2) an increase in the concentration of $CO_2(g)$
   (3) a decrease in the concentration of $O_2(g)$
   (4) a decrease in the concentration of $H_2O(g)$

20 Solution equilibrium always exists in a solution that is
   (1) unsaturated   (2) saturated   (3) dilute   (4) concentrated

21 When a catalyst is added to a system at equilibrium, the concentration of the products
   (1) decreases   (2) increases   (3) remains the same

22 The addition of concentrated hydrochloric acid to the system
   $$AgCl(s) \leftrightarrow Ag^+(aq) + Cl^-(aq)$$

   at equilibrium will result in
   (1) a decrease in the amount of $AgCl(s)$

   (2) a decrease in the concentration of $H_3O^+(aq)$

   (3) an increase in the concentration of $Ag^+(aq)$

   (4) an increase in the concentration of $Cl^-(aq)$

23 The diagram at the right shows a bottle containing $NH_3(g)$ dissolved in water. How can the equilibrium $NH_3(g) \leftrightarrow NH_3(aq)$ be reached?
   (1) Add more water.
   (2) Add more $NH_3(g)$.
   (3) Cool the contents.
   (4) Stopper the bottle.

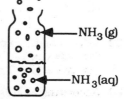

24 Which factors must be equal in a reversible chemical reaction at equilibrium?
   (1) the concentrations of the reactants and products
   (2) the potential energies of the reactants and products
   (3) the activation energies of the forward and reverse reactions
   (4) the rates of reaction of the forward and reverse reactions

III. Spontaneous reactions; including energy and entropy change, FREE energy change, predicting spontaneous reactions and solubility product constant

25 As products are formed in the reaction

$$NH_4Cl(s) + 3.5 \text{ kcal} \xrightarrow{H_2O} NH_4^+(aq) + Cl^-(aq),$$

the entropy of the system
(1) decreases and heat is absorbed (3) increases and heat is absorbed
(2) decreases and heat is released (4) increases and heat is released

26 According to Reference Table G, which substance will form spontaneously from its elements in their standard states at 1 atmosphere and 298 K?
(1) ethene   (3) hydrogen iodide
(2) ethyne   (4) hydrogen fluoride

27 Based on Reference Table E, which of the following compounds would most likely have the *smallest* K_{sp}?
(1) barium carbonate   (3) magnesium nitrate
(2) calcium sulfate   (4) silver acetate

28 A nonreversible chemical reaction is exothermic and occurs with an increase in entropy. The ΔG for this reaction
(1) must be negative   (3) could be negative or positive
(2) must be positive   (4) could be zero

29 Based on Reference Table M, which of the following compounds is *least* soluble at 298 K?
(1) $PbCl_2$ (2) PbI_2 (3) $AgCl$ (4) AgI

30 Which compound shows the *least* increase in solubility in water from 50°C to 60°C?
(1) KCl (2) $NaCl$ (3) KNO_3 (4) $NaNO_3$

31 What is the solubility product constant expression (K_{sp}) for the reaction below? $Ag_2S(s) \leftrightarrow 2Ag^+(aq) + S^{2-}(aq)$

(1) $K_{sp} = \dfrac{[Ag^+]^2[S^{2-}]}{[Ag_2S]}$ (3) $K_{sp} = [Ag^+]^2[S^{2-}]$

(2) $K_{sp} = \dfrac{[Ag_2S]}{[Ag^+]^2[S^{2-}]}$ (4) $K_{sp} = [Ag_2S]$

32 Which is the expression for the free energy change of a chemical reaction?
(1) $\Delta H = \Delta G - T\Delta S$ (3) $\Delta G = \Delta H - T\Delta S$
(2) $\Delta G = \Delta S - T\Delta H$ (4) $\Delta S = \Delta G - T\Delta H$

33 According to Reference Table I, the dissolving of which salt is accompanied by the release of energy?
(1) $LiBr$ (2) NH_4Cl (3) $NaCl$ (4) KNO_3

34 As $NaCl(s)$ dissolves according to the equation
$NaCl(s) \rightarrow Na^+(aq) + Cl^-(aq)$, the entropy of the system
(1) decreases   (2) increases   (3) remains the same

7 Acids and Bases

Important Terms and Concepts

Electrolytes
Acids and Bases
   Indicators
   Arrhenius' Theory
   Brönsted–Lowry Theory
Neutralization
   Acid–Base Titration

Amphoteric Substances
Normality
Salts
Hydrolysis
Conjugate Acid – Base Pair
Ionization Constants
pH

I. Electrolytes

An electrolyte will dissolve in water to form a solution that will conduct an electric current. The ability of a solution to conduct an electric current is due to the presence of ions that are free to move. Therefore, all ionic compounds are electrolytes. Also, some polar covalent compounds form ions and conduct electricity when dissolved in water. For example, HCl and HBr. However, nonelectrolytes, such as organic solvents, do not conduct electricity.

Weak electrolytes in aqueous solution attain an equilibrium between ions and the undisassociated compound. The equilibrium constant for such systems is called the **dissociation constant**. The value of the dissociation constant changes with a change of temperature.

II. Acids And Bases

One means of defining a substance is to list its properties and reactions. This form of definition is called an **operational definition** and is based on **experimental observations**, which include a set of conditions.

As the understanding of acid-base reactions has grown, **conceptual definitions** (those that try to **answer why(?) and how(?) statements**) of acids and bases have been extended. These conceptual definitions have also been applied to reactions that do not necessarily take place in aqueous solutions.

A. Acids

Acids may be defined in terms of their characteristic properties. These properties can be observed experimentally and form the basis of the **operational definition** of an acid. They include:

Operational Definition of Acids:
1. **Aqueous solutions of acids conduct electricity.** They conduct electricity in relation to the degree of their ionization. A few acids ionize almost completely in aqueous solution and are strong electrolytes (strong acids). Others ionize only to a slight degree and are weak electrolytes (weak acids).

2. Acids will react with certain active metals to liberate hydrogen gas. Those metals below hydrogen in the Table N of Standard Electrode Potentials, given in *Reference Tables for Chemistry*, will react with acids to produce a salt of the metal and hydrogen gas.

Note: Some acids, in addition to their acid properties, have strong oxidizing ability. Therefore, except in very dilute solution, they do not release hydrogen gas on reaction with metals. For example, nitric acid and concentrated sulfuric acid have strong oxidizing properties.

3. Acids cause color changes in acid-base indicators. Acid-base indicators are substances that have different colors in acid and basic solutions. Two common indicators are **litmus** which is blue in basic solution and red in acid solution, and **phenolphthalein** which is pink in basic solution and colorless in acid solution.

Note: These indicators do not change color exactly at pH 7. Different indicators change color at different concentrations of hydrogen ions.

4. Acids react with hydroxides to form water and a salt. When hydrogen ions react with hydroxide ions, water is formed. This reaction is called **neutralization**.

5. Dilute aqueous solutions of acids have a sour taste, such as vinegar and acetic acid.

6. Acids react with metallic oxides to form salts and water.

Conceptual Definitions of Acids:
1. Arrhenius' theory. An acid is a substance that yields hydrogen ions in aqueous solutions. This conceptual definition is adequate when considering reactions in aqueous solutions. As knowledge of the mechanism of chemical reactions has increased, more inclusive definitions have been advanced. The characteristic properties of acids in aqueous solution are due to an excess of **hydrogen ions** (hydronium ions).

2. Brönsted-Lowry theory. An acid is any species (molecule or ion) that can donate a proton to another species. The Brönsted-Lowry theory does not replace the Arrhenius' theory, but extends it. The Brönsted-Lowry definition of an acid includes all substances that are acids according to the Arrhenius definition.

In addition, some molecules and ions are classified as acids under the Brönsted-Lowry definition that are not acids in the Arrhenius sense. For example, in the reaction:

$$NH_3 + H_2O \rightleftarrows NH_4^+ + OH^-$$

The water molecule donates a proton to the ammonia and is considered an **acid** in the Brönsted-Lowry sense. In the reverse reaction, the ammonium ion will donate a proton and act as the acid, while the hydroxide ion accepts the proton and acts as a base.

B. Bases.

Bases may be defined in terms of their characteristic properties.

Operational Definitions of Bases:

1. **Aqueous solutions of bases conduct electricity.** Some examples of the relative degree of ionization are shown in the *Reference Tables for Chemistry.*

2. **Bases cause color changes in acid-base indicators.** They cause red litmus to turn blue and phenolphthalein (a weak, colorless acid) to turn pink.

3. **Bases react with acids to form water and a salt.**

4. **Aqueous solutions of bases feel slippery.**

5. **Strong bases have a caustic action on the skin.**

Conceptual Definition of Bases:

1. **Arrhenius' theory. A base is a substance that yields hydroxide ions as the only negative ions in aqueous solution.** According to the Arrhenius definition, the only bases are hydroxides. The characteristic properties of bases in an ion aqueous solution are due to the hydroxide ion.

2. **Brönsted-Lowry theory. A base is any species (molecule or ion) that can accept a proton.** The Brönsted-Lowry definition extends the Arrhenius' definition to include many species in addition to the OH$^-$ that can accept a proton. For example, in the reaction:

$$H_2O \ + \ HCl \ \longrightarrow \ H_3O^+ \ + \ Cl^-$$

The water molecule combines with a proton to form the **hydronium ion** and is here considered a **base** in the Brönsted-Lowry sense. In the reverse reaction, the hydronium ion will donate a proton and act as an acid. While the chloride ion, which accepts the proton, acts as a base.

C. Amphoteric (Amphiprotic) Substances

An amphoteric (amphiprotic) substance is one that **can act either as an acid or as a base,** depending on its chemical environment. By referring to Table *L* in the *Reference Tables for Chemistry,* one can easily find out which items are amphiprotic. If they appear on **either side of the reactions,** it means that they can act as either acids or bases. For example, NH_3, HSO_4^-, HCO_3^-. Therefore, they are considered amphiprotic. Also, as shown in the above equation, water (H_2O) can act as either a base or an acid. Therefore, it is considered amphiprotic.

Questions VII-1

1   Which Brönsted base can also act as an acid?
    (1) Cl$^-$ (2) OH$^-$ (3) S^{2-} (4) SO$_4^{2-}$

2   According to Reference Table *L*, which particle is amphiprotic?
    (1) HNO$_3$ (2) NO$_3^-$ (3) NH$_3$ (4) NH$_4^+$

3   According to Reference Table L, which ion is amphiprotic?

    (1) $Br^-(aq)$ (2) $HS^-(aq)$ (3) $S^{2-}(aq)$ (4) $CO_3{}^{2-}(aq)$

4   Which reaction best illustrates amphoterism?

    (1) $H_2O + HCl \rightarrow H_3O^+ + Cl^-$ (3) $H_2O + H_2SO_4 \rightarrow H_3O^+ + HSO_4{}^-$

    (2) $NH_3 + H_2O \rightarrow NH_4{}^+ + OH^-$ (4) $H_2O + H_2O \rightarrow H_3O^+ + OH^-$

5   According to Reference Table L, which substance is amphiprotic?

    (1) H_2SO_4 (2) HNO_3 (3) NH_3 (4) HBr

6   When tested, a solution turns red litmus to blue. This indicates that the solution contains more:

    (1) H^+ ions than OH^- ions

    (2) H_3O^+ ions than OH^- ions

    (3) OH^- ions than H_3O^+ ions

    (4) H^+ and OH^- ions than H_2O molecules

7   Which compound is an electrolyte?

    (1) $C_6H_{12}O_6$ (2) $C_{12}H_{22}O_{11}$ (3) C_2H_5OH (4) CH_3COOH

8   In the reaction $H_2O + H_2O \leftrightarrow H_3O^+ + OH^-$, the water is acting as:

    (1) a proton acceptor, *only*   (3) *both* a proton acceptor *and* donor

    (2) a proton donor, *only*   (4) *neither* a proton acceptor *nor* donor

9   Which solution will turn litmus from red to blue?

    (1) $H_2S(aq)$ (2) $NH_3(aq)$ (3) $SO_2(aq)$ (4) $CO_2(aq)$

10 According to Reference Table L, which is an amphiprotic ion?

    (1) $HSO_4{}^-$ (2) $NH_4{}^+$ (3) $NO_3{}^-$ (4) Cl^-

III. Acid — Base Reactions

A. Neutralization

Acid-base neutralization pertains to the reaction that occurs when equivalent quantities of an acid and a hydroxide are mixed. One mole of hydrogen ions will react with one mole of hydroxide ions to form water.

$$H^+ + OH^- \longrightarrow H_2O$$

Neutralization reactions are used for the following:

1. Acid-base titration. The molarity of an acid (or base) of unknown concentration can be determined by slowly combining it with a base (or acid) of known molarity. The acid or base solution of a known molarity is called the **standard solution**. This process of metering a standard solution into a solution of unknown concentration is called **titration**. During titration, when the molar quantities of acid and base mixed are equal, **neutralization** has occurred. This point of neutralization is called the "**equivalence point.**"

The molarity of a solution of unknown concentration can be calculated from an understanding of the molar relationship involved. By knowing the concentration of the standard solution and the volumes of both solutions needed to reach the equivalence point, we can use the following procedure to find the molarity of the unknown solution:

a. After writing the balanced equation, we can determine from the coefficients, the molar ratio of the reactants and products.

b. By using the following equation, you can determine the moles of standard solution required to neutralize the solution of unknown concentration.

moles of known solute = volume of solution in liters x molarity

c. By molar ratios, determine the moles of the solution of unknown concentration that were used.

$$\frac{\text{moles of known}}{\text{moles of unknown}} = \frac{\text{liters of known x molarity of known}}{\text{liters of known x molarity of unknown}}$$

d. By substituting in the above equation, find the molarity of the solution of unknown concentration. When monoprotic acids, which provide one proton per molecule, or bases, which can accept one proton per molecule are used, you can use the following equation:

liters of acid x molarity of acid = liters of base x molarity of base

Sample Problem:
How many milliliters of 0.5 M NaOH solution are required to neutralize 50 ml of 0.2 M HCl solution?

Solution:
Since both the acid and base are monoprotic, we can convert the volumes given in milliliters to liters and use the equation.

liters of acid x molarity of acid = vol. of base x molarity of base

$$0.050\ \text{L x } 0.2\ \text{M} = V_B \times 0.5\ \text{M}$$

$$\frac{\overset{0.1}{\cancel{0.050\ \text{L x } 0.2\ \text{M}}}}{\underset{1}{\cancel{0.5\ \text{M}}}} = V_B$$

$$V_B = 0.002\ \text{liters (answer)}$$

Normality

In solving titration problems, as long as the acid and base are monoprotic, (by using the above equation), we can easily determine the concentration of either, if we know the concentration of one of them. However, if one of the two (either the acid or base) are diprotic (H_2SO_4), or triprotic (H_3PO_4), a complication arises.

To overcome this complication, chemists have devised an expression of concentration, which is concerned with the amount of hydrogen ions given off by an acid or accepted by a base. For example, since a mole of HCl molecules can donate a mole of H^+ (recall that one mole of H^+ weighs 1 g.), it is called a monoprotic acid. It is capable of supplying 1 gram-equivalent of H^+. A solution of one mole of a monoprotic acid would be called a 1 molar solution (1M) or a 1 normal solution (1N). A mole of a diprotic acid, such as H_2SO_4, is considered 2 normal (2N) and can supply 2 gram equivalents of H^+. H_3PO_4 is a triprotic acid. It can supply 3 gram equivalents of H^+ per mole of acid molecules and is considered 3N.

Question: What is the normality of 1 liter of a 3 molar solution of H_3PO_4?

Answer: _____

The normality of bases is classified in the same manner. For example, NaOH will break up so that one OH^- ion is produced for every molecule broken up. Therefore, NaOH is called a monoprotic base, and a mole of this base will be capable of accepting 1 mole of H^+, so it is considered a 1 gram-equivalent base. A 1M solution of $Mg(OH)_2$ is 2 normal (2N). Mathematically, the relationship between reacting acids and bases includes their volumes and their concentrations expressed in normality, similar to monoprotic acids and bases. For example,

Acid Vol. x Acid Normality = Base Vol. x Base Normality

Expressing the concentration in normality allows us to use diprotic and triprotic acids or bases in the equation. Although normality is no longer a suggested way to express concentration, it will probably be in common usage for some time. A limitation of the normality scale is that a given solution may be more than one normality, depending on the reaction for which it is used and how much it ionizes in solution.

2. Salts. A salt is an ionic compound containing positive ions other than hydrogen and negative ions other than hydroxide. Most salts are strong electrolytes and are considered to be completely dissociated in aqueous solution. Some salts in aqueous solution react with the water to form solutions that are acidic or basic.

This process is called **hydrolysis,** and it is considered to be the **opposite of a neutralization reaction.** For example, in a neutralization reaction, an acid and base react to form a salt and water. In hydrolysis, the salt is added to the water to form the acid and base, which originally formed the salt. Of the various reactions that occur when salts react with water, four cases may be distinguished. They are:

a. A salt of a strong acid and a strong base forms a neutral solution with a pH of about 7. An example is NaCl, formed from the strong acid, HCl, and the strong base NaOH.

b. A salt of a weak acid and a strong base forms a basic solution with a pH of greater than 7. An example is $NaC_2H_3O_2$, which is formed from acetic acid, $HC_2H_3O_2$ (acetic acid is also shown as CH_3COOH), a weak acid, and the strong base NaOH.

c. A salt of a strong acid and a weak base forms an acidic solution with a pH of less than 7. An example is NH_4Cl, which is formed from the strong acid, HCl, and the weak base, NH_4OH.

d. A salt of a weak acid and base forms a solution that may be acidic or basic, or neutral, depending on the ionization constants of the products. An example is $NH_4C_2H_3O_2$ which is formed from the weak acid, acetic acid, $HC_2H_3O_2$, and the weak base ammonium hydroxide, NH_4OH.

B. Conjugate Acid — Base Pair

According to the Brönsted-Lowry theory, acid-base reactions involve a transfer of protons from the acid to the base.

To accept a proton, a base must have at least one pair of unshared electrons. The proton (H^+) will share a pair of electrons belonging to the base, forming a coordinate covalent bond.

In an acid-base reaction, an acid transfers a proton to become a conjugate base. This acid and its newly formed base form a conjugate acid-base pair. A base gains a proton to become a conjugate acid, forming a second acid- base pair. Each pair, made up of an acid and its base, is related by the transfer of a proton.

In the following reactions the two **conjugate pairs** are identified by subscripts.

$$Base_1 + Acid_2 \rightleftharpoons Acid_1 + Base_2$$

$$H_2O + HCl \rightleftharpoons H_3O^+ + Cl^-$$

$$NH_3 + H_2O \rightleftharpoons NH_4^+ + OH^-$$

The same form of identification of acid and base pairs can be done using brackets.

$$H_2O + H_2SO_4 \rightleftharpoons H_3O^+ + HSO_4^-$$

Base $_1$ Acid $_2$ Acid $_1$ Base $_2$

Note: The strongest acids have the weakest conjugate bases, and the strongest bases have the weakest conjugate acids.

Refer to Table *L* in the *Reference Tables for Chemistry*.

Questions VII-2

1 If HCl and H_2O react together in an acid-base reaction to form their Brönsted-Lowry conjugates, the products would be:
   (1) HCl and H_3O^+ (3) Cl_2 and H_2
   (2) Cl^- and OH^- (4) Cl^- and H_3O^+

2 The conjugate acid of the HS^- ion is
   (1) H^+ (3) H_2O
   (2) S^{2-} (4) H_2S

3   A solution containing more H_3O^+ ions than OH^- ions is:
     (1) basic and turns litmus paper red
     (2) basic and turns litmus paper blue
     (3) acidic and turns litmus paper red
     (4) acidic and turns litmus paper blue

4   What is a conjugate acid-base pair in the reaction
     $H_2O + HI \rightarrow H_3O^+ + I^-$?
     (1) H_2O and HI   (3) HI and I^-
     (2) H_2O and I^- (4) HI and H_3O^+

5   In the reaction: $NH_2^- + H_2O \leftrightarrow NH_3 + OH^-$, two Brönsted acids are:
     (1) NH_2^- and H_2O (3) H_2O and NH_3
     (2) NH_2^- and NH_3 (4) H_2O and OH^-

6   In the reaction $CH_3COOH + H_2O \leftrightarrow CH_3COO^- + H_3O^+$, a conjugate acid-base pair is:
     (1) CH_3COOH and H_3O^+ (3) CH_3COOH and CH_3COO^-
     (2) CH_3COO^- and H_2O (4) CH_3COO^- and H_3O^+

7   In the reaction: $H_3PO_4 + H_2O \leftrightarrow H_2PO_4^- + H_3O^+$, a conjugate acid–base pair is
     (1) H_3PO_4 and H_2O (3) $H_2PO_4^-$ and H_2O
     (2) H_3PO_4 and $H_2PO_4^-$ (4) $H_2PO_4^-$ and H_3O^+

8   A sample of pure water contains:
     (1) neither OH^- ions nor H_3O^+ ions
     (2) equal concentrations of OH^- and H_3O^+ ions
     (3) a larger concentration of H_3O^+ ions than OH^- ions
     (4) a smaller concentration of H_3O^+ ions than OH^- ions

9   Which of the following acids has the weakest conjugate base?
     (1) HCl   (2) H_3PO_4 (3) HF   (4) H_2S

10   Which 0.1 M solution is the best conductor of electricity?
     (1) HCl(aq)   (3) $C_6H_{12}O_6$(aq)
     (2) HF(aq)   (4) CH_3OH(aq)

IV. Ionization Constant

The equilibrium constant for the ionization of acids (K_a) is a convenient method for comparing the relative strength of acids. For the reaction:

$$HB(aq) \longrightarrow H^+(aq) + B^-(aq)$$

$$K_a = \frac{[H^+][B^-]}{[HB]}$$

Ionization constants can be calculated for all acids that are not completely ionized. For acids that are completely ionized there is no equilibrium. The denominator (HB) approaches zero and K_a approaches infinity.

Ionization constants such as those given in the *Reference Tables for Chemistry* can be used in comparing the relative strength of acids. *Note: The magnitude of the number indicates the strength of the acid.*

For example, an acid with a $K_a = 1.8$ x 10^{-5} (although a weak acid) is stronger than an acid with $K_a = 1.0$ x 10^{-7}.

A. The Ionization Constant For Water — K_w

Only 1 x 10^{-7} moles per liter of water molecules ionize at room temperature and standard pressure. When that small amount of water molecules do ionize, they break up into equal portions of hydrogen and hydroxide ions:

$$H_2O \longrightarrow H^+ + OH^- \quad \text{also} \quad 2H_2O \longrightarrow H_3O^+ + OH^-$$

The molar proportions of the equation (preceding page) are: $1 \rightarrow 1 + 1$. Therefore, if 1 x 10^{-7} moles per liter of water molecules ionizes, it will form 1 x 10^{-7} moles per liter H^+ and 1 x 10^{-7} moles per liter OH^-. Since water is such a weak electrolyte, its concentration is considered a constant. The equilibrium constant expression for water would be:

$$K_{eq} = \frac{[H^+]\ [OH^-]}{[H_2O]}$$

However, as stated above, for practical purposes, the concentration of water is a constant. Since the product of two constants is another constant, (in this case called K_w) the equilibrium expression for water is written as follows:

$$K_w = [H^+]\ [OH^-]$$

Therefore, as shown above, it follows that in water and aqueous solutions the product of the hydrogen ion concentration and the hydroxide ion concentration is a constant at constant temperature. This constant, K_w, is useful in solving problems involving hydrogen ion and hydroxide ion concentrations.

Substituting the molar concentrations of hydrogen and hydroxide ions into the formula, we can derive a numerical constant at 25°C for the ionization of water per liter.

$$K_w = [H^+]\ [OH^-]$$

$$K_w = [1 \text{ x } 10^{-7}]\ [1 \text{ x } 10^{-7}] = 1.0 \text{ x } 10^{-14}$$

Note: in pure water, $[H^+] = [OH^-] = 1.0$ x 10^{-7} at 25° C

B. pH

pH is the logarithm (exponent) of the reciprocal (negative logarithm) of the hydrogen ion concentration. Therefore, if the hydrogen ion concentration is 1x 10^{-5}, its reciprocal would be 1 x 10^5, and its logarithm (exponent) is 5. Its pH is expressed as 5. The pH of a solution indicates the concentration of hydrogen ions (acid strength) in a solution.

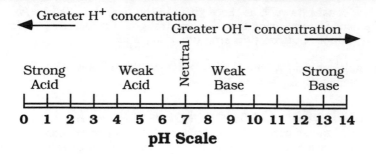

pH Scale

A pH of 7 is neutral.
A pH of less than 7 is acidic.
A pH greater than 7 is basic. Pure water has a pH of 7 at 25°C.

Questions VII-3

1 As 50. milliliters of 0.1 M HCl is added to 100 milliliters of 0.1 M NaOH, the pH of the NaOH solution:
   (1) decreases   (2) increases   (3) remains the same

2 Which salt hydrolyzes in water to form a solution that is acidic?
   (1) KCl   (2) NaCl   (3) NH_4Cl (4) LiCl

3 In a solution with a pH of 3, the color of:
   (1) litmus is red   (3) phenolphthalein is red
   (2) litmus is blue   (4) phenolphthalein is blue

4 Based on Reference Table *L*, which 0.1 M solution would have the lowest concentration of H+ ions:
   (1) HI   (2) HCl   (3) HNO_2 (4) HNO_3

5 What is the ionization constant, K_a, for H_2S at 1 atmosphere and 298 K?
   (1) 1.0×10^{-4} (2) 1.0×10^{-5} (3) 9.5×10^{-8} (4) 1.0×10^{-17}

6 OH- ion concentration is greater than H_3O^+ ion concentration in a 0.1 M solution of:
   (1) NaOH   (2) CH_3OH (3) HNO_3 (4) H_2SO_4

7 What is the pH of a solution whose H_3O^+ ion concentration is 0.0001 mole per liter?
   (1) 1   (2) 10   (3) 14   (4) 4

8 The [H+] of a solution is 1×10^{-2} at 298 K. What is the [OH-] of this solution?
   (1) 1×10^{-14} (2) 1×10^{-12} (3) 1×10^{-7} (4) 1×10^{-2}

9 What is the total number of moles of H+ ions that will neutralize 2.0 moles of OH- ions?
   (1) 1.0   (2) 2.0   (3) 17   (4) 34

10 Phenolphthalein has a pink color in solution which has a pH of:
   (1) 1   (2) 5   (3) 7   (4) 11

11 If 50. milliliters of a 1.0 M NaOH solution is needed to exactly neutralize 10. milliliters of an HCl solution, the molarity of the HCl solution is
   (1) 1.0 M   (2) 0.20 M   (3) 5.0 M   (4) 10. M

12 What is the K_w of water at 1 atm and 298 K?
   (1) 1.0×10^{-14} (2) 1.0×10^{-7} (3) 1.0×10^{14} (4) 1.0×10^{7}

13 If a solution has a hydronium ion concentration of 1×10^{-9} M, the solution is:
(1) basic and has a pH of 9   (3) acidic and has a pH of 9
(2) basic and has a pH of 5   (4) acidic and has a pH of 5

14 If 50 milliliters (mL) of a 0.01 M HCl solution is required to neutralize exactly 25 milliliters (mL) of NaOH, what is the concentration of the base?
(1) 0.01 M   (2) 0.02 M   (3) 0.0005 M   (4) 0.04 M

15 A 0.1 M acid solution at 298 K would conduct electricity best if the acid had a K_a value of:

(1) 1.0×10^{-7} (2) 1.8×10^{-5} (3) 6.7×10^{-4} (4) 1.78×10^{2}

16 In any water solution at 25°C, the product of the H_3O^+ ion concentration times the hydroxide ion concentration is equal to.

(1) 1×10^{-1} (2) 1×10^{-7} (3) 1×10^{-9} (4) 1×10^{-14}

17 What is the pH of a 0.001 M solution of HCl?
(1) 1   (2) 7   (3) 3   (4) 11

18 What is the hydronium ion concentration of a solution at 298 K whose hydroxide ion concentration is 1×10^{-8}?

(1) 1×10^{-6} (2) 1×10^{-7} (3) 1×10^{-8} (4) 1×10^{-14}

19 How many moles of NaOH are required to exactly neutralize 0.50 liter of 2.0 M HCl?
(1) 1.0   (2) 2.0   (3) 0.50   (4) 4.0

20 How many milliliters of 2.0 M HCl are required to exactly neutralize 10. milliliters of 3.0 M NaOH?
(1) 5.0   (2) 15   (3) 20   (4) 30

21 As 0.1 M HCl is added to 0.1 M KOH, the pH of the basic solution:
(1) decreases and basicity decreases
(2) increases and basicity decreases
(3) decreases and basicity increases
(4) increases and basicity increases

22 Which of the following acids ionizes to the least extent at 298 K?

(1) HF   (2) HNO_2 (3) H_2S (4) H_2SO_4

23 What is the pH of an aqueous solution of $C_6H_{12}O_6$?
(1) 1   (2) 7   (3) 11   (4) 14

24 A solution that has a pH of 11:
(1) is neutral   (3) will turn litmus red
(2) is acidic   (4) will turn phenolphthalein pink

25 What is the hydroxide ion concentration in moles per liter of a solution whose $[H_3O^+]$ is equal to 1×10^{-5}?

(1) 1×10^{-5} (2) 5×10^{-1} (3) 1×10^{-9} (4) 9×10^{-1}

26 The $[OH^-]$ of a solution is 1×10^{-6}. At 1 atmosphere and 298 K, the product of the $[H_3O^+][OH^-]$ is:

(1) 1×10^{-2} (2) 1×10^{-6} (3) 1×10^{-8} (4) 1×10^{-14}

27 Which solution listed below is the best conductor of electricity?
(1) 0.1 M C_2H_5OH (3) 0.1 M $C_6H_{12}O_6$
(2) 0.1 M CH_3COOH (4) 0.1 M HNO_3

28 Pure water at 25°C has a pH of
(1) 1×10^{-7} (2) 1×10^{-14} (3) 7   (4) 14

29 As the H_3O^+ ion concentration of a solution increases, the pH of the
    solution:
    (1) decreases   (2) increases   (3) remains the same
30 What is the ionization constant for water at 298 K?
    (1) 1.0×10^{-14} (3) 1.0×10^7
    (2) 1.0×10^{-7} (4) 1.0×10^{14}
31 The pH of a 0.1 M CH_3COOH solution is
    (1) less than 1   (3) equal to 7
    (2) greater than 1 but less than 7   (4) greater than 7 but less than 14
32 Which solution can turn phenolphthalein pink?
    (1) $CH_3OH(aq)$ (3) $HCl(aq)$
    (2) $CH_3COOH(aq)$ (4) $NaOH(aq)$
33 The H_3O^+ ion concentration of a solution is 1×10^{-4} mole per liter. This
    solution is
    (1) acidic and has a pH of 4   (3) basic and has a pH of 4
    (2) acidic and has a pH of 10   (4) basic and has a pH of 10

Self-Help Questions VII

Unit VII - Acids And Bases

I. Electrolytes and dissociation constant,
Acids - operational and conceptual definitions;
Bases - operational and conceptual definitions,
Amphoteric (Amphoprotic) substances.

1   According to Reference Table *L*, which species is amphiprotic?
    (1) H_3PO_4 (2) $PO_4{}^{3-}$ (3) NH_3 (4) $NH_4{}^+$

2   Which solution can turn phenolphthalein pink?
    (1) $CH_3OH(aq)$ (3) $HCl(aq)$
    (2) $CH_3COOH(aq)$ (4) $NaOH(aq)$

3   Water containing dissolved electrolyte conducts electricity because the
    solution contains mobile
    (1) electrons   (2) molecules   (3) atoms   (4) ions

4   According to Reference Table *L*, which substance is amphiprotic?
    (1) HI   (2) $NH_4{}^+$ (3) HNO_3 (4) HS^-

5   Which compound is an electrolyte?
    (1) $C_6H_{12}O_6$ (3) CH_3CH_2OH
    (2) $C_{12}H_{22}O_{11}$ (4) CH_3COOH

6   According to the Brönsted-Lowry theory, an acid is any species that can
    (1) donate a proton   (3) donate an electron
    (2) accept a proton   (4) accept an electron

7   According to the Arrhenius theory, when an acid substance is dissolved in
    water it will produce a solution containing only one kind of positive ion.
    To which ion does the theory refer?
    (1) acetate   (2) hydrogen   (3) chloride   (4) sodium

8   According to Reference Table L, ammonia can act as
(1) a Brönsted acid, only
(2) a Brönsted base, only
(3) either a Brönsted acid or a Brönsted base
(4) neither a Brönsted acid nor a Brönsted base

9   A substance is added to a water solution containing phenolphthalein, causing the solution to turn pink. Which substance would produce this result?

(1) $HC_2H_3O_2$ (2) H_2CO_3 (3) KOH   (4) CH_3OH

10   Which compound is an electrolyte?
(1) $C_6H_{12}O_6$ (2) C_6H_6 (3) CH_3OH (4) CH_3COOH

11   According to Reference Table L, which species is amphoteric (amphiprotic)?

(1) HS^- (2) HCl   (3) NH_4^+ (4) HBr

12   At 298 K, which metal will release $H_2(g)$ when reacted with HCl(aq)?
(1) Au(s)   (2) Zn(s)   (3) Hg(l)   (4) Ag(s)

13   When Arrhenius acid is dissolved in water, it produces
(1) H^+ as the only positive ion in solution
(2) NH_4^+ as the only positive ion in solution
(3) OH^- as the only negative ion in solution
(4) HCO_3^- as the only negative ion in solution

II. Acid - Base Reactions; inc. neutralization, and titration problems Salts, including hydrolysis, Conjugate acid - base pairs

14   In the reaction $HBr + H_2O \leftrightarrow H_3O^+ + Br^-$, which is a conjugate acid-base pair?
(1) HBr and Br^- (3) H_3O^+ and Br^-
(2) HBr and H_2O (4) H_3O^+ and HBr

15   How many milliliters of 2.5 M HCl are required to exactly neutralize 15 milliliters of 5.0 M NaOH?
(1) 10.   (2) 20.   (3) 30.   (4) 40.

16   Based on Reference Table L, which species is amphoteric?
(1) NH_2^- (2) NH_3 (3) I^- (4) HI

17   Which equation represents a neutralization reaction?
(1) $CaO + H_2O \rightarrow Ca(OH)_2$
(2) $2HCl + Zn \rightarrow ZnCl_2 + H_2$
(3) $H_2SO_4 + CaCO_3 \rightarrow CaSO_4 + H_2O + CO_2$
(4) $HNO_3 + KOH \rightarrow KNO_3 + H_2O$

18 How much water is formed when 1.0 mole of HCl reacts completely with 1.0 mole of NaOH?
(1) 1.0 mole   (2) 2.0 moles   (3) 0.50 mole   (4) 0.25 mole

19 When the salt $NaHCO_3$ is dissolved in water, the solution becomes

(1) basic due to the production of H_3O^+ ions

(2) acidic due to the production of H_3O^+ ions

(3) basic due to the production of OH^- ions

(4) acidic due to the production of OH^- ions

20 As a solution of NaOH is added to a solution of HCl at constant temperature, the product of $[H_3O^+]$ $[OH^-]$
(1) decreases   (2) increases   (3) remains the same

21 Which substance is classified as a salt?
(1) $Ca(OH)_2$ (3) CCl_4
(2) $C_2H_4(OH)_2$ (4) $CaCl_2$

22 What volume of a 0.200 M NaOH solution is needed to exactly neutralize 40.0 milliliters of a 0.100 M HCl solution?
(1) 10.0 mL   (2) 20.0 mL   (3) 40.0 mL   (4) 80.0 mL

23 In the reaction $NH_3(g)$ + $H_2O(l)$ ↔ $NH_4^+(aq)$ + $OH^-(aq)$, which pair are Brönsted bases?
(1) NH_3 and H_2O (3) NH_4^+ and H_2O
(2) NH_3 and OH^- (4) NH_4^+ and OH^-

24 Which is a net ionic equation for a neutralization reaction?
(1) H^+ + HCO_3^- → H_2CO_3 (3) Ag^+ + Cl^- → AgCl
(2) NH_4^+ + OH^- → NH_4OH (4) H^+ + OH^- → H_2O

25 As water solution contains 0.50 mole of HCl. How much KOH should be added to the HCl solution to exactly neutralize it?
(1) 1.0 mole   (3) 0.25 mole
(2) 2.0 moles   (4) 0.50 mole

26 Which ion is the conjugate base of H_2SO_4?
(1) SO_3^{2-} (2) S^{2-} (3) HSO_3^- (4) HSO_4^-

27 Which of the following is the best conductor of electricity?
(1) NaCl(s)   (3) $C_6H_{12}O_6(s)$
(2) NaCl(aq)   (4) $C_6H_{12}O_6(aq)$

28 Given the reaction: HCl(aq) + $H_2O(l)$ → X(aq) + Y(aq)
Which pair does X(aq) + Y(aq) represent?
(1) H_3O^+ + Cl^- (3) HClO + H_2
(2) H_2Cl^+ + OH^- (4) H^+ + H_2OCl^-

29 Which compound reacts with an acid to form a salt and water?
(1) CH_3Cl (3) KCl
(2) CH_3COOH (4) KOH

30 In the reaction $HSO_4^- + H_2O \leftrightarrow H_3O^+ + SO_4^{2-}$, an acid-base conjugate pair is
(1) HSO_4^- and SO_4^{2-}
(2) HSO_4^- and H_2O
(3) SO_4^{2-} and H_3O^+
(4) SO_4^{2-} and H_2O

31 Which type of reaction is represented by the following equation?
$$Al_2S_3 + 6H_2O \rightarrow 2Al(OH)_3 + 3H_2S$$
(1) neutralization
(2) dehydration
(3) electrolysis
(4) hydrolysis

32 If 50. milliliters of 0.50 M HCl is used to completely neutralize 25 milliliters of KOH solution, what is the molarity of the base?
(1) 1.0 M   (2) 0.25 M   (3) 0.50 M   (4) 2.5 M

33 When an acid solution exactly neutralizes a base solution, which acid-base combination always produces a mixture with a pH less than 7?
(1) a strong acid and a strong base
(2) a strong acid and a weak base
(3) a weak acid and a strong base
(4) a weak acid and a weak base

III. Ionization Constant and pH

34 What is the ionization constant for water at 298 K?
(1) 1.0×10^{-14} (2) 1.0×10^{-7} (3) 1.0×10^7 (4) 1.0×10^{14}

35 The pH of a 0.1 M CH_3COOH solution is
(1) less than 1
(2) greater than 1 but less than 7
(3) equal to 7
(4) greater than 7 but less than 14

36 According to Reference Table *L*, which of the following is the strongest Brönsted-Lowry base?
(1) HS^- (2) S^{2-} (3) HSO_4^- (4) SO_4^{2-}

37 Which relationship between ion concentrations always exists in an aqueous solution that is basic?
(1) $[H^+]$ equals zero.
(2) $[H^+]$ equals $[OH^-]$.
(3) $[H^+]$ is less than $[OH^-]$.
(4) $[H^+]$ is greater than $[OH^-]$.

38 Which statement best describes a solution with a pH of 3?
(1) It has an H_3O^+ ion concentration of 1×10^3 mol/L and is acidic.
(2) It has an H_3O^+ ion concentration of 1×10^{-3} mol/L and is acidic.
(3) It has an H_3O^+ ion concentration of 1×10^3 mol/L and is basic.
(4) It has an H_3O^+ ion concentration of 1×10^{-3} mol/L and is basic.

39 Based on Reference Table *L*, which of the following aqueous solutions is the *poorest* conductor of electricity?
(1) 0.1 M H_2S
(2) 0.1 M HF
(3) 0.1 M H_3PO_4
(4) 0.1 M H_2SO_4

40 Which pH value represents a solution with the *lowest* OH⁻ ion concentration?
    (1) 1   (2) 7   (3) 10   (4) 14

41 Given K_w = [H⁺][OH⁻] = 1×10^{-14} at 298 K. What is the concentration of H⁺ in pure water at 298 K?
    (1) 1×10^{-7} mole per liter   (3) 1×10^{-14} mole per liter
    (2) 1×10^{7} moles per liter   (4) 1×10^{14} moles per liter

42 Which 0.1 M solution has the highest concentration of H_3O^+ ions?
    (1) CH_3COOH (2) NaCl   (3) KBr   (4) $Ba(OH)_2$

43 When equal volumes of 0.5 M HCl and 0.5 M NaOH are mixed, the pH of the resulting solution is
    (1) 1   (2) 2   (3) 7   (4) 4

44 What is the hydroxide ion concentration of a solution that has a hydronium ion concentration of 1×10^{-9} mole per liter at 298 K?
    (1) 1×10^{-5} mole per liter   (3) 1×10^{-9} mole per liter
    (2) 1×10^{-7} mole per liter   (4) 1×10^{-14} mole per liter

45 According to Reference Table *L*, which of the following acids will ionize to the greatest degree in a water solution?
    (1) H_3PO_4 (2) HF   (3) H_2S (4) HNO_2

Important Terms and Concepts

Redox – Oxidation and Reduction
   Oxidation Number
Electrochemistry
   Half–Reactions, Half–Cells
   Chemical Cells
   Electrolytic Cells

Standard Electrode Potentials
Electrochemical Cells
Electrodes
   Cathode and Anode
Electroplating
Redox Equations

I. Redox: Oxidation — Reduction

Many reactions result from the competition for electrons between atoms, and the term used for this oxidation-reduction competition is redox.

A. Oxidation

Oxygen, having the second highest electronegativity rating of 3.5, after fluorine (4.0), was isolated over 100 years before fluorine, and its properties were well known. Chemists found that whenever an element combined with oxygen, it had a tendency of losing electrons. This tendency of losing electrons was associated with the element oxygen and was called oxidation.

Oxidation represents a loss or an apparent loss of electrons. Any chemical change in which there is an increase in oxidation number, due to a loss of negative charge (electrons), is called oxidation.

The particle that increases in oxidation number is said to be oxidized. Since it is the agent that causes the reduction of another, it is referred to as the **reducing agent**.

B. Reduction

Reduction represents a gain, or apparent gain, of electrons. Any chemical change in which there is a decrease of oxidation number is called reduction.

The particle that decreases in oxidation number is said to be reduced. Since it is the agent that causes the oxidation of another, it is referred to as the **oxidizing agent**.

C. Oxidation Number

The oxidation number (oxidation state) of an atom is the charge which an atom has, or appears to have, when electrons are counted according to certain arbitrary rules. This oxidation number, although arbitrary, is a convenient notation for keeping track of the number of electrons involved in a chemical reaction. In assigning oxidation numbers, electrons shared between two unlike atoms are counted as belonging to the more electronegative atom. The electrons shared between two like atoms are divided equally between the sharing atoms.

Operational Rules For Determining Oxidation Number

Applying the general rules above, has resulted in the following operational rules:

1. In the free elements, each atom has an oxidation number of zero. For example, hydrogen in H_2, sodium in Na, and sulfur in S_8; all have oxidation numbers of zero.

2. In simple ions (ions containing one atom) the oxidation number is equal to the charge on the ion. These common ionic charges can be found in the Reference Tables for Chemistry, for example, Na^+, Zn^{++}, Cl^-.

3. When monatomic ions make up an ionic compound, the oxidation number of each ion is equal to its ionic charge. The algebraic sum of these charges is equal to zero. For example, in $CaCl_2$, calcium has an oxidation number of +2 and each chlorine -1, giving chlorine a total charge of -2; therefore, the total charge of the compound adds up to zero. Iron in $FeCl_2$ has an oxidation number of +2 , and each chlorine -1, which gives chlorine a total oxidation state of -2, so that the total sum of the positive and negative charges is again zero. In $FeCl_3$, iron has an oxidation number of +3, and each chlorine a -1, giving chlorine a total negative charge of -3, and the compound a total charge of zero.

4. All metals in **Group 1** form only **1$^+$** ions and their oxidation number is +1 in all compounds.

5. All metals in **Group 2** form only **2$^+$** ions and their oxidation number is +2 in all compounds.

6. **Oxygen** has an oxidation number of **-2** in all its compounds **except in peroxides** (such as H_2O_2) when it is **-1** and in compounds with fluorine (OF and OF_2) when it may be **+1** or **+2**. For example, in H_2SO_4, oxygen has an oxidation number of -2.

7. **Hydrogen** has an oxidation number of **+1** in all its compounds (such as HCl and H_2SO_4) **except in the metal hydrides** (such as LiH and CaH_2) when it is **-1**.

8. For polyatomic ions (charged particles that contain more than one atom) the oxidation numbers of all the atoms must add up to the charge on the ion. For example, in $SO_4{}^{2-}$ the four oxygen atoms contribute a total oxidation number of -8. Therefore, the sulfur must contribute an oxidation number of +6 to give the ion a charge of 2-.

All oxidation numbers must be consistent with the conservation of charge.

For neutral molecules, the algebraic sum of the oxidation number of all the atoms must add up to zero. For example, in H_2SO_4, the two hydrogens contribute a total of +2, and the four oxygens contribute a total of -8. Therefore, the sulfur must contribute an oxidation number of +6.

D. Redox Reactions

Oxidation and reduction occur simultaneously, *one cannot occur without the other.*

In oxidation and reduction, the increase and decrease of oxidation number results from a shift of electrons. The only way by which electrons can be shifted away from an atom (oxidation) is for them to be pulled toward another atom or ion (reduction). There is a conservation of charge as well as a conservation of mass in a redox reaction.

Redox reactions fall into categories which include:

1. Composition Reactions:

$$\overset{0}{2Na} + \overset{0}{Cl_2} \longrightarrow \overset{1+ \; 1-}{2NaCl}$$

In this reaction (above):
    a. Sodium's oxidation state changes from 0 to +1, and it is oxidized.
    b. Chlorine's oxidation state changes from 0 to -1, and it is reduced.

2. Decomposition or analysis reactions:

$$\overset{1+ \; 1-}{2HCl} \longrightarrow \overset{0}{H_2} + \overset{0}{Cl_2}$$

In this reaction:
    a. Hydrogen's oxidation state changes from +1 to 0, and it is reduced.
    b. Chlorine's oxidation state changes from -1 to 0, and it is oxidized.

3. Single replacement reactions:

$$\overset{0}{Mg} + \overset{1+ \; 1-}{2HCl} \longrightarrow \overset{2+ \; 1-}{MgCl_2} + \overset{0}{H_2}$$

In this reaction:
    a. Magnesium's oxidation state changes from 0 to +2, and it is oxidized.
    b. Hydrogen's oxidation state changes from +1 to 0, and it is reduced.

Note: Ionic or double replacement reactions are not usually redox reactions.

Questions VIII-1

1   If element X forms the oxides XO and X_2O_3, the oxidation numbers of element X are:
    (1) +1 and +2   (3) +1 and +3
    (2) +2 and +3   (4) +2 and + 4

2   Given the reaction: $3Ca^0 + 2Al^{3+} \rightarrow 3Ca^{2+} + 2Al^0$, the ratio of the total number of electrons lost to the total number of electrons gained is:
    (1) 1:1   (2) 2:1   (3) 3:1   (4) 1:2

3   Oxygen has a positive oxidation number in the compound:
    (1) H_2O (2) H_2O_2 (3) OF_2 (4) IO_2

4 Which is an oxidation-reduction reaction?
   (1) $4Na + O_2 \rightarrow 2Na_2O$
   (2) $3O_2 \rightarrow 2O_3$
   (3) $AgNO_3 + NaCl \rightarrow AgCl + NaNO_3$
   (4) $KI \rightarrow K^+ + I^-$

5 Which is a redox reaction?
   (1) $CaCO_3 \rightarrow CaO + CO_2$
   (2) $NaOH + HCl \rightarrow NaCl + H_2O$
   (3) $2NH_4Cl + Ca(OH)_2 \rightarrow 2NH_3 + 2H_2O + CaCl_2$
   (4) $2H_2O \rightarrow 2H_2 + O_2$

6 Which is a redox reaction?
   (1) $H_2O + CO_2 \rightarrow H^+ + HCO_3^-$ (3) $HSO_4^- \rightarrow H^+ + SO_4^{2-}$
   (2) $H^+ + OH^- \rightarrow H_2O$ (4) $2Br^- + Cl_2 \rightarrow 2Cl^- + Br_2$

7 According to Reference Table *N*, which is the strongest reducing agent?
   (1) $Li(s)$ (2) $Na(s)$ (3) $F_2(g)$ (4) $Br_2(l)$

8 Which of the following elements is the strongest oxidizing agent?
   (1) F_2 (2) Cl_2 (3) Br_2 (4) I_2

9 Hydrogen would have a -1 oxidation state if it formed a compound with:
   (1) N   (2) O   (3) F   (4) K

10 The oxidation number of nitrogen is highest in:
   (1) N_2 (2) NH_3 (3) NO_2 (4) N_2O

11 An oxide ion is oxidized to an oxygen atom by:
   (1) gaining electrons   (3) gaining protons
   (2) losing electrons   (4) losing protons

12 Which metal will react spontaneously with HCl (aq)?
   (1) Au   (2) Ag   (3) Cu   (4) Mg

13 What is the oxidation number of sulfur in H_2SO_4?
   (1) 0   (2) -2   (3) +6   (4) +4.

14 In the reaction, $3Zn^0 + 2Fe^{3+} \rightarrow 3Zn^{2+} + 2Fe^0$, the oxidizing agent is:
   (1) Fe^{3+} (2) Zn^0 (3) Zn^{2+} (4) Fe^0

15 Which is not an oxidation and reduction reaction?
   (1) $KOH + HCl \rightarrow KCl + H_2O$ (3) $2KClO_3 \rightarrow 2KCl + 3O_2$
   (2) $2K + Cl_2 \rightarrow 2KCl$ (4) $2K + 2H_2O \rightarrow 2KOH + H_2$

16 In the reaction of $Zn + Cu^{2+} \rightarrow Zn^{2+} + Cu$, the reducing agent:
   (1) gains protons   (3) gains electrons
   (2) loses protons   (4) loses electrons

17 In the reaction: $2KMnO_4 + 5SO_2 + 2H_2O \rightarrow K_2SO_4 + 2MnSO_4 + 2H_2SO_4$, the oxidation number of manganese changes from:
   (1) +5 to +2   (3) +7 to +2
   (2) +6 to +3   (4) +4 to +3

18 How many moles of electrons would be required to completely reduce 1.5 moles of Al^{3+} to Al?
   (1) 0.50   (2) 1.5   (3) 3.0   (4) 4.5

19 Which half-reaction correctly represents reduction?
   (1) $S^{2-} + 2e^- \rightarrow S^0$ (3) $Mn^{7+} + 3e^- \rightarrow Mn^{4+}$
   (2) $S^{2-} \rightarrow S^0 + 2e^-$ (4) $Mn^{7+} \rightarrow Mn^{4+} + 3e^-$

20 Given the reaction: $Sn^{2+}(aq) + 2Fe^{3+}(aq) \rightarrow Sn^{4+}(aq) + 2Fe^{2+}(aq)$, the oxidizing agent in this reaction is:
(1) Sn^{2+} (2) Fe^{3+} (3) Sn^{4+} (4) Fe^{2+}

21 What is the oxidizing agent in the reaction
$Zn^0 + 2Ag^+ \rightarrow Zn^{2+} + 2Ag^0$?
(1) Zn^0 (2) Ag^0 (3) Zn^{2+} (4) Ag^+

22 In the reaction $3Ag + Au^{3+} \rightarrow 3Ag^+ + Au$, what is the total number of moles of electrons gained by 1 mole of Au^{3+}?
(1) 1   (2) 6   (3) 3   (4) 9

23 Given the reaction: $Cu^{2+}(aq) + Zn(s) \rightarrow Cu(s) + Zn^{2+}(aq)$, which is the reducing agent?
(1) $Cu^{2+}(aq)$ (2) $Zn^{2+}(aq)$ (3) $Zn(s)$ (4) $Cu(s)$

24 In the reaction: $4HCl + MnO_2 \rightarrow MnCl_2 + 2H_2O + Cl_2$, the manganese is:
(1) reduced and the oxidation number changes from +4 to +2
(2) oxidized and the oxidation number changes from +4 to +2
(3) reduced and the oxidation number changes from +2 to +4.
(4) oxidized and the oxidation number changes from +2 to +4.

25 In the reaction: $2Na + 2H_2O \rightarrow 2Na^+ + 2OH^- + H_2^0$, the substance oxidized is:
(1) H_2 (2) H^+ (3) Na (4) Na^+

26 Given the reaction: $3Cu + 8HNO_3 \rightarrow 3Cu(NO_3)_2 + 2NO + 4H_2O$, the reducing agent is:
(1) Cu^0 (2) N^{+5} (3) Cu^{+5} (4) N^{+2}

27 How many moles of electrons are needed to reduce one mole of Cu^{2+} to Cu^{1+}?
(1) 1   (2) 2   (3) 3   (4) 4

28 In the reaction $2H_2S + 3O_2 \rightarrow 2SO_2 + 2H_2O$, the oxidation number of oxygen changes from:
(1) -4 to 0   (2) -2 to 0   (3) 0 to -4   (4) 0 to -2

29 In the reaction $2K + Cl_2 \rightarrow 2KCl$, the substance oxidized is:
(1) K^+ (2) K^0 (3) Cl^- (4) Cl_2^0

II. Electrochemistry

A. Half — Reactions

A redox reaction may be considered in two parts, one representing a loss of electrons (oxidation) and the other representing a gain of electrons (reduction). Each reaction is known as a half-reaction.

A separate equation showing gain or loss of electrons (electronic equation) can be written for each half-reaction. For example:

$$\overset{0}{Mg} + \overset{0}{Cl_2} \longrightarrow \overset{2+ \ 1-}{MgCl_2}$$

can be represented as

$$Mg^0 \longrightarrow Mg^{2+} + 2e^- \text{ (oxidation)}$$

$$Cl_2 + 2e^- \longrightarrow 2Cl^- \text{ (reduction)}$$

In the above reaction, Mg^o is supplying the electrons and is considered the agent that causes the chlorine to be reduced. So, we call it the **reducing agent**. Substances with low ionization energies and low electronegativities are easily oxidized so they are called **strong reducing agents**.

Question: Which group of elements are strong reducing agents?

Answer: _____

On the other hand the Cl_2 is considered the agent that causes Mg to be oxidized. It is called the **oxidizing agent**. Elements with high electronegativities and ionization energies are more easily reduced and, therefore, are called **strong oxidizing agents**.

Question: Which group of elements include strong oxidizing agents?

Answer: _____

In other words, when attempting to identify the oxidizing or reducing agents, a simple rule could be applied. The item oxidized is the reducing agent, and the item reduced is the oxidizing agent.

B. Half — Cells

It is possible to set up reactions so that each half of a redox reaction takes place in a separate vessel. This occurs if the vessels are connected by an external conductor and a salt bridge or porous partition. This permits the migration of ions but does not allow the solutions to mix.

C. Chemical Cells (Electrochemical or Daniell Cells)

Redox reactions that occur spontaneously may be employed to provide a source of electrical energy.

When the two half-cells of a redox reaction are connected by an external conductor and a **salt bridge** or a porous cup that allows the migration of ions, a flow of electrons (electric current) is produced .

In an electrochemical cell, a chemical reaction is used to produce an electric current.

$$\overset{0}{Al} \xrightarrow[\text{(at anode)}]{\text{oxidation}} \overset{3+}{Al} + 3e^- \qquad \overset{2+}{Zn} + 2e^- \xrightarrow[\text{(at cathode)}]{\text{reduction}} \overset{0}{Zn}$$

In the figure (next page), when the voltmeter is connected to allow the flow of electrons, the aluminum metal strip which is immersed in a solution of $Al_2(SO_4)_3$ which contains both Al^{+++} and SO_4^{--} ions will supply the electrons which flow through the voltmeter to the zinc electrode. The excess of electrons allows the zinc ions in the solution of $Zn^{++}SO_4^{--}$ to pick up electrons at the zinc electrode and become zinc metal atoms Zn^o.

The aluminum metal atoms Al^0 each lost three electrons to become Al^{3+} ions. Since the loss of electrons takes place at the aluminum electrode it is called the **anode**, and the zinc electrode, where reduction takes place, is called the **cathode**.

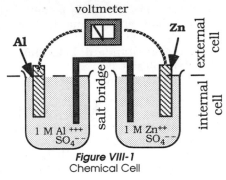

Figure VIII-1
Chemical Cell

The E^o values are established after setting up the two half-cell reactions

$$Al \overset{0}{\longrightarrow} Al^{3+} + 3e^- \qquad + 1.66$$

$$Zn^{2+} + 2e^- \longrightarrow Zn^0 \qquad - 0.76$$

$$+ 0.90$$

The +0.90 E^o value indicated:
   a. the reaction is spontaneous
   b. the maximum voltage that can be registered on the voltmeter at 1 M concentration.

As the current continues, the flow of electrons diminishes until the cell reaches a state of equilibrium when the voltmeter will read zero.

D. Electrolytic Cells
Redox reactions that do not occur spontaneously can be forced to take place by supplying energy with an externally applied electric current.

The use of an electric current to bring about a chemical reaction is called electrolysis. In an electrolytic cell, an electric current is used to produce a chemical reaction.

Additional Materials In Redox And Electrochemistry
AM 1. Standard Electrode Potentials
Half-Cell Potential
Comparison of the driving force of a half-reaction with that of the hydrogen standard establishes a scale of voltages.

It is impossible to measure the absolute reducing tendency of a half-reaction except by comparison with an attendant half-reaction. For purposes of measuring the relative reducing tendency of a half-reaction, it has been found convenient to adopt the half reaction for the reduction of the hydrogen ion as an arbitrary standard:

$$2H^+ + 2e^- \longrightarrow H_2 \qquad E^0 = 0.00$$

When each half-reaction is compared to the standard under specified conditions of concentration, temperature, and pressure, standard electrode potentials can be obtained.

The standard electrode potential (E^0) gives the potential difference in the specified half-reaction and the hydrogen half-reaction.

A table of Standard Potentials is included in the Reference Tables for Chemistry (Table N).

AM 2. Electrochemical cells

Use of Standard Electrode Potentials

Standard electrode potentials are useful in determining whether or not a specific redox reaction will take place. Any pair of half-reactions can be combined to give the complete reaction for a cell whose potential difference can be calculated by adding the appropriate half-cell potentials.

In combining half-reactions, it must be remembered that, in any redox reaction there must be an oxidation half-reaction and a reduction half-reaction. In the table of Standard Electrode Potentials (Table N), half-reactions are written as reductions. If read from left to right, all half-reactions result in the gaining of electrons.

To write the equation and the potential for the oxidation half-reaction correctly, the equation as written in the table must be reversed, and the sign E^0 changed. For example, in the reaction:

$$\overset{0}{Mg} + \overset{0}{Cl_2} \longrightarrow \overset{2+ \quad 1-}{MgCl_2}$$

The magnesium is oxidized. Therefore, to obtain the equation and the potential for this oxidation half-reaction, the equation for magnesium as written in the table must be reversed, and the sign of E^0 changed. Chlorine is reduced in this reaction, and, therefore, its reduction potential is obtained directly from Table N.

The two half-reactions may now be combined:

$$
\begin{array}{ll}
Mg \longrightarrow Mg^{2+} + 2e^- & E^0 = +2.37 \\
\underline{Cl_2 + 2e^- \longrightarrow 2Cl^-} & \underline{E^0 = +1.36} \\
Mg + Cl_2 \longrightarrow Mg^{2+} + 2Cl^- & E^0 = +3.73
\end{array}
$$

In combining half-reactions, if the potential (E^0) for the overall reaction is positive, the reaction is spontaneous. Such a reaction is called an **electrochemical cell reaction**.

When combining half-reactions, the electron transfer must be balanced. For example, in the reaction:

$$2Na + Cl_2 \longrightarrow 2NaCl$$

The two half-reactions would be combined as follows:

$$2 \, (Na \longrightarrow Na^+ + e^-) \qquad E^0 = +2.71$$

$$Cl_2 + 2e^- \longrightarrow 2Cl^- \qquad E^0 = +1.36$$

$$2Na + Cl_2 \longrightarrow 2Na^+ + 2Cl^- \qquad E^0 = +4.07$$

Metals with negative reduction potentials will produce hydrogen upon reaction with an acid. For example, if magnesium metal is added to a solution of an acid, the net reaction, if one occurs, could be represented by the equation:

$$Mg + 2H^+ \longrightarrow Mg^{2+} + H_2$$

Combining the half-reactions:

$$Mg \longrightarrow Mg^{2+} + 2e^- \qquad E^0 = +2.37$$

$$2H^+ + 2e^- \longrightarrow H_2 \qquad E^0 = 0.00$$

$$Mg + 2H^+ \longrightarrow Mg^{2+} + H_2 \qquad E^0 = +2.37$$

Since this value is positive, we know that the reaction will take place spontaneously.

Note: The Standard Electrode Potentials (E^0) are *not* multiplied by the coefficients in calculating the E^0 for the reaction!

AM 3. Electrolytic cells

In combining half-reactions, if the potential (E^0) for the overall reaction is negative, a reaction will not take place spontaneously. For example, metals with positive reduction potentials will not produce hydrogen on reaction with an acid. If copper metal is added to a solution of an acid, the reaction would not occur spontaneously, for example,

$$Cu + 2H^+ \longrightarrow Cu^{2+} + H_2$$

or combining the half-reactions:

$$Cu \longrightarrow Cu^{2+} + 2e^- \qquad E^0 = -0.34$$

$$2H^+ + 2e^- \longrightarrow H_2 \qquad E^0 = 0.00$$

$$Cu + 2H^+ \longrightarrow Cu^{2+} + H_2 \qquad E^0 = -0.34$$

Since the resulting E^0 is negative, we know the reaction will not take place spontaneously.

Electrolysis is an application of an electrolytic cell. Examples include:

1. Electrolysis of water:

$$2H_2O + \text{electricity} \longrightarrow 2H_2 + O_2$$

2. The electrolysis of concentrated brine:

$$2NaCl + 2H_2O + \text{electricity} \longrightarrow 2NaOH + H_2 + Cl_2$$

3. The electrolysis of molten salts:

$$2KCl \text{ (fused)} + \text{electricity} \longrightarrow 2K + Cl_2$$

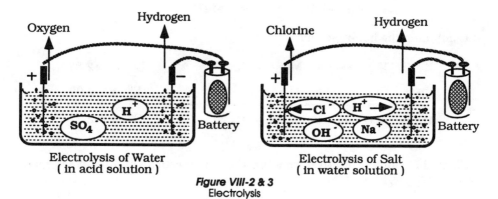

Figure VIII-2 & 3
Electrolysis

Equilibrium

Equilibrium is attained in chemical cells when the voltage measured is equal to **ZERO**.

The E^0 values given in the table of standard electrode potentials are for definite concentrations. As a reaction proceeds, these concentrations change, and the measured value falls off until at equilibrium the measured voltage is equal to zero.

AM 4. Electrodes

Cathode

The electrode at which reduction occurs in a cell is called a cathode. We identify the cathode as follows:

1. In an **electrochemical** cell, the cathode is the **positive** electrode.

2. In an **electrolytic** cell, the cathode is the **negative** electrode

Anode

The electrode at which oxidation occurs in a cell is called the anode. We identify the anode as follows:

1. In an **electrochemical** cell, the anode is the **negative** electrode.

2. In an **electrolytic** cell, the anode is the **positive** electrode.

AM 5. Electroplating

Silver, chrome, and stainless steel plating are some processes that make use of this principle. In this process an electric current is used to produce a chemical reaction. This results in the covering of a surface (usually a metallic item such as a spoon, car bumper, or trim) with a metal plating.

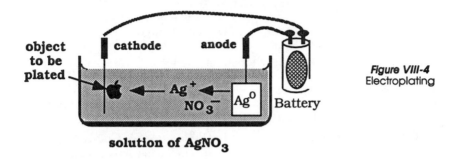

Figure VIII-4
Electroplating

In the illustration shown, the passage of one mole of electrons (6.02×10^{23} electrons) through the cathode will allow one mole of Ag^+ ions to be plated onto the object to be plated.

The procedure used is to make a solution of the salt that contains the plating metal ions and immerse the object to be plated (which is attached to the negative pole of a power pack) into the solution; then switch on the flow of electric current. The concentration of the solution and the amount of time that the current is allowed to continue determines how thick the layer of plated material will be.

In this process reduction takes place at the cathode, which is negative in charge and to which the object to be plated is placed. Oxidation takes place at the anode, which is composed of the metal to be plated.

Calculations to perform concerning chemical cells:

1. Predict the direction of electron flow. By referring to the Standard Electrode Potential Table in the Reference Tables for Chemistry (Table *N*), you can deduce which item will be oxidized and which will be reduced. The element closest to the top of the table will be reduced, and that which is closest to the bottom will be oxidized.

2. Predict the direction of ion movement. By first identifying the anode and cathode and determining the charge on the ion, you can easily predict the direction of ion movement.

3. Calculate the net potential for the redox reaction. This is done by writing the two half-cell reactions, along with their E^0 values. The algebraic sum of the E^0 values will give the net potential of the reaction.

Note: **Do not forget to change the value of the species which is oxidized.**

Questions VIII-2 and AM 1—5

1 Given the reaction: $Mg + Fe^{+2} \rightarrow Mg^{+2} + Fe$, the potential difference E^0 of this cell is:
   (1) 2.82 volts   (3) 1.92 volts
   (2) 2.37 volts   (4) 0 .44 volt

2 In a chemical cell composed of two half-cells, ions are allowed to flow from one-cell to another by means of:
   (1) electrodes   (3) a voltmeter
   (2) an external conductor   (4) a salt bridge

3 Given the reaction: $2Au^{3+}(aq) + 3Ni^0 \rightarrow 2Au^0 + 3Ni^{2+}(aq)$, the cell potential (E^0) for the overall reaction is:
   (1) 3.75 volts   (3) 1.75 volts
   (2) 2.25 volts   (4) 1.25 volts

4 Which half-reaction occurs at the cathode in an electrolytic cell in which an object is being plated with copper?
   (1) $Cu(s) \rightarrow Cu^{2+} + 2e^-$ (3) $Cu^{2+} + 2e^- \rightarrow Cu(s)$
   (2) $Cu(s) + 2e^- \rightarrow Cu^{2+}$ (4) no choice is correct

5 In order for a redox reaction to be spontaneous, the potential (E^0) for the overall reaction must be:
   (1) greater than zero   (3) between zero and -1
   (2) zero   (4) less than -1

6 According to Reference Table N, what is the standard electrode potential (E^0) for the oxidation of $Cu(s)$ to Cu^{2+}?
   (1) +0.52   (3) -0.52
   (2) +0.34   (4) -0.34

7 Given the equation: $Cu^{2+} + 2e^- \rightarrow Cu$, the reduction potential (E^0) for this half-reaction is:
   (1) +0.52 v   (3) -0.54 v
   (2) +0.34 v   (4) -0.34v

8 What is the oxidation potential (E^0) of the half-reaction $Cu(s) \rightarrow Cu^+ + e$?
   (1) +0.34 volt   (3) -0.34 volt
   (2) +0.52 volt   (4) -0.52 volt

9 According to Reference Table N, which metal will react spontaneously with $Cr^{3+}(aq)$ but not with $Ca^{2+}(aq)$?
   (1) Mg   (2) Co   (3) Ba   (4) Pb

10 Given the reaction: $Mg^0 + Pb^{2+} \rightarrow Mg^{2+} + Pb^0$, what is the cell potential (E^0) for the overall reaction?
   (1) -2.24 volts   (3) -2.50 volts
   (2) +2.24 volts   (4) +2.50 volts

11 In the electrolytic process used to plate copper onto a material, the material is the:
   (1) cathode which is positive   (3) anode which is positive
   (2) cathode which is negative   (4) anode which is negative

12 Given the overall cell reaction: $Zn(s) + 2Ag^+(aq) \rightarrow Zn^{+2}(aq) + 2\,Ag(s)$, which will occur as the cell operates?
   (1) the amount of $Zn(s)$ will increase
   (2) the amount of $Ag(s)$ will decrease
   (3) the concentration of $Zn^{2+}(aq)$ will increase
   (4) the concentration of $Ag^{1+}(aq)$ will increase

13 What is the voltage for a chemical cell that has reached equilibrium?
    (1) 1   (3) between 0 and 1
    (2) greater than 1   (4) 0

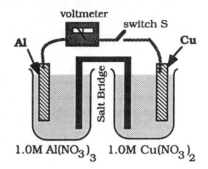

Base your answer to question 14 on the diagram which represents a chemical cell at 298 K. The equation which accompanies the diagram represents the net cell reaction.

$$2Al^0(s) + 3Cu^{2+}(aq) \rightarrow 2Al^{3+}(aq) + 3Cu^0(s)$$

14 When switch S is closed, the maximum potential (E^0) for the cell will be:
    (1) 1.32 volts   (3) -1.32 volts
    (2) 2.00 volts   (4) -2.00 volts

15 Which atom-ion pair will react spontaneously under standard conditions:
    (1) $Mg + Li^+$ (3) $Mg + Ag^+$

    (2) $Mg + Ba^{2+}$ (4) $Mg + Sr^{2+}$

16 During the electrolysis of fused KBr, which reaction occurs at the positive electrode:
    (1) Br^- ions are oxidized   (3) K^+ ions are reduced
    (2) Br^- ions are reduced   (4) K^+ ions are oxidized

17 According to Reference Table *N*, which halogen will react spontaneously with Au(s) to produce Au^{3+}?
    (1) Br_2 (2) F_2 (3) I_2 (4) Cl_2

18 Based on Reference Table *N*, which of the following elements will replace Pb from $Pb(NO_3)_2(aq)$?
    (1) Mg(s)   (3) Cu(s)
    (2) Au(s)   (4) Ag(s)

19 Which occurs in the half-reaction $Na^0(s) \rightarrow Na^+ + e^-$?
    (1) Na(s) is reduced   (3) Na(s) gains electrons
    (2) Na(s) is oxidized   (4) Na^+ is oxidized

20 Which half-reaction correctly represents reduction?
    (1) $Cu^0 \rightarrow Cu^{2+} + 2e^-$

    (2) $Zn^0 + 2e^- \rightarrow Zn^{2+}$

    (3) $Cr^{3+} + 3e^- \rightarrow Cr^0$

    (4) $2I^- \rightarrow I_2^0 + 2e^-$

21 According to Reference Table *N*, which pair will react spontaneously:
    (1) $I_2 + 2Br^-$ (3) $F_2 + 2 Cl^-$

    (2) $I_2 + 2 Cl^-$ (4) $Cl_2 + 2 F^-$

III. Balancing
Simple Redox Equations

In any reaction the loss of electrons by the species oxidized must be equal to the gain of electrons by the species reduced. In other words, there is a conservation of charge as well as a conservation of mass in a redox reaction. For example:

$$Cu^0 + 2Ag^+ \longrightarrow Cu^{2+} + 2Ag^0$$

In the above reaction, the oxidation state of copper has changed from 0 to 2+. This means that the copper atom has lost two of its electrons and has become a positive 2 ion. Meanwhile, two silver ions have each picked up one electron and have changed their oxidation state from positive one (+1) to zero (0).

Another example includes the following reaction:

$$2Al^0 + 6H^+ \longrightarrow 2Al^{3+} + 3H_2^0$$

In the above reaction two aluminum atoms give up 3 electrons each to become 2 aluminum ions with a charge of positive 3. The six hydrogen ions have picked up the 6 electrons given up by the 2 aluminum atoms and have become 3 molecules of hydrogen.

AM 6. Balancing Redox Reactions

In any reaction the loss of electrons by the species oxidized must be equal to the gain of electrons by the species reduced. There is a conservation of charge as well as a conservation of mass in a redox reaction.

One method is illustrated below for balancing the reaction between aqueous nitric acid and solid iodine.

Given the unbalanced equation:

$$HNO_3 + I_2 \longrightarrow HIO_3 + NO_2 + H_2O$$

Proceed with the following steps:

1. Assign oxidation numbers to each element.

$$\overset{+1\ +5\ -2}{H\,N\,O_3} + \overset{0}{I_2} \longrightarrow \overset{+1\ +5\ -2}{H\,I\,O_3} + \overset{+4\ -2}{N\,O_2} + \overset{+1\ \ -2}{H_2\,O}$$

2. Determine the change in oxidation number (transfer of electrons) of the elements.

For nitrogen, $+5 \rightarrow +4$ (this shows a gain of 1 electron)
For iodine, $0 \rightarrow +5$ (this shows a loss of 5 electrons)

3. Write partial electronic equations for the materials oxidized and reduced.

$$N^{+5} + 1e^- \longrightarrow N^{+4}$$

$$I_2^0 \longrightarrow 2I^{+5} + 10e^-$$

4. Balance the electrons gained and lost by writing appropriate coefficients for the two half-reactions and cancel out the electrons gained and lost.

$$10N^{+5} + 10e^- \longrightarrow 10N^{+4}$$

$$2I^0 \longrightarrow 2I^{+5} + 10e^-$$

$$\overline{\phantom{10N^{+5} + I_2^0 \longrightarrow 10N^{+4} + 2I^{+5}}}$$

$$10N^{+5} + I_2^0 \longrightarrow 10N^{+4} + 2I^{+5}$$

This equation is the net equation and does not include any spectator ions or atoms that are not involved in the redox reaction.

5. Insert the coefficients from the net equation into the skeletal equation:

$$10\ HNO_3 + I_2 \longrightarrow 2\ HIO_3 + 10\ NO_2 + H_2O$$

6. Insert other coefficients consistent with the conservation of matter, and balance by inspection.

$$10\ HNO_3 + I_2 \longrightarrow 2\ HIO_3 + 10\ NO_2 + 4\ H_2O$$

Questions VIII-3 and AM 6

1   Given the reaction:
    __Cu(s) + __HNO$_3$(aq) → __Cu(NO$_3$)$_2$(aq) + __NO$_2$(g) + __H$_2$O(l)
    When the reaction is completely balanced using smallest whole numbers, the coefficient of HNO$_3$(aq) will be
    (1) 1   (2) 2   (3) 3   (4) 4

2   What is the coefficient of the Cl$_2$ when the equation below is correctly balanced:
    MnO$_2$ + 4H$^+$ + 4 Cl$^-$ → 2H$_2$O + Mn^{2+} + 2Cl$^-$ + Cl$_2$:
    (1) 1   (2) 2   (3) 3   (4) 4

3   Given the unbalanced equation: Ca^0 + Al^{3+} → Ca^{2+} + Al^0, when the equation is completely balanced with the smallest whole-number coefficients, what is the coefficient of Ca^0?
    (1) 1   (2) 2   (3) 3   (4) 4

4   When the equation NH$_3$ + O$_2$ → N$_2$ + H$_2$O is completely balanced using the smallest whole numbers, the coefficient of the O$_2$ will be:
    (1) 1   (2) 2   (3) 3   (4) 4

Self-Help Questions VIII

Unit VIII - Redox And Electrochemistry
I. Oxidation- Reduction, oxidizing and reducing agents defined,
rules for determining oxidation number, types of redox reactions.

1 What is the oxidation number of iodine in KIO_4?
   (1) +1   (2) -1   (3) +7   (4) -7

2 A redox reaction always involves
   (1) a change in oxidation number (3) the transfer of protons
   (2) a change of phase   (4) the formation of ions

3 Which species is produced when a hydrogen atom is oxidized?
   (1) H:H   (2) H :   (3) H ·   (4) H^+

4 Given the reaction: $Mg(s) + Cl_2(g) \rightarrow MgCl_2(s)$ Which half-reaction
   correctly represents the reduction that occurs?
   (1) $Mg(s) + 2e^- \rightarrow Mg^{2+}$ (3) $Mg^{2+} \rightarrow Mg(s) + 2e^-$
   (2) $Cl_2(g) + 2e^- \rightarrow 2Cl^-$ (4) $2Cl^- \rightarrow Cl_2(g) + 2e^-$

5 In the equation $Cu(s) + 2Ag^+(aq) \rightarrow Cu^{2+}(aq) + 2Ag(s)$, the oxidizing
   agent is
   (1) Cu^0 (2) Ag^+ (3) Cu^{2+} (4) Ag^0

6 In the reaction $3Cu^0 + 2NO_3^- + 8H^+ \rightarrow 3Cu^{2+} + 2NO + 4H_2O$, the
   substance oxidized is
   (1) H^+ (2) O^{2-} (3) N^{2+} (4) Cu^0

7 When Fe^{3+} is reduced to Fe^{2+}, the Fe^{3+} ion
   (1) gains 1 proton   (3) gains 1 electron
   (2) loses 1 proton   (4) loses 1 electron

8 The reaction $BaCO_3 \rightarrow BaO + CO_2$ involves
   (1) oxidation, only   (3) both oxidation and reduction
   (2) reduction, only   (4) neither oxidation nor reduction

9 In the reaction $Al + Cr^{3+} \rightarrow Al^{3+} + Cr$, the reducing agent is
   (1) Al   (2) Cr^{3+} (3) Al^{3+} (4) Cr

10 In the reaction $MnO_2 + 4HCl \rightarrow MnCl_2 + 2H_2O + Cl_2$, the oxidation
   number of the manganese
   (1) decreases
   (2) increases
   (3) remains the same

11 In the compound Na_2HPO_4, which element has a negative oxidation
   number?
   (1) H   (2) O   (3) P   (4) Na

12 In the reaction $Pb + Cu^{2+} \rightarrow Pb^{2+} + Cu$, the Cu^{2+}
   (1) gains protons   (3) gains electrons
   (2) loses protons   (4) loses electrons

13 Given the reaction:

$$Zn(s) + 2H^+(aq) + 2Cl^-(aq) \rightarrow Zn^{2+}(aq) + 2Cl^-(aq) + H_2(g)$$

Which species is oxidized?
(1) $Zn(s)$ (2) $H^+(aq)$ (3) $Cl^-(aq)$ (4) $H_2(g)$

14 Given the reaction: $Mg + 2H^+ \rightarrow Mg^{2+} + H_2$ Which particle is the reducing agent?
(1) H^+ (2) Mg (3) Mg^{2+} (4) H_2

15 The oxidation number of hydrogen in sodium hydride (NaH) is
(1) +1   (2) +2   (3) -1   (4) -2

16 In which compound does chlorine have the highest oxidation number?
(1) $KClO$ (3) $KClO_3$
(2) $KClO_2$ (4) $KClO_4$

17 All redox reactions involve
(1) the gain of electrons, only
(2) the loss of electrons, only
(3) both the gain and the loss of electrons
(4) neither the gain nor the loss of electrons

18 Which is the oxidizing agent in the reaction:
$2Fe^{2+} + Cl_2 \rightarrow Fe^{3+} + 2Cl^-$?
(1) Fe^{2+} (2) Cl_2 (3) Fe^{3+} (4) Cl^-

19 Which ion can be both an oxidizing agent and a reducing agent?
(1) Sn^{2+} (2) Cu^{2+} (3) Al^{3+} (4) Fe^{3+}

20 In the reaction: $Mg(s) + 2Ag^+(aq) \rightarrow Mg^{2+}(aq) + 2Ag(s)$
Which species undergoes a loss of electrons?
(1) $Mg(s)$ (3) $Mg^{2+}(aq)$
(2) $Ag^+(aq)$ (4) $Ag(s)$

II. Electrochemistry; including half cell potentials, electrochemical, and electrolytic cells

21 Based on Reference Table *N*, which of the following elements will replace Pb from $Pb(NO_3)_2(aq)$?
(1) $Mg(s)$ (2) $Au(s)$ (3) $Cu(s)$ (4) $Ag(s)$

22 In an electrolytic cell, Cu^{2+} ions will
(1) migrate to the positive electrode
(2) migrate to the negative electrode
(3) be reduced at the positive electrode
(4) be oxidized at the negative electrode

23 If redox reactions are forced to occur by use of an externally applied electric current, the procedure is called
(1) neutralization   (3) electrolysis
(2) esterification   (4) hydrolysis

Base your answers to questions 24 through 26 on
the diagram of the chemical cell at 298 K and on
the equation below.

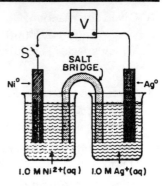

$$Ni^0(s) + 2Ag^+(aq) \rightarrow Ni^{2+}(aq) + 2Ag^0(s)$$

24 In the given reaction, the Ag^+ ions
    (1) gain electrons
    (2) lose electrons
    (3) gain protons
    (4) lose protons

25 When the switch is closed, the potential (E^0) of this cell is
    (1) 0.55 V   (2) 1.05 V   (3) 1.35 V   (4) 1.85 V

26 As the reaction in this cell takes place, the concentration of Ni^{2+} ions
    (1) decreases and the concentration of Ag^+ ions decreases
    (2) decreases and the concentration of Ag^+ ions increases
    (3) increases and the concentration of Ag^+ ions decreases
    (4) increases and the concentration of Ag^+ ions increases

27 According to Ref. Table *N*, which ion will react spontaneously with Ag?
    (1) Mn^{2+} (2) Mg^{2+} (3) Al^{3+} (4) Au^{3+}

28 In the chemical cell reaction $2Al + 3Ni^{2+} \rightarrow 2Al^{3+} + 3Ni$, which
    species is reduced?
    (1) Al   (2) Ni^{2+} (3) Al^{3+} (4) Ni

29 Given the reaction: $2Cr(s) + 3Cu^{2+} \rightarrow 3Cu(s) + 2Cr^{3+}$, the cell
    potential (E^0) for the overall reaction is
    (1) +0.40 V   (2) -0.40 V   (3) +1.08 V   (4) -1.08 V

30 According to the Reference Table *N*, which molecule is most easily
    reduced?
    (1) Br_2 (2) Cl_2 (3) F_2 (4) I_2

31 What is the purpose of the salt bridge in an electrochemical cell?
    (1) It allows ion migration.   (3) It prevents ion migration.
    (2) It allows electron flow.   (4) It prevents electron flow.

32 Which half-reaction is the arbitrary standard used in the measurement of
    the Standard Electrode Potentials in Reference Table *N*?
    (1) $2H^+ + 2e^- \rightarrow H_2(g)$
    (2) $2H_2O + 2e^- \rightarrow 2OH^- + H_2(g)$
    (3) $F_2(g) + 2e^- \rightarrow 2F^-$
    (4) $Li^+ + e^- \rightarrow Li(s)$

33 Given the reaction: $2Fe^{3+} + 2I^- \rightarrow 2Fe^{2+} + I_2$, the net potential ($E^0$)
    for the overall reaction is
    (1) 1.00 V   (2) 1.31 V   (3) 2.08 V   (4) 0.23 V

34 The diagram represents an electroplating arrangement.
In the setup shown, an object to be plated with metal would be the
(1) anode at A
(2) anode at B
(3) cathode at A
(4) cathode at B

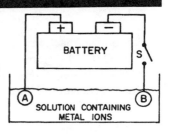

35 In the reaction $4Zn + 10HNO_3 \rightarrow 4Zn(NO_3)_2 + NH_4NO_3 + 3H_2O$, the zinc is
(1) reduced and the oxidation number changes from 0 to +2
(2) oxidized and the oxidation number changes from 0 to +2
(3) reduced and the oxidation number changes from +2 to 0
(4) oxidized and the oxidation number changes from +2 to 0

36 Based on Reference Table N, which half-cell has a greater reduction potential than the standard hydrogen half-cell?
(1) $Na^+ + e^- \rightarrow Na(s)$
(2) $Ni^{2+} + 2e^- \rightarrow Ni(s)$
(3) $Pb^{2+} + 2e^- \rightarrow Pb(s)$
(4) $Sn^{4+} + 2e^- \rightarrow Sn^{2+}$

37 Given the reaction: $2H_2O + electricity \rightarrow 2H_2 + O_2$, in which type of cell would this reaction most likely occur?
(1) a chemical cell, because it is exothermic
(2) an electrolytic cell, because it is exothermic
(3) a chemical cell, because it is endothermic
(4) an electrolytic cell, because it is endothermic

38 The diagram at the right represents an electrochemical cell.

When switch S is closed, which particles undergo reduction?
(1) Zn^{2+} ions
(2) Zn atoms
(3) Cu^{2+} ions
(4) Cu atoms

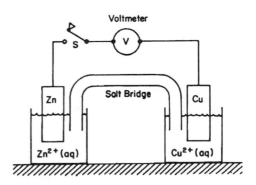

39 In the following reaction: $Mg(s) + 2Ag^+(aq) \rightarrow Mg^{2+}(aq) + 2Ag(s)$
What is the cell voltage (E°) for the overall reaction?
(1) +1.57 V
(2) +2.37 V
(3) +3.17 V
(4) +3.97 V

40 Which reaction will take place spontaneously?
(1) $Cu + 2H^+ \rightarrow Cu^{2+} + H_2$
(2) $2Au + 6H^+ \rightarrow 2Au^{3+} + 3H_2$
(3) $Pb + 2H^+ \rightarrow Pb^{2+} + H_2$
(4) $2Ag + 2H^+ \rightarrow 2Ag^+ + H_2$

III. Balancing redox reactions.

41 Given the reaction:

$$\_Cu(s) + \_HNO_3(aq) \rightarrow \_Cu(NO_3)_2(aq) + \_NO_2(g) + \_H_2O(l)$$

When the reaction is completely balanced using smallest whole numbers, the coefficient of $HNO_3(aq)$ will be

(1) 1   (2) 2   (3) 3   (4) 4

42 Given the equation:

$$\_Cr_2O_7^{2-} + \_H_2SO_3 + \_H^+ \rightarrow \_Cr^{3+} + \_HSO_4^- + \_H_2O$$

When the equation is completely balanced using the smallest whole numbers, the coefficient of the $Cr_2O_7^{2-}$ will be

(1) 1   (2) 5   (3) 3   (4) 4

43 Which reduction half-reaction is correctly balanced?

(1) $Cl_2(g) + e^- \rightarrow Cl^-$ (3) $Cl_2(g) + e^- \rightarrow 2Cl^-$

(2) $Cl_2(g) + 2e^- \rightarrow Cl^-$ (4) $Cl_2(g) + 2e^- \rightarrow 2Cl^-$

44 When the equation: $\_NO_3^- + 4H^+ + \_Cu \rightarrow \_Cu^{2+} + \_NO_2 + 2H_2O$

is correctly balanced, the coefficient of NO_3^- is

(1) 1   (2) 2   (3) 3   (4) 4

9 Organic Chemistry

Important Terms and Concepts

Organic Compounds – Carbon Bonding
   Tetrahedral Model
   Spatial Arrangement
Structural Formulas
Isomers
Saturated and Unsaturated Hydrocarbons
   Alkanes and Alkenes
   Alkynes and Alkadienes

Benzene and Alkyl Radicals
Organic Reactions
   Substitution and Addition
   Fermentation and Esterification
   Saponification and Oxidation
   Polymerization
Organic Acids and Alcohols
Aldehydes and Ketones
Ethers and Polymers
   Condensation and Addition

Organic Chemistry is the chemistry of the compounds of carbon. Organic compounds occur extensively in nature. All living things are composed predominantly of organic compounds.

Carbon is able to form four covalent bonds not only with other kinds of atoms but also with other carbon atoms. This makes possible a very large number of compounds. Consequently, organic compounds are much more numerous than inorganic compounds.

The major sources of raw materials from which organic chemicals are obtained are petroleum, coal, wood, and other plant products, and animal sources.

I. Characteristics Of Organic Compounds

1. Solubility. Organic compounds are generally nonpolar and tend to dissolve in nonpolar solvents.

Question: Can you name a few nonpolar solvents?

        Answer: _____

Those organic compounds that are somewhat polar, such as acetic acid, are soluble in water.

Generally organic compounds are insoluble in water and soluble in nonaqueous solvents.

2. Conductivity. Organic compounds are generally nonelectrolytes. Those organic acids that are electrolytes are very weak.

3. Melting Points. Organic compounds generally have low melting points.

Since most organic compounds are essentially nonpolar, the intermolecular forces are weak. Therefore, the compounds have relatively low melting points (under 300°C).

4. Organic Compound Reactions. Most reactions involving organic compounds are slower than those involving inorganic compounds.

Because of strong covalent bonding within the molecule, organic compounds do not readily form activated complexes (intermediates), and thus, reactions take place slowly. The activation energy required for organic reactions is generally high.

A. Bonding
Carbon atoms bond with the following characteristics:

- The four valence electrons of the carbon atom allow it to form four covalent bonds.
- These four single bonds of the carbon atom are spatially directed toward the corners of a regular tetrahedron.
- The carbon atom can share electrons with other carbon atoms.
- Two adjacent carbon atoms can share one to three pairs of electrons.
- The covalent bonding results in compounds that have molecular characteristics.

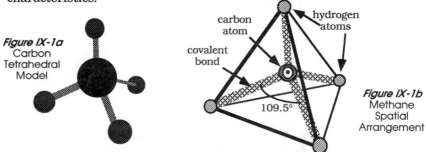

Figure IX-1a
Carbon
Tetrahedral
Model

carbon atom   hydrogen atoms

covalent bond

109.5°

Figure IX-1b
Methane
Spatial
Arrangement

Ball and stick model (*Figure IX-1a*) showing the spatial arrangement of the tetrahedral model of the carbon atom. Spatial arrangement (*Figure IX-1b*) showing bond angles and overlap covalent bonds of methane.

B. Structural Formulas
The covalent bond is usually represented by a short line (or dash) representing one pair of shared electrons. A formula showing the bonding in this manner is known as a structural formula.

The illustration shows molecular formulas and structural formulas of a single, double and triple bonded organic molecule. It is important to remember that molecules or organic compounds are three-dimensional in nature.

C. Isomers

Compounds that have the same molecular formula, but different structures are called **isomers**. The compounds CH_3CH_2CHO (propanal) and CH_3COCH_3 (acetone) are isomers, both having the molecular formula C_3H_6O.

Note: You are expected to identify isomers both in molecular formulas and structural formulas.

As the number of atoms in the molecule increases, the possibilities of more spatial arrangements (therefore, the number of isomers) increase.

propanal

propanone
(acetone)

D. Saturated And Unsaturated Compounds

A bond, formed between carbon atoms by the sharing of one pair of electrons, is referred to as a single bond. Organic compounds, where carbon atoms are bonded by the sharing of a **single pair** of electrons, are said to be **saturated** compounds.

Organic compounds, containing two adjacent carbon atoms bonded by the sharing of **more than one pair** of electrons, are said to be **unsaturated** compounds. A bond between carbon atoms by the sharing of two pairs of electrons is referred to as a **double bond**. A bond formed between carbon atoms by the sharing of three pairs of electrons is referred to as a **triple bond**.

II. Homologous
Series Of Hydrocarbons

Compounds containing only carbon and hydrogen are known as **hydrocarbons**. The study of organic chemistry is simplified by the fact that organic compounds can be classified into groups having related structures and properties. Such groups are called homologous series. Each member of a homologous series differs from the one before it by a common **increment**. The increment CH_2 is what distinguishes one member of a series from another in each of the homologous series that you will study.

Most carbon compounds are named from, and can be considered as related to, corresponding hydrocarbons. **As the members of a series increase in molecular size, the boiling points and freezing points increase due to the increased number of van der Waals forces.**

A. Alkanes

The series of **saturated** hydrocarbons having the general formula C_nH_{2n+2} is called the alkane series. This general formula allows us to deduce the molecular formula; once we know what **n** (number of carbon atoms) equals. Therefore, if we know that a molecule contains 16 carbon atoms, we simply substitute **16** for **n** and find the molecular formula to be $C_{16}H_{34}$. The **alkane series** is also called the **methane series** or the **paraffin series**.

In naming organic compounds the **IUPAC** (International Union of Pure and Applied Chemists) rules of nomenclature are followed. For the hydrocarbon series these rules are based on the number of carbon atoms in the molecule. The number of carbon atoms is indicated in the prefix of the name of each molecule, followed by the suffix, which for the **alkanes** is "**-ane.**"

The table at the right describes the naming of the first ten members of the alkane series. The alkane series shows isomerism beginning with the fourth member (butane, C_4H_{10}).

Number of Carbon Atoms	Prefix	Full Name	Molecular Formula
1	meth–	methane	CH_4
2	eth–	ethane	C_2H_6
3	prop–	propane	C_3H_8
4	but–	butane	C_4H_{10}
5	pent–	pentane	C_5H_{12}
6	hex–	hexane	C_6H_{14}
7	hept–	heptane	C_7H_{16}
8	oct–	octane	C_8H_{18}
9	non–	nonane	C_9H_{20}
10	dec–	decane	$C_{10}H_{22}$

```
      H
      |
  H – C – H
      |
      H
   methane
```

```
    H   H
    |   |
H – C – C – H
    |   |
    H   H
    ethane
```

```
    H   H   H
    |   |   |
H – C – C – C – H
    |   |   |
    H   H   H
      propane
```

Isomers of Butane C_4H_{10}. In alkanes, when adding a side chain or group, number the parent carbon chain by *starting at whichever end* results in the use of the *lowest number*.

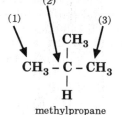

methylpropane

```
      (2)
(4)  (3)          (1)
       \    CH₃    /
        CH₃ C CH₂ CH₃
             |
             H
       2 - methylbutane
```

Isomers of Pentane C_5H_{12}. Numbering the carbon atoms left to right, this molecule is called 3 - methylbutane. Numbering from right to left, this molecule is called 2 - methylbutane

If the same group occurs more than once as a side chain, indicate this by the prefix di–, tri–, tetra– to show how many of these groups there are. Indicate by various numbers the positions of each group as illustrated by the second isomer of pentane: 2,2 dimethylpropane.

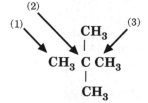

2,2 - dimethylpropane

B. Alkenes

The series of **unsaturated** hydrocarbons containing one double bond and having the general formula C_nH_{2n} is called the alkene series.

In the **IUPAC** system of nomenclature the **alkenes** are named from the corresponding alkane by changing the ending "-ane" to "**-ene.**"

The **alkene series** is also called the **ethylene series** or the **olefin series**.

First member:	**ethene — C_2H_4**
Second member:	**propene — C_3H_6**
Third member:	**butene — C_4H_8**
Fourth member:	**pentene — C_5H_{10}**

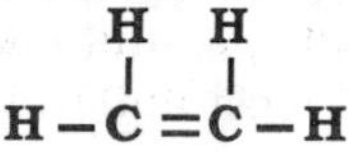

$$H-C=C-H$$

ethene

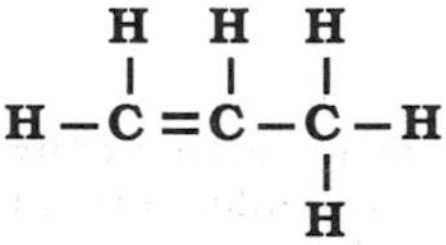

propene 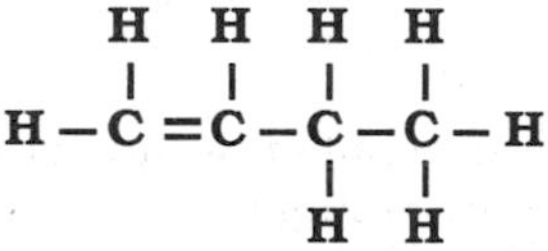   butene 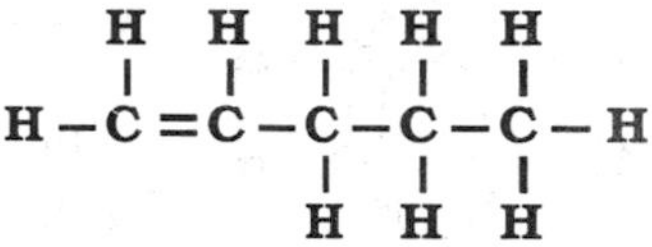   pentene

C. Alkynes

The series of **unsaturated** hydrocarbons containing one triple bond and having the general formula C_nH_{2n-2} is called the alkyne series. In the **IUPAC** system of nomenclature the **alkynes** are named from the corresponding alkane by changing the ending "-ane" to "-yne."

The common name of the first member of this series C_2H_2, is "acetylene," and the common name of the series is the **acetylene series**.

Note: In naming alkyne compounds, the prefix is the same as the alkanes; However, the suffix is "yne" instead of "ane":

ethyne	C_2H_2
propyne	C_3H_4
butyne	C_4H_6
pentyne	C_5H_8

$$H-C\equiv C-H$$

ethyne

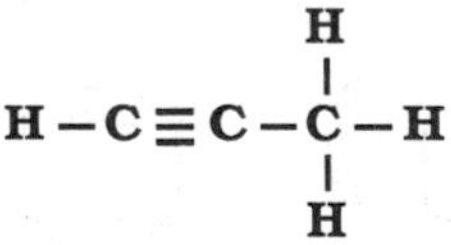

$$H-C\equiv C-C-H$$

propyne

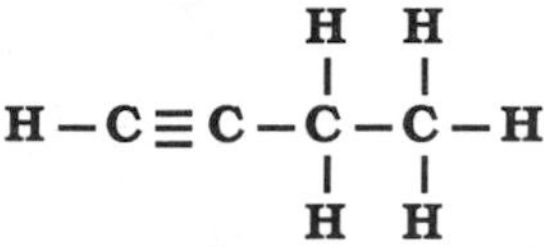

$$H-C\equiv C-C-C-H$$

butyne

$$H-C\equiv C-C-C-C-H$$

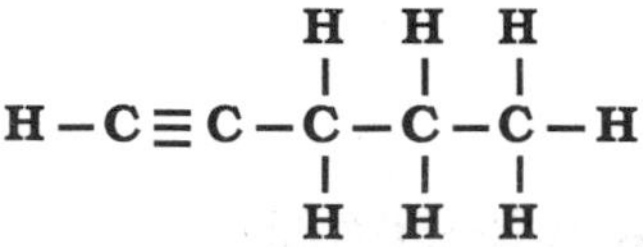

pentyne

D. Alkadienes

There is an homologous series of hydrocarbons containing **more than one** double bond, such as the **dienes**. They have the same general formula as the alkynes C_nH_{2n-2}.

They are not members of the alkene series. An example of an alkadiene is butadiene, $CH_2CHCHCH_2$.

$$H-C=C-C=C-H$$

butadiene

E. Benzene Series

Cyclic hydrocarbons differ from **open chain** hydrocarbons (includes the alkane, alkene, and alkyne series), in that they are arranged in a ring structure. The most important **cyclic series** is called the **BENZENE series**.

The benzene series is a group of aromatic hydrocarbons having the general formula C_nH_{2n-6}. Members of the **aromatic** hydrocarbons can be easily detected by their odors.

The simplest member of the benzene series is **benzene**, C_6H_6. The second member is **toluene**; its molecular formula is C_7H_8 (toluene is also called methylbenzene and is represented by the formula $C_6H_5CH_3$).

benzene

All of the carbon-carbon bonds in the benzene ring are the same, and they have structure and properties intermediate between single bonds and double bonds. Benzene is rather unreactive and in many of its reactions behaves like a saturated hydrocarbon, rather than unsaturated hydrocarbons.

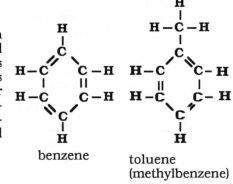

benzene

toluene
(methylbenzene)

benzene

CH₃

toluene
(methylbenzene)

This represents a **"super position,"** average of single and double bonds. For simplicity, the chemist often uses either one of the structures shown on left.

F. Alkyl Radicals

When an alkane molecule loses a hydrogen atom, it has an open bond and is called an **alkyl radical**. Its formula will be the same as the alkane molecule except that it will have one less hydrogen in the formula, for example:

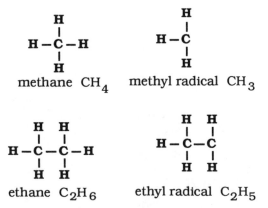

methane CH_4 methyl radical CH_3

ethane C_2H_6 ethyl radical C_2H_5

Questions IX-1

1 Which is a saturated hydrocarbon?
   (1) C_2H_6 (2) C_3H_6 (3) C_4H_8 (4) C_5H_{10}

2 What is the number of hydrogen atoms in a molecule of ethyne?
   (1) 6   (2) 2   (3) 8   (4) 4

3 Which compound is a member of the alkyne series?
   (1) C_2H_2 (3) C_5H_{10}
   (2) C_3H_6 (4) C_6H_6

4 Which compound contains a triple bond?
   (1) CH_4 (3) C_3H_6
   (2) C_2H_2 (4) C_4H_{10}

5 Which hydrocarbon has more than one possible structural formula?
   (1) CH_4 (3) C_3H_8
   (2) C_2H_6 (4) C_4H_{10}

6 The members of the alkane series of hydrocarbons are similar in that each member has the same:
   (1) empirical formula   (3) structural formula
   (2) general formula   (4) molecular formula

7 Which compound is a member of the alkene series of hydrocarbons?
   (1) benzene   (3) toluene
   (2) propene   (4) butadiene

8 Organic compounds that are essentially nonpolar and exhibit weak intermolecular forces have:
   (1) low melting points   (3) high conductivity in solution
   (2) low vapor pressure   (4) high boiling points

9 How many carbon atoms are contained in a methyl group?
   (1) 1   (2) 2   (3) 3   (4) 4

10 A compound with the formula C_6H_6 is:
   (1) toluene   (3) butene
   (2) benzene   (4) pentene

11 Each member of the alkane series differs from the preceding member by one additional carbon atom and:
   (1) 1 hydrogen atom   (3) 3 hydrogen atoms
   (2) 2 hydrogen atoms   (4) 4 hydrogen atoms

12 Which formula represents a saturated hydrocarbon?
   (1) C_2H_2 (2) C_2H_4 (3) C_3H_6 (4) C_3H_8

13 Given the compounds: $CH_3CH_2CHCH_2$ and CH_3CHCH_2 these compounds are both:
   (1) alkynes   (3) isomers of butane
   (2) alkenes   (4) isomers of propane

14 Molecules of 2-methylpentane and 3-methylpentane have different:
   (1) percentage compositions   (3) molecular formulas
   (2) molecular masses   (4) structural formulas

15 Which is the structure for 1,2 - dibromoethane?

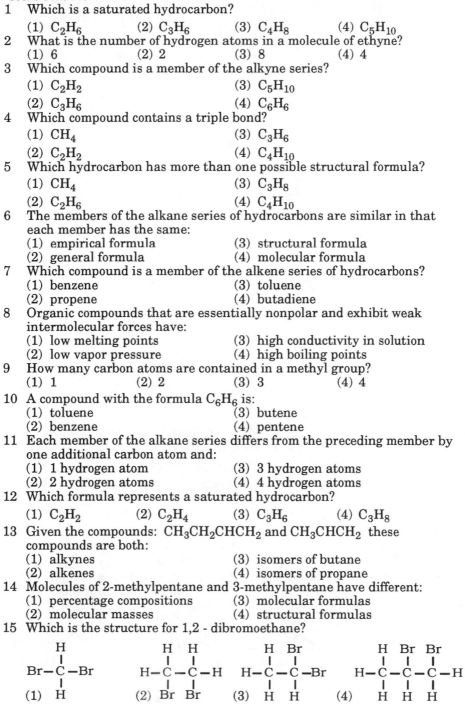

16 Which is the correct structural formula of propene?

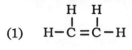

(1) H–C=C–H (with H, H above the two carbons)

(3) H–C≡C–C–H (with H above and H below the last carbon)

(2) H–C≡C–H

(4) H–C=C–C–H (with H, H, H above and H below the last carbon)

17 Which compound is an isomer of n-butane?

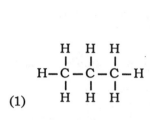

(1) H–C–C–C–H (with H, H, H above and H, H, H below)

(3) branched structure with H–C–H top and H–C–H bottom

(2) branched structure with H–C–H below

(4) H–C–C–C–C–H with H–C–H branch below

18 Which compound is an isomer of CH_3COOCH_3?
    (1) CH_3OCH_3 (3) CH_3COOCH_3
    (2) CH_3CH_2COOH (4) $CH_3CH_2CH_2OH$

19 Which compound is a saturated hydrocarbon?
    (1) ethene   (2) ethane   (3) ethylene   (4) ethyne

20 The compound C_2H_2 belongs to the series of hydrocarbons with the general formula:
    (1) C_nH_{2n+2} (3) C_nH_{2n-2}
    (2) C_nH_{2n} (4) $C_{2n}H_{2n-2}$

21 Which of the following compounds has the lowest normal boiling point?
    (1) butane   (2) ethane   (3) methane   (4) propane

22 Which of the following represents toluene?

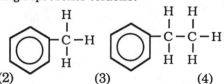

(1)   (2)   (3)   (4)

III. Organic Reactions

Organic reactions generally take place more slowly than inorganic reactions. These reactions frequently involve only the functional groups of the reacting species, leaving the greater part of the reacting molecules relatively unchanged during the course of the reaction. The reactions include:

A. Substitution

Substitution means replacement of one kind of atom or group by another kind of atom or group. For saturated hydrocarbons, reactions (except for combustion and thermal decomposition) necessarily involve replacement of one or more hydrogen atoms. The hydrogen atoms of saturated hydrocarbons can be replaced by active halogen family atoms. The general term for these reactions is **halogen substitution** (or **halogenation**) and the products are called **halogen derivatives**. For example:

$$CH_4 + Cl_2 \longrightarrow CH_3Cl + HCl$$

B. Addition

Addition usually involves adding one or more atoms at a double or triple bond of an unsaturated molecule, resulting in saturation of the compound. Addition is characteristic of unsaturated compounds.

Because addition reactions take place more easily than substitution reactions, unsaturated compounds tend to be more reactive than saturated compounds. Some addition reactions are about as fast as the reactions between ions. Alkynes are more reactive than alkenes. The addition of hydrogen to an unsaturated substance is called **hydrogenation**. This reaction usually requires the presence of a catalyst and a raised temperature.

The addition of chlorine and bromine (iodine usually cannot be added) takes place at room temperature. The compounds formed are also called halogen derivatives, and the reaction is referred to as **halogenation**. For example:

$$\begin{array}{c} \text{H} \quad \text{H} \\ | \quad\ | \\ \text{H} - \text{C} = \text{C} - \text{H} \end{array} + \text{Cl}_2 \longrightarrow \begin{array}{c} \text{H} \quad \text{H} \\ | \quad\ | \\ \text{H} - \text{C} - \text{C} - \text{H} \\ | \quad\ | \\ \text{Cl} \quad \text{Cl} \end{array}$$

C. Fermentation

In the fermentation process, enzymes produced by living organisms act as catalysts. A common fermentation product, ethanol, results from the fermentation of sugar. For example:

$$C_6H_{12}O_6 \xrightarrow[\text{(from yeast)}]{\text{zymase}} 2C_2H_5OH + 2CO_2$$

D. Esterification

Esterification is the reaction of an acid with an alcohol to give an ester and water:

$$\text{acid + alcohol} \rightarrow \text{ester + water}$$

$$HCOOH + CH_3CH_2CH_2CH_2OH \longrightarrow HCOOCH_2CH_2CH_2CH_3 + HOH$$

Esterification is not an ionic reaction, proceeds slowly, and is reversible. To increase the yield of the ester, concentrated sulfuric acid is added decreasing the concentration of the water and favoring the forward reaction.

Esters have certain characteristics which include:

a. They are covalent compounds.

b. Esters usually have pleasant odors. The aromas of many fruits, flowers, and perfumes are due to esters. **Fats** are esters derived from glycerol and long-chain organic acids.

E. Saponification

The hydrolysis of fats by bases is called **saponification**. To make **soap**, fat (a glycerol ester) is saponified by hot alkali (base). The products are soap (a salt of an organic fatty acid) and glycerol.

F. Oxidation (Combustion)

Saturated hydrocarbons react readily with oxygen under conditions of combustion. In an excess of oxygen, hydrocarbons burn completely to form carbon dioxide and water.

$$CH_4 + 2O_2 \longrightarrow CO_2 + 2H_2O$$

Burning in a limited supply of oxygen may produce carbon monoxide and carbon as well.

$$2CH_4 + 3O_2 \longrightarrow 2CO + 4H_2O$$

$$CH_4 + O_2 \longrightarrow C + 2H_2O$$

G. Polymerization

Polymerization involves the formation of a large molecule from smaller molecules called **monomers**. Synthetic rubbers, plastics such as polyethylene, and other chain molecules synthesized by persons are polymers. In nature polymerization occurs in the production of proteins, starches, and other chemicals by living organisms.

$$C_6H_{10}O_4 + C_6H_{16}N_2 \longrightarrow (C_{12}H_{22}O_2N_2)_n + nH_2O$$

Adipic Acid + Hexamethylene $\longrightarrow$ Nylon   +   Water

Questions IX-2

1   The organic reaction:

$$HCOOH + CH_3CH_2CH_2CH_2OH \rightarrow HCOOCH_2CH_2CH_2CH_3 + HOH$$

is an example of:

(1) fermentation   (3) polymerization
(2) esterification   (4) saponification

2   Which structural formula represents the product of the reaction between ethene and bromine (Br_2)?

3   Which is the product of the reaction between ethene and chlorine?

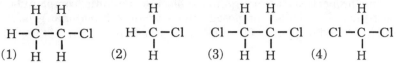

$$(1)\quad H\!-\!\underset{\underset{H}{|}}{\overset{\overset{H}{|}}{C}}\!-\!\underset{\underset{H}{|}}{\overset{\overset{H}{|}}{C}}\!-\!Cl \qquad (2)\quad H\!-\!\underset{\underset{H}{|}}{\overset{\overset{H}{|}}{C}}\!-\!Cl \qquad (3)\quad Cl\!-\!\underset{\underset{H}{|}}{\overset{\overset{H}{|}}{C}}\!-\!\underset{\underset{H}{|}}{\overset{\overset{H}{|}}{C}}\!-\!Cl \qquad (4)\quad Cl\!-\!\underset{\underset{H}{|}}{\overset{\overset{H}{|}}{C}}\!-\!Cl$$

4   The reaction $C_3H_8 + Cl_2 \rightarrow C_3H_7Cl + HCl$ is an example of:
(1) substitution   (3) esterification
(2) addition   (4) hydrogenation

5   Which equation represents a fermentation reaction?

(1) $C_2H_4 + H_2 \rightarrow C_2H_6$

(2) $CH_4 + 2Cl_2 \rightarrow CH_2Cl_2 + 2HCl$

(3) $HCOOH + CH_3OH \rightarrow HCOOCH_3 + HOH$

(4) $C_6H_{12}O_6 \rightarrow 2C_2H_5OH + 2CO_2$

6   Which is the formula of 2,2 – dichloropropane?

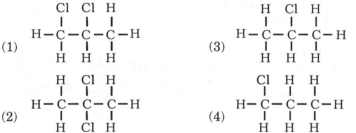

7   What is the structural formula for 2-chloropropane?

8   The reaction $CH_4 + Br_2 \rightarrow CH_3Br + HBr$ is one example of:
(1) substitution   (3) addition
(2) hydrogenation   (4) polymerization

IV. Other Organic Compounds

Other series of organic compounds occur when one or more hydrogen atoms of a hydrocarbon have been replaced by other elements. These compounds are usually named for their corresponding hydrocarbons but are not necessarily prepared directly from the original hydrocarbon itself.

A **functional group** is a particular arrangement of a few atoms that gives characteristic properties to an organic molecule. Organic compounds can often be considered as being composed of one or more functional groups attached to a hydrocarbon group. The functional groups include the following:

A. Alcohols

In alcohols, one or more hydrogens of a hydrocarbon have been replaced by an **-OH** group. No more than one **-OH** group can be attached to one carbon atom under ordinary conditions. The alcohols are not bases. The **-OH** group of an alcohol does not form a hydroxide ion in aqueous solution.

1. Primary alcohols. In primary alcohols, one **-OH** group is attached to the end carbon of a hydrocarbon. Since the functional group can be the end group of any hydrocarbon, the typical alcohol is frequently represented as **R-OH**, where "**R**" represents the rest of the molecule. The end group of a primary alcohol has the structure seen at the right.

$$R_1 - \overset{\overset{\displaystyle H}{|}}{\underset{\underset{\displaystyle H}{|}}{C}} - OH$$

It is frequently written as **-CH$_2$OH**

Primary alcohols contain the functional group **-CH$_2$OH**.

In the IUPAC system, primary alcohols are named from the corresponding hydrocarbon by replacing the final "-e" with the ending "**-ol**."

The common names of the alcohols were formerly derived from the name of the corresponding hydrocarbon by changing the ending "-ane" to "**-yl**" and adding the name "**alcohol**." Thus, **CH$_3$OH**, methanol, was called methyl alcohol.

$$H - \overset{\overset{\displaystyle H}{|}}{\underset{\underset{\displaystyle H}{|}}{C}} - OH$$
methanol

$$H - \overset{\overset{\displaystyle H}{|}}{\underset{\underset{\displaystyle H}{|}}{C}} - \overset{\overset{\displaystyle H}{|}}{\underset{\underset{\displaystyle H}{|}}{C}} - OH$$
ethanol

B. Organic Acids

Organic acids contain the functional group **-COOH**. Acids are represented by the general formula **R-COOH**, except for the first member. The **structural formula** of the acid group is shown on the left.

$$R' - C \overset{\displaystyle O}{\underset{\displaystyle OH}{\big\backslash}}$$

In the IUPAC system of nomenclature, organic acids are named from the corresponding hydrocarbons by replacing the final "-e" with the ending "**-oic**" and adding the name "**acid**."

The first two members of this series, methanoic acid, HCOOH, and ethanoic acid, CH$_3$COOH, are more familiarly known as their common names, **formic acid** and **acetic acid**.

Additional Materials In Organic Chemistry
AM 1. Alcohols

Alcohols contain the functional group of **-O-H**. Alcohols can be classified according to the number of **-O-H** groups in the molecule. They may also be classified according to the number of carbon chains attached to the carbon having the -O-H group. "R" is an abbreviation and represents any hydrocarbon group.

A. Monohydroxy Alcohols

Monohydroxy alcohols have one -O-H group.

$$H-\overset{\overset{\displaystyle H}{|}}{\underset{\underset{\displaystyle H}{|}}{C}}-OH$$

methanol

1. Primary alcohols.

The primary alcohol group is:

$$-\overset{\overset{\displaystyle H}{|}}{\underset{\underset{\displaystyle H}{|}}{C}}-OH$$

Methanol and ethanol are common primary alcohols.

$$H-\overset{\overset{\displaystyle H}{|}}{\underset{\underset{\displaystyle H}{|}}{C}}-\overset{\overset{\displaystyle H}{|}}{\underset{\underset{\displaystyle H}{|}}{C}}-OH$$

ethanol

2. Secondary alcohols.

The secondary alcohol group is:

$$R_1-\overset{\overset{\displaystyle H}{|}}{\underset{\underset{\displaystyle OH}{|}}{C}}-R_2$$

An example is 2-propanol (isopropanol). In this molecule, the carbon atom, to which the hydroxyl group is attached, is bonded to the other two carbon atoms.

$$H-\overset{\overset{\displaystyle H}{|}}{\underset{\underset{\displaystyle H}{|}}{C}}-\overset{\overset{\displaystyle H}{|}}{\underset{\underset{\displaystyle OH}{|}}{C}}-\overset{\overset{\displaystyle H}{|}}{\underset{\underset{\displaystyle H}{|}}{C}}-H$$

2-propanol

3. Tertiary alcohols.

The tertiary alcohol group is:

$$R_1-\overset{\overset{\displaystyle R_2}{|}}{\underset{\underset{\displaystyle OH}{|}}{C}}-R_3$$

An example is 2-methyl-2-propanol (tertiary butanol). In this molecule, the carbon atom, to which the hydroxyl group is attached, is bonded to three other carbon atoms.

2-methyl 2-propanol

B. Dihydroxy Alcohols

ethylene glycol

Compounds containing two -O-H groups are known as dihydroxy (dihydric) alcohols, or **glycols**. The most important glycol is **1,2-ethanediol**, commonly called ethylene glycol. It has the structural formula (left):

C. Trihydroxy Alcohols

Compounds containing three -O-H groups are known as trihydroxy (trihydric) alcohols. The most important trihydroxy alcohol is **1,2,3 - propanetriol** (glycerol), having the structural formula (right):

glycerol (glycerine)

AM 2. Aldehydes

Aldehydes contain the functional group: $-\overset{\overset{\displaystyle H}{|}}{C}=O$

In the IUPAC system of nomenclature, aldehydes are named from the corresponding hydrocarbons by replacing the final "-e" with the ending "**al**." The first member of the aldehydes is called methanal and is commonly called **formaldehyde**. Its molecular formula is **HCHO**, and its structural formula is represented at the right.

$$H-\overset{\overset{\displaystyle }{}}{\underset{\underset{\displaystyle H}{|}}{C}}=O$$

All other aldehydes are represented by the general formula **R-CHO** where "**R**" is any hydrocarbon group.

Primary alcohols can be oxidized to aldehydes:

$$R-\overset{\overset{\displaystyle H}{|}}{\underset{\underset{\displaystyle H}{|}}{C}}-OH \;+\; \textbf{oxidizing agent} \;\longrightarrow\; R-\overset{\overset{\displaystyle H}{|}}{C}=O \;+\; H_2O$$

Aldehyde groups are easily oxidized to acids:

$$R-\overset{\overset{\displaystyle H}{|}}{C}=O \;+\; \textbf{oxidizing agent} \;\longrightarrow\; R-C\overset{\displaystyle \diagup O}{\diagdown OH}$$

AM 3. Ketones

Ketones contain the functional group: $R_1-\overset{\overset{\displaystyle O}{||}}{C}-R_2$

R_1 and R_2 are hydrocarbon groups.

An important ketone, widely used as a solvent, is propanone. It is generally referred to by its common name, **acetone**. Secondary alcohols can be oxidized to ketones:

$$H-\overset{\overset{\displaystyle H}{|}}{\underset{\underset{\displaystyle H}{|}}{C}}-\overset{\overset{\displaystyle O}{||}}{C}-\overset{\overset{\displaystyle H}{|}}{\underset{\underset{\displaystyle H}{|}}{C}}-H$$

acetone

$$R_1-\overset{\overset{\displaystyle OH}{|}}{\underset{\underset{\displaystyle H}{|}}{C}}-R_2 \;+\; \textbf{oxidizing agent} \;\longrightarrow\; R_1-\overset{\overset{\displaystyle O}{||}}{C}-R_2 \;+\; H_2O$$

AM 4. Ethers

The functional group of an ether is $R_1 - O - R_2$

Diethyl ether, $C_2H_5OC_2H_5$, is used as an anesthetic. Primary alcohols can be dehydrated to give ethers.

$$R_1-OH \;+\; R_2-OH \;\longrightarrow\; R_1-O-R_2 \;+\; H_2O$$

AM 5. Polymers

A polymer is composed of many repeating units called monomers. Starch, cellulose, and proteins are natural polymers. **Nylon** and **polyethylene** are synthetic polymers.

Polymerization is the process of joining monomers. Polymers may be formed by condensation or by additional polymerization.

A. Condensation Polymers

Condensation polymerization results from the bonding of monomers by a dehydration reaction. Water is the usual by-product. A condensation process may be illustrated by:

$$
\underset{\text{monomer}}{HO-\overset{\displaystyle H}{\underset{\displaystyle H}{C}}-\overset{\displaystyle H}{\underset{\displaystyle H}{C}}-OH} \; + \; \underset{\text{monomer}}{HO-\overset{\displaystyle H}{\underset{\displaystyle H}{C}}-\overset{\displaystyle H}{\underset{\displaystyle H}{C}}-OH} \; \rightarrow \; \underset{\text{dimer}}{HO-\overset{\displaystyle H}{\underset{\displaystyle H}{C}}-\overset{\displaystyle H}{\underset{\displaystyle H}{C}}-O-\overset{\displaystyle H}{\underset{\displaystyle H}{C}}-\overset{\displaystyle H}{\underset{\displaystyle H}{C}}-OH} \; + \; H_2O
$$

This process may be repeated to give a long-chain polymer. The prerequisite for this is that the starting material (monomer) has at least two functional groups, one at each end, so that an attachment can be accomplished. Silicone, polyester, polyamides, phenolic plastics, and nylons are all examples of condensation polymers.

B. Addition Polymers

An addition polymerization results from the joining of monomers of unsaturated compounds by "opening" double or triple bonds in the carbon chain. An addition process may be illustrated as:

$$nC_2H_4 \longrightarrow (\text{-}C_2H_4\text{-})_n$$

Vinyl plastics such as polyethylene and polystyrene are examples of addition polymers.

$$n\left(CH_2 = CH_2\right) \longrightarrow \left(-CH_2 - CH_2 -\right)_n$$

polyethylene

$$n\left(\overset{\displaystyle H}{\underset{\displaystyle \bighexagon}{C}} = CH_2\right) \longrightarrow \left(-\overset{\displaystyle H}{\underset{\displaystyle \bighexagon}{C}} - CH_2 -\right)_n$$

polystyrene

Questions IX-3 and AM 1—5

1 What is the formula of ethanol?

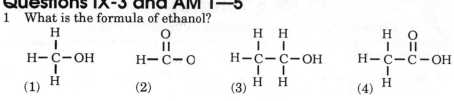

(1)   (2)   (3)   (4)

2 Which structural formula represents an aldehyde?

(1)   (2)   (3)   (4)

3 Which is the correct structural formula for glycerol?

(1)   (2)   (3)   (4)

4 Which is the general formula for an ether?

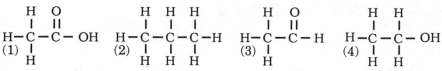

R − OH   R − C ⟋O ⟍H   R₁ − O − R₂   R₁ − C − R₂

(1)   (2)   (3)   (4)

$$R - OH \quad R - C{\overset{O}{\diagup}}{\underset{H}{\diagdown}} \quad R_1 - O - R_2 \quad R_1 - \overset{O}{\overset{\|}{C}} - R_2$$

5 In an aqueous solution, which compound will be acidic?
   (1) CH_3COOH (3) $C_3H_5(OH)_3$
   (2) CH_3CH_2OH (4) CH_3OH

6 Which alcohol has three hydroxyl groups per molecule?
   (1) butanol   (2) ethanol   (3) glycerol   (4) methanol

7 Which is the correct formula of methanoic acid?

(1)   (2)   (3)   (4)

8 The functional group − C ⟋O ⟍OH is always found in an organic:

   (1) aldehyde   (2) ester   (3) acid   (4) ether

9 Which is represented by
   this structural formula?   H − C ⟋O ⟍H

   (1) an aldehyde   (3) an alkane
   (2) an alcohol   (4) an acid

10 What could be the name of a compound that has the general formula of R-OH?
   (1) methanol   (3) methyl methanoate
   (2) methane   (4) methanoic acid

11 Which organic reaction involves the bonding of monomers by a dehydration process?
   (1) substitution   (3) addition polymerization
   (2) oxidation   (4) condensation polymerization

12 Which structural formula represents 2-propanol?

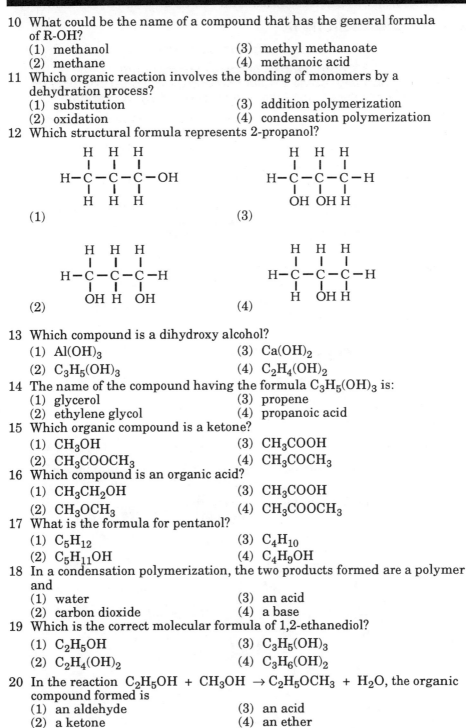

(1)   (3)

(2)   (4)

13 Which compound is a dihydroxy alcohol?
   (1) $Al(OH)_3$ (3) $Ca(OH)_2$
   (2) $C_3H_5(OH)_3$ (4) $C_2H_4(OH)_2$

14 The name of the compound having the formula $C_3H_5(OH)_3$ is:
   (1) glycerol   (3) propene
   (2) ethylene glycol   (4) propanoic acid

15 Which organic compound is a ketone?
   (1) CH_3OH (3) CH_3COOH
   (2) CH_3COOCH_3 (4) CH_3COCH_3

16 Which compound is an organic acid?
   (1) CH_3CH_2OH (3) CH_3COOH
   (2) CH_3OCH_3 (4) CH_3COOCH_3

17 What is the formula for pentanol?
   (1) C_5H_{12} (3) C_4H_{10}
   (2) $C_5H_{11}OH$ (4) C_4H_9OH

18 In a condensation polymerization, the two products formed are a polymer and
   (1) water   (3) an acid
   (2) carbon dioxide   (4) a base

19 Which is the correct molecular formula of 1,2-ethanediol?
   (1) C_2H_5OH (3) $C_3H_5(OH)_3$
   (2) $C_2H_4(OH)_2$ (4) $C_3H_6(OH)_2$

20 In the reaction $C_2H_5OH + CH_3OH \rightarrow C_2H_5OCH_3 + H_2O$, the organic compound formed is
   (1) an aldehyde   (3) an acid
   (2) a ketone   (4) an ether

Self-Help Questions IX

Unit IX - Organic Chemistry

I. Characteristics of organic compounds, including bonding, isomers, saturated and unsaturated compounds nomenclature of the homologous series of hydrocarbons.

1 Which of the following compounds has the greatest possible number of isomers?
(1) butane   (2) ethane   (3) pentane   (4) propane

2 Which is the correct molecular formula of pentene?
(1) C_5H_8 (2) C_5H_{10} (3) C_5H_{12} (4) C_5H_{14}

3 The bonds between the atoms in an organic molecule are generally
(1) ionic   (3) covalent
(2) coordinate covalent   (4) hydrogen

4 As the length of the chain of carbon atoms in molecules of the alkene series increases, the number of double bonds per molecule
(1) decreases   (2) increases   (3) remains the same

5 Which element is composed of atoms that can form more than one covalent bond with each other?
(1) hydrogen   (2) helium   (3) carbon   (4) calcium

6 A hydrocarbon molecule containing one triple covalent bond is classified as an
(1) alkene   (2) alkane   (3) alkyne   (4) alkadiene

7 The structure ⬡ represents a molecule of

(1) cyclopentane   (3) toluene
(2) cyclopropane   (4) benzene

8 What is the total number of hydrogen atoms in a molecule of butene?
(1) 10   (2) 8   (3) 6   (4) 4

9 Which compound belongs to the alkene series?
(1) C_2H_2 (2) C_2H_4 (3) C_6H_6 (4) C_6H_{14}

10 Which is the structural formula of methane?

(1)
```
    H   H
    |   |
H — C — C — H
    |   |
    H   H
```

(3)
```
              H
              |
      H   H — C — H   H
      |       |       |
H — C ——— C ——— C — H
      |       |       |
      H       H       H
```

(2)
```
   H       H
    \     /
     C = C
    /     \
   H       H
```

(4)
```
    H
    |
H — C — H
    |
    H
```

11 All organic compounds must contain the element
   (1) hydrogen   (3) carbon
   (2) nitrogen   (4) oxygen

12 Which structural formula represents a compound that is an isomer of

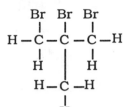

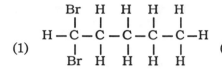

(1)

(3)

(2)

(4)

13 Which hydrocarbon is a member of the series with the general formula C_nH_{2n-2}?
   (1) ethyne   (2) ethene   (3) butane   (4) benzene

14 Which structural formula represents an aromatic hydrocarbon?

(1)

(3)

(2)

(4)

15 In the alkane series, each molecule contains
   (1) only one double bond   (3) one triple bond
   (2) two double bonds   (4) all single bonds

II. Organic reactions

16 In the reaction at the right, what is the formula of the product represented by the *X*?

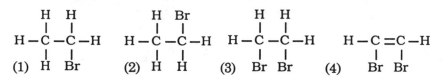

$$\text{(1)} \quad H-\underset{\underset{H}{|}}{\overset{\overset{H}{|}}{C}}-\underset{\underset{Br}{|}}{\overset{\overset{H}{|}}{C}}-H$$

$$\text{(2)} \quad H-\underset{\underset{H}{|}}{\overset{\overset{H}{|}}{C}}-\underset{\underset{H}{|}}{\overset{\overset{Br}{|}}{C}}-H$$

$$\text{(3)} \quad H-\underset{\underset{Br}{|}}{\overset{\overset{H}{|}}{C}}-\underset{\underset{Br}{|}}{\overset{\overset{H}{|}}{C}}-H$$

$$\text{(4)} \quad H-\underset{\underset{Br}{|}}{C}=\underset{\underset{Br}{|}}{C}-H$$

17 In a condensation polymerization, the two products formed are a polymer and
(1) water   (3) an acid
(2) carbon dioxide   (4) a base

18 Given the equation: $C_6H_{12}O_6 \rightarrow 2C_2H_5OH + 2CO_2$ The reaction represented by the equation is called
(1) polymerization   (3) esterification
(2) fermentation   (4) saponification

19 Which are the products of a fermentation reaction?
(1) an ester and water   (3) an alcohol and carbon dioxide
(2) a salt and water   (4) a soap and glycerol

20 Which type of reaction is represented by $C_2H_4 + H_2 \rightarrow C_2H_6$?
(1) addition   (3) condensation
(2) substitution   (4) polymerization

21 Which hydrocarbon will undergo a substitution reaction with chlorine?
(1) methane   (2) ethyne   (3) propene   (4) butene

22 In the reaction $\quad H-\underset{\underset{H}{|}}{\overset{\overset{H}{|}}{C}}-\overset{\overset{H}{|}}{C}=\overset{\overset{H}{|}}{C}-\underset{\underset{H}{|}}{\overset{\overset{H}{|}}{C}}-H + Cl_2 \longrightarrow X$

which structural formula correctly represents *X*?

$$\text{(1)} \quad H-\underset{\underset{H}{|}}{\overset{\overset{H}{|}}{C}}-\underset{\underset{Cl}{|}}{\overset{\overset{H}{|}}{C}}=\underset{\underset{Cl}{|}}{\overset{\overset{H}{|}}{C}}-\underset{\underset{H}{|}}{\overset{\overset{H}{|}}{C}}-H$$

$$\text{(3)} \quad H-\underset{\underset{H}{|}}{\overset{\overset{H}{|}}{C}}-\underset{\underset{H}{|}}{\overset{\overset{H}{|}}{C}}-\underset{\underset{Cl}{|}}{\overset{\overset{H}{|}}{C}}-\underset{\underset{H}{|}}{\overset{\overset{H}{|}}{C}}-H$$

$$\text{(2)} \quad H-\underset{\underset{Cl}{|}}{\overset{\overset{H}{|}}{C}}-\underset{\underset{H}{|}}{\overset{\overset{H}{|}}{C}}-\underset{\underset{H}{|}}{\overset{\overset{H}{|}}{C}}-\underset{\underset{Cl}{|}}{\overset{\overset{H}{|}}{C}}-H$$

$$\text{(4)} \quad H-\underset{\underset{H}{|}}{\overset{\overset{H}{|}}{C}}-\underset{\underset{Cl}{|}}{\overset{\overset{H}{|}}{C}}-\underset{\underset{Cl}{|}}{\overset{\overset{H}{|}}{C}}-\underset{\underset{H}{|}}{\overset{\overset{H}{|}}{C}}-H$$

III. Other organic compounds
(functional groups)

23 Which is the correct IUPAC name for the hydro-carbon with the structural formula shown at the right?

(1) 1-methyl-2-ethylethane
(2) 1-propylethane
(3) n-propane
(4) n-pentanol

24 Which is the correct molecular formula of 1,2-ethanediol?
(1) C_2H_5OH (2) $C_2H_4(OH)_2$ (3) $C_3H_5(OH)_3$ (4) $C_3H_6(OH)_2$

25 Which structural formula represents a secondary alcohol?

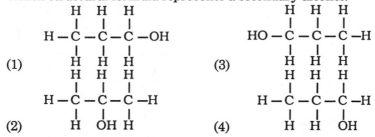

26 Which structural formula represents a ketone?

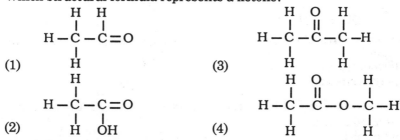

27 In the reaction C_2H_5OH + CH_3OH → $C_2H_5OCH_3$ + H_2O, the organic compound formed is
(1) an aldehyde   (2) a ketone   (3) an acid   (4) an ether

28 Both cellulose and proteins are classified as
(1) aldehydes   (2) esters   (3) polymers   (4) ketones

29 Which structural formula represents the trihydroxy alcohol, glycerol?

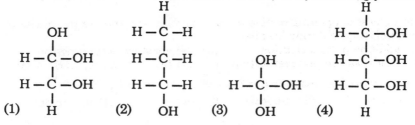

30 The compound 2-propanol is classified as a
   (1) primary alcohol   (3) tertiary alcohol
   (2) secondary alcohol   (4) dihydroxy alcohol

31 Which is the general formula for an aldehyde?

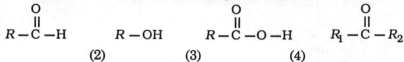

   (1)   (2)   (3)   (4)

32 How many hydroxy groups (—OH) does a primary alcohol molecule contain?
   (1) 1   (2) 2   (3) 3   (4) 4

33 Which structural formula represents a tertiary alcohol?

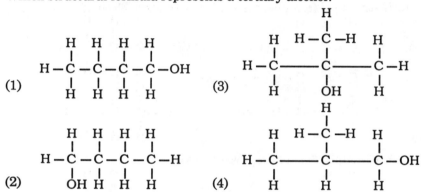

34 What is the name of the process that begins with the joining of monomer molecules?
   (1) fermentation   (3) esterification
   (2) polymerization   (4) hydrogenation

35 In the equation CH_3OH + oxidizing agent → X + H_2O, the product represented by X is
   (1) an alcohol   (3) an ether
   (2) an aldehyde   (4) a ketone

36 Which structural formula represents an isomer of dimethyl ether?

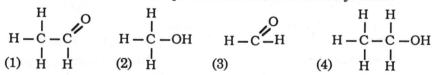

37 The process of opening double bonds and joining monomer molecules to form polyvinyl chloride is called
   (1) addition polymerization   (3) dehydration polymerization
   (2) condensation polymerization (4) neutralization polymerization

38 Which class of compounds has the general formula R_1–O–R_2?
   (1) esters   (2) alcohols   (3) ethers   (4) aldehyde

Important Terms and Concepts

Pure and Applied Research
Equilibrium and Reaction Rates
  Haber Process
  Temperature and Pressure
  Catalyst

Redox – Reduction and Crrosion
Lead–Acid Battery
Nickel oxide–cadmium Battery
Petroleum
  Fractional Distillation, Cracking

I. Chemical Theory And Industry

Pure research is directed toward the acquiring of knowledge without the consideration of the immediate practical application of that knowledge. The scientist's search for truth often leads to the benefit of mankind.

Industrial processes are concerned with the application of pure research and obtaining the maximum yield of products with maximum economic efficiency. An understanding and proper use of the principles involved in an industrial process can materially effect the quality, yield, and cost of the product.

These understandings and the principles derived from them are the result of **applied research.**

II. Industrial Applications
A. Equilibrium And Reaction Rates

An application of the factors (such as concentration, temperature, pressure, and catalysts) affecting the rates of reaction, as well as equilibrium conditions, make many industrial processes practical. They include:

1. Haber Process. Ammonia is produced by the reaction:

$$N_2(g) + 3H_2(g) \longrightarrow 2NH_3(g) + heat$$

The above reaction, if allowed to take place under ordinary conditions, would not be practical. The reaction is slow in reaching equilibrium, and, once there, there is the problem of small yield. Once the above reaction is at equilibrium; however, we can use Le Chatelier's Principle to increase the rate of production of ammonia. This is done by the manipulation of the following factors:

Temperature. The rate of formation of ammonia is slow at low temperatures, so that the temperature must be raised to increase this rate. However, since this is an exothermic reaction, high temperatures decrease the yield of ammonia. In practice a compromise temperature of 400-600°C is used.

Pressure. Note that the formula on the previous page shows 4 moles of gases on the left and 2 on the right of the equation. If we increase the pressure, a stress would occur on the left causing a shift in the equilibrium to the right and the production of more ammonia would occur. Therefore, a pressure of 200-1000 atmospheres is used to secure reasonable yields.

Equilibrium Shift. As the ammonia is drawn out of the reaction container, the concentration of ammonia decreases. This causes an equilibrium shift towards the making of more ammonia in order to offset this decreased concentration.

Catalyst. To increase the rate of the formation of the ammonia, a catalyst is used in the reaction. The catalyst is iron with traces of aluminum oxide or silicon oxide, and potassium oxide.

2. Contact Process. This process is used in the manufacture of sulfuric acid, and proceeds in the following steps:

a. Sulfur or sulfide ores, such as iron pyrites, are burned in order to produce sulfur dioxide:

$$S + O_2 \longrightarrow SO_2 + heat$$

b. The sulfur dioxide is then oxidized to sulfur trioxide by the reaction:

$$2SO_2 + O_2 \longrightarrow 2SO_3 + heat$$

Platinum or vanadium pentoxide is used as a catalyst in SO_2 oxidation.

c. From the equation (above), we can deduce that, like the Haber process, an increase in pressure and a decrease in temperature favor the formation of SO_3. Since the rate of formation of SO_3 is slow at low temperatures, the temperature must be raised to increase the rate of formation. However, since this is an exothermic reaction, high temperatures decrease the yield of SO_3. In practice, the temperatures are adjusted between 450-575°C to attain reasonable yields of SO_3.

It was found that SO_3 is absorbed very slowly in water but very readily in concentrated sulfuric acid, forming pyrosulfuric acid. The absorption of the sulfur trioxide by sulfuric acid with the resulting formation of pyrosulfuric acid is illustrated below:

$$H_2SO_4 + SO_3 \longrightarrow H_2SO_4SO_3 \text{ or } H_2S_2O_7$$

The pyrosulfuric acid ($H_2S_2O_7$) is then diluted with water to produce sulfuric acid (commonly called **oil of vitriol**).

$$H_2S_2O_7 + H_2O \longrightarrow 2H_2SO_4$$

B. Redox

Principles involving the "competition" for electrons by atoms help to explain many processes used by industry. Many basic redox reactions are employed to reduce or oxidize elements from compounds. Redox reactions are also used in refining and protecting metals.

1. Reduction of metals. The ores of most metals contain the metal in an oxidized state (with a positive ionic charge). The form in which a metal occurs is related to the chemical activity of the metal and the stability of its compounds. In general, compounds of metals that occur in nature have high stability and low solubility in water.

Reduction (the taking on of electrons by the positively charged metal ion so that it may become a metal atom with a zero charge of the ore) is necessary to obtain the metal. The method of reduction depends on the activity of the metal and the type of ore.

The most active metals are obtained from their fused compounds by electrolytic reduction. These include Groups 1 and 2 metals, which are obtained by the electrolysis of their fused (melted) salts:

$$2NaCl \text{ (fused)} + \text{electricity} \longrightarrow 2Na + Cl_2$$

Metals that form relatively stable compounds can be liberated from their compounds by stronger reducing agents. For example, in the production of chromium, aluminum reduces the chromium oxide:

$$2Al + Cr_2O_3 \longrightarrow Al_2O_3 + 2Cr$$

Many metals are found as sulfides and carbonates (ZnS, Cu_2S). These ores are converted to the oxide form by reacting the ore with oxygen through a process called **"roasting."** After that, the oxide ores are reduced by carbon (coke) or carbon monoxide:

$$ZnO + C + \text{heat} \longrightarrow Zn + CO$$

$$Fe_2O_3 + 3CO + \text{heat} \longrightarrow 3CO_2 + 2Fe$$

2. Corrosion. Corrosion is a gradual attack on a metal by its surroundings. When the metal returns to its ionic form, the usefulness of the metal may be destroyed. Corrosion is a redox reaction. Moisture, some gases in the air, and some chemicals contribute to corrosion.

Some metals like aluminum and zinc form self-protective coatings. Aluminum is more susceptible to corrosion than iron. However, the corrosion of aluminum is not a serious problem, since the aluminum oxide formed can adhere tightly to the uncorroded aluminum beneath it and provide a protective layer that prevents further corrosion.

In the case of iron, the oxide formed from corrosion lacks the ability to adhere to the metal. It constantly flakes off, exposing fresh iron to corrosion.

Metals that corrode easily, like iron, may be protected by a variety of methods. Some methods include:

a. Plating with self-protective metals like aluminum and zinc, or plating with corrosion resistant metals like chromium and nickel.

b. Sometimes, a more active metal like magnesium corrodes preferentially when it is connected to iron. This is called **cathode protection,** and magnesium plates are added to the hulls of sea going ships so that the magnesium corrodes and not the hull.

c. The alloying of iron with corrosion resistant metals like nickel and chromium has produced stainless steel.

d. Coating iron with paints, oils, or glass (porcelain) has proven effective against corrosion.

Questions X-1

1 Which type of chemical reaction is the corrosion of iron?
(1) reduction-oxidation   (3) polymerization
(2) substitution   (4) decomposition

2 The Haber process is used to produce:
(1) sulfur dioxide   (3) sulfuric acid
(2) ammonia   (4) sodium chloride

3 Which reaction occurs during the contact process?

(1) $2SO_2 + O_2 \rightarrow 2SO_3$ (3) $Al + Cr_2O_3 \rightarrow AlO_3 + 2Cr$

(2) $N_2 + 3H_2 \rightarrow 2NH_3$ (4) $2NaCl \rightarrow 2Na + Cl_2$

4 The self-protective coating which forms on aluminum metal is:
(1) an oxide   (3) an oxalate
(2) a sulfide   (4) a chloride

5 Which of the following metals forms a self-protective coating when exposed to air and moisture?
(1) zinc   (2) calcium   (3) iron   (4) sodium

6 Which of the following atmospheric gases is least likely to cause iron to corrode?
(1) $O_2(g)$ (2) $CO_2(g)$ (3) $Ar(g)$ (4) $H_2O(g)$

7 Which metal must be combined with chromium to produce stainless steel?
(1) radium   (2) iron   (3) copper   (4) zinc

8 Which oxide will react with carbon (coke) to produce a free metal?
(1) MgO   (2) ZnO   (3) Na_2O (4) Li_2O

9 Which metal is usually obtained by electrolytic reduction?
(1) iron   (2) tin   (3) lead   (4) potassium

10 Which reaction correctly represents the step in the contact process that produces the compound SO_3?

(1) $2S + 3O_2 \rightarrow 2SO_3$ (3) $2SO_2 + O_2 \rightarrow 2SO_3$

(2) $S + O_3 \rightarrow SO_3$ (4) $3SO_2 + O_3 \rightarrow 3SO$

3. Batteries. Batteries are chemical cells in which spontaneous redox reactions are used to provide a source of electrical energy. The two half-cells in a battery are separated by an open switch. When the switch is closed, the two half-cells react spontaneously to produce electrical energy from chemical energy.

1. Lead-acid battery. The change in oxidation states of lead is used to produce electrical energy. When fully charged, the positive and negative electrodes of this widely used battery are PbO_2 and Pb respectively. The electrolyte is a sulfuric acid solution and is also a reactant. Therefore, the concentration of sulfuric acid decreases as the battery discharges.

The overall reaction is:

$$Pb + PbO_2 + 2\,H_2SO_4 \; \underset{\text{charge}}{\overset{\text{discharge}}{\rightleftarrows}} \; 2PbSO_4 + 2\,H_2O$$

As noted in the reaction, reversal of the chemical process recharges the storage battery for further use.

2. Nickel oxide-cadmium battery. The changes in oxidation states of nickel and cadmium are used to produce a source of electric energy.

The positive and negative electrodes of this compact rechargeable battery are NiO_2 and Cd, respectively. The electrolyte is a solution of KOH, and its concentration remains the same, whether discharging or charging. The probable reaction is:

$$NiO_2(s) + Cd(s) + H_2O \; \underset{\text{charge}}{\overset{\overset{\displaystyle KOH}{\text{discharge}}}{\rightleftarrows}} \; Ni(OH)_2(s) + Cd(OH)_2(s)$$

At the anode cadmium is oxidized in the basic electrolyte:

$$Cd(s) + 2\,OH^-(aq) \longrightarrow Cd(OH)_2(s) + 2e^-$$

At the cathode $NiO_2(s)$ is reduced:

$$NiO_2(s) + 2\,H_2O + 2e^- \longrightarrow Ni(OH)_2(s) + 2\,OH(aq)$$

C. Petroleum

Petroleum is a complex natural mixture of many hydrocarbons. Many petroleum products are used as fuels, and they are the important starting material for many chemical products like plastics, textiles, rubber, and detergents.

Natural gas (mostly methane), a common fuel, is often found with petroleum. Common bottled gases like propane and butane are obtained from petroleum. Many products that we derive from petroleum are obtained by the following processes:

1. Fractional distillation. Because hydrocarbons differ in boiling points, petroleum (called crude oil) can be separated into fractions of simpler mixtures by distillation. The important fractions, which are boiled off at different temperatures, include gasoline, kerosene, fuel oil, lubricating oils and greases, paraffin wax, and asphalt. The uses of these different fractions are various.

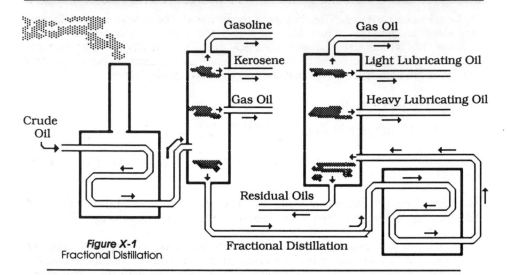

Figure X-1
Fractional Distillation

Fractional Distillation

2. Cracking. In cracking, some large molecules of hydrocarbons (higher boiling fractions) are broken down into smaller molecules of lower boiling points.

The cracking process is utilized to increase the yield of valuable fractions. Two examples are gasoline and fuel oil. Generally, cracking requires the use of a catalyst (for example, oxides of aluminum and silicon) and increased temperature.

Questions X-2

1 A process in which large molecules are broken down into smaller molecules is used commercially to increase the yield of gasoline from petroleum. This process is called:
   (1) polymerization   (3) esterification
   (2) hydrogenation   (4) cracking

2 Which equation represents a simple example of cracking?
   (1) $N_2 + 3H_2 \rightarrow 2NH_3$
   (2) $S + O_2 \rightarrow SO_2$
   (3) $C_3H_8 + 5O_2 \rightarrow 3CO_2 + 4H_2O$
   (4) $C_{14}H_{30} \rightarrow C_7H_{16} + C_7H_{14}$

3 Which process increases the yield of gasoline and kerosene from crude oil?
   (1) oxidation   (3) Haber
   (2) cracking   (4) contact

4 The components of petroleum are separated from each other by fractional distillation. The process depends on differences in the components'
   (1) solubilities in water   (3) freezing points
   (2) adsorption on charcoal   (4) boiling points

Self-Help Questions X

Unit X - Applications Of Chemical Principles

I. Industrial applications - including the Haber reaction, reduction of metal, and corrosion.

1 Which substance is produced by the Haber process?
(1) ammonia
(3) sulfuric acid
(2) aluminum
(4) sodium

2 Which metal is *not* obtained from its ore by electrolytic reduction?
(1) Na
(2) Li
(3) Au
(4) K

3 Given the reaction at equilibrium:

$$N_2(g) + 3H_2(g) \leftrightarrow 2NH_3(g) + heat$$

Which change would shift the equilibrium to the right?
(1) Increase the temperature.
(3) Decrease the $[N_2]$.
(2) Increase the pressure.
(4) Decrease the $[H_2]$.

4 Which metal will undergo the greatest degree of corrosion if left unprotected from its surroundings?
(1) nickel
(3) chromium
(2) zinc
(4) iron

5 The contact process is used to produce
(1) ammonia
(3) sodium
(2) sulfuric acid
(4) ethanol

6 Which group contains metals that are obtained from their fused compounds by electrolytic reduction?
(1) 1 (IA)
(2) 6 (VIB)
(3) 11 (IB)
(4) 16 (VIA)

7 Which compound forms on the surface of aluminum and protects it from further corrosion?
(1) $Al_2(CO_3)_3$
(3) $Al_2(SO_3)_3$
(2) Al_2O_3
(4) Al_2S_3

8 Given the reaction: $2X + Cr_2O_3 \rightarrow X_2O_3 + 2Cr$ The metal represented by X in the equation could be
(1) Pb
(2) Sn
(3) Ni
(4) Al

9 Which group contains elements that are obtained by the electrolytic reduction of their fused salts?
(1) 1 (IA)
(2) 12 (IIB)
(3) 14 (IVA)
(4) 16 (VIA)

10 The reaction $CuO + CO \rightarrow CO_2 + Cu$ is an example of
(1) reduction, only
(3) both oxidation and reduction
(2) oxidation, only
(4) neither oxidation nor reduction

11 Which group of metals is normally obtained by the electrolysis of their fused salts?
(1) Group 17 (VIIA)
(3) Group 7 (VIIB)
(2) Group 2 (IIA)
(4) Group 4 (IVB)

12 Given the Haber reaction at equilibrium:

$N_2(g) + 3H_2(g) \leftrightarrow 2NH_3(g)$ + heat   Which stress on the system at
equilibrium favors the production of $NH_3(g)$?
(1) decreasing the concentration of $N_2(g)$
(2) decreasing the concentration of $H_2(g)$
(3) increasing the pressure on the system
(4) increasing the temperature of the system

13 Which type of reaction is the Haber process,

$$N_2(g) + 3H_2(g) \rightarrow 2NH_3(g) + \text{heat?}$$

(1) exothermic, with an increase in entropy
(2) exothermic, with a decrease in entropy
(3) endothermic, with an increase in entropy
(4) endothermic, with a decrease in entropy

II. Batteries, including lead-acid, and nickel oxide - cadmium batteries. petroleum, including fractional distillation and cracking.

14 The components of petroleum are separated from each other by fractional
distillation. The process depends on differences in the components'
(1) solubilities in water   (3) freezing points
(2) adsorption on charcoal   (4) boiling points

15 Which type of reaction in a battery produces electrical energy?
(1) decomposition   (3) redox
(2) neutralization   (4) hydrolysis

16 Petroleum is an important starting material for manufacturing
(1) sulfuric acid   (3) ammonia
(2) methyl alcohol   (4) rubber

17 Which process is used to separate the components of a mixture based on
the differences in their boiling points?
(1) cracking   (3) fractional distillation
(2) polymerization   (4) fractional crystallization

18 When a battery is in use, stored chemical energy is first changed to
(1) electrical energy   (3) light energy
(2) heat energy   (4) mechanical energy

19 Molecules in the less valued fractions of petroleum may undergo cracking, which results in
(1) fewer molecules with larger molecular masses
(2) fewer molecules with smaller molecular masses
(3) more molecules with larger molecular masses
(4) more molecules with smaller molecular masses

20 Kerosene is a mixture of compounds called
(1) esters   (2) alcohols   (3) aldehydes   (4) hydrocarbons

11 Nuclear Chemistry

Important Terms and Concepts

Artificial Radioactivity
    Artificial Transmutation
    Radioisotopes
        Tracers, Half–Life
    Accelerators
Nuclear Energy

Mass Defect
    Binding Energy
    Fission and Fusion Reactions
        Fuels, Moderators, Coolants
        Control Rods, Shielding
Radioactive Wastes

I. Artificial Radioactivity

In Unit II, we studied natural radioactivity, in that it occurs under natural conditions and results in natural transmutation. The products of natural transmutation are radioactive isotopes, which decay finally into a stable isotope of lead. However, elements can be made radioactive by bombarding their nuclei with high energy particles such as protons, neutrons, and alpha particles. When the element nitrogen, for example, is bombarded by alpha particles, the nuclear reaction that takes place is:

$$\text{alpha particle} \; + \; \text{nitrogen} \; \longrightarrow \; \text{proton} \; + \; \text{oxygen}$$

$$_2^4\text{He} + \; _7^{14}\text{N} \; \longrightarrow \; _1^1\text{H} \; + \; _8^{17}\text{O}$$

As noted from the above reaction, isotopes of two new elements are formed. This is called **Artificial Transmutation**. The process includes the bombardment of nuclei by accelerated particles which cause nuclei to become unstable and may result in the formation of radioactive isotopes, or radioactive isotopes of new elements called **radioisotopes**. Another example follows:

sum of atomic masses on left = 31   sum of atomic masses on right = 31

$$_{13}^{27}\text{Al} + \; _2^4\text{He} \; \longrightarrow \; _{15}^{30}\text{P} + \; _0^1\text{n}$$

sum of atomic numbers on left = 15   sum of atomic numbers on right = 15

In the above equations you should note the following:

a. The sum of the atomic masses (the superscripts) on the left must equal the total of atomic masses (the superscripts) on the right.

b. The sum of the atomic numbers (the subscripts) on the left must equal the total of the atomic numbers (the subscripts) on the right.

c. In an equation depicting artificial radioactivity, there are at least two reactants on the left side of the equation. In natural radioactivity equations, there appears only one reactant on the left side of the equation. The following example points out how the above suggestions work:

$$X + {}_1^1H \longrightarrow {}_3^6Li + {}_2^4He$$

Problem: In the previous reaction what nucleus is represented by X?

Solution: Noting that the sums of both superscripts of the completed right side of the equation are 10 and 5, respectively, and the sums on the incomplete left side are 1 and 1, respectively, we now can easily ascertain the atomic mass and number of element X.

Since the atomic number identifies the element, this element with an atomic number of 4 is Beryllium, which has an atomic mass of 9. Therefore, the answer is written (at right):

$${}_4^9X$$

$${}_4^9Be$$

Accelerators. Accelerators are used to give charged particles sufficient kinetic energy to overcome electrostatic forces and penetrate the nucleus. Electric and magnetic fields are used to accelerate these charged particles.

II. Nuclear Energy

In nuclear reactions, mass is converted to energy. Nuclear reactions involve energies a million or more times greater than ordinary chemical reactions. The energy changes are due to the the changes in binding energy as a result of what is called **mass defect**. In order to understand these reactions, it is better that we clarify these terms now.

Mass Defect. The mass of a free proton (1.6725×10^{-24}g) and free neutron (1.6748×10^{-24}g) is known. Knowing that the nucleus of elements represents the total number of protons and neutrons, one would think that we can easily predict what the mass of an element will be by adding together the total mass of all neutrons and protons in the nucleus.

In the case of Helium with 2 neutrons and 2 protons in its nucleus, consider the following:

Mass of 2 free neutrons = 1.6748×10^{-24}g x 2	=	3.3496×10^{-24}g
Mass of 2 free protons = 1.6725×10^{-24}g x 2	=	3.3450×10^{-24}g
Total Mass of 2 free protons and neutrons	=	6.6946×10^{-24}g

However, the sum of the masses of the nucleons in the Helium nucleus was found to be 6.641236×10^{-24}g. If we subtract one total from the other, we find a total mass deficiency of 0.053364×10^{-24}g for every Helium atom.

If we substitute this value for the mass factor in Einstein's equation, $E=mc^2$, we find, that when nuclear particles merge to make up a nucleus, a great deal of energy is released as this small amount of mass is changed into energy. The amount of energy released is called the **Binding Energy** and is a measure of the stability of the atom formed. The greater the amount of energy released in the formation of the nucleus, the greater will be the amount of energy required to separate the nucleus into separate particles.

A. Fission Reaction

A fission reaction results in the "splitting" of heavier nuclei into lighter ones. Fission is brought about by a nucleus capturing slow moving neutrons, which results in the nucleus becoming very unstable. The unstable nucleus splits to form fission fragments of elements of lighter weight, liberation of energy, and release of two or more neutrons. The liberation of energy is the result of conversion of mass into energy.

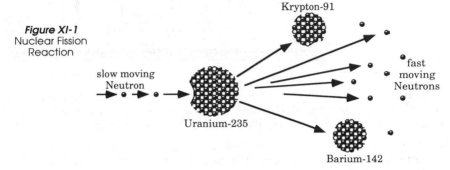

Figure XI-1
Nuclear Fission
Reaction

Krypton-91

slow moving
Neutron

fast
moving
Neutrons

Uranium-235

Barium-142

Only unstable elements of high atomic numbers can be fissioned. When a heavy element fissions, the new elements formed have a more stable configuration because of the greater binding energy per nucleon. In a nuclear reactor, the chain reaction can be controlled with control rods which limit the amount of interacting neutrons. In an atomic bomb, the chain reaction is not controlled.

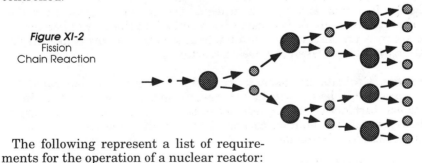

Figure XI-2
Fission
Chain Reaction

The following represent a list of requirements for the operation of a nuclear reactor:

1. Fuels. Uranium-233, Uranium-235, and plutonium-239 are fissionable. Natural uranium (99.3% uranium-238 and 0.7% uranium-235), and enriched uranium (3 to 4 percent enrichment with uranium -235) are commonly used as fuels. Uranium-233 produced from thorium-232 and plutonium-239 produced from uranium-238, are obtained as fuels in breeder reactors. Breeder reactors produce more fuel than is consumed. Through a complex process, these fuels are used to make the uranium oxide which is packed in the fuel rods made of stainless steel.

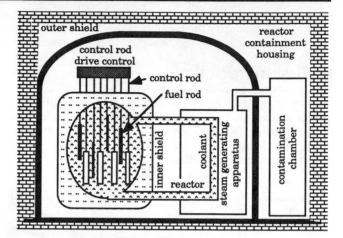

Figure XI-3
General Nuclear
Reactor and
Power Generating
System

2. Moderators. For efficient nuclear fission, it is necessary to slow down the speed of the neutrons. Moderators are materials that have the ability to slow down neutrons quickly with little tendency to absorb them.

Particles of similar mass such as hydrogen and its isotope deuterium have been found effective as moderators. The neutrons are slowed down most effectively by a head-on collision with a particle of similar mass. Water, heavy water, beryllium, and graphite are commonly used as moderators, which slow down the neutrons without capturing them.

3. Control rods. The fission process in a reactor can be controlled by adjusting the number of neutrons available. Boron and cadmium are commonly used in control rods because they absorb neutrons very well. They are placed alongside the fuel rods and by withdrawing and inserting these rods the amount of neutrons available for fission is controlled. In the event of an emergency, these rods are inserted to completely absorb the neutrons needed for fission, thereby shutting down the fission reaction.

4. Coolants. Coolants are used to keep the temperatures generated by fission at reasonable levels within the reactor and to carry heat to heat exchanges and turbines so that it can be utilized in the production of energy. Water, heavy water, air, helium, carbon dioxide, molten sodium and molten lithium are examples of coolants.

In some reactors, the coolants also serve as a help to the moderator in the removal of heat from the reactor core, where the fuel rods are located, and help prevent a core melt down.

5. Shielding. There are two shields used in nuclear reactors.

The internal shield, made of a steel lining, protects the walls of the reactor from radiation damage.

The external shield, made of high density concrete, acts as a radiation containment vessel in the event of a nuclear accident.

B. Fusion Reaction

When two light nuclei fuse into a heavier nucleus at high temperatures and pressures, an element of more stable configuration (with greater binding energy per nucleon) is formed. The mass of the heavier nucleus formed is less than the sum of the masses of the lighter nuclei. The difference in mass is converted into energy.

Fusion is the process of combining two light nuclei to form a heavier one. The energy released in a fusion reaction is much greater than in a fission reaction. Solar energy is probably the result of the fusion of ordinary hydrogen atoms to form helium. The nuclear reaction of a hydrogen bomb utilizes fission as a trigger for fusion.

Nuclear fusion presents the most appealing method of producing great amounts of energy for many reasons. They include:

1. Production. The production process is safe and does not present a threat. That is, it can be much more easily controlled than other forms of energy producing forms and shuts down automatically. Also, the isotopes produced are not radioactive but "clean", stable isotopes, that reduce the pollution threat to life.

2. Fuels. The isotopes of hydrogen, $^{2}_{1}H$ (deuterium) and $^{3}_{1}H$ (tritium), are used as fuels. Heavy water (deuterium oxide) is obtained by concentrating the trace qualities present in water. Tritium is made by the nuclear reaction.

$$^{6}_{3}Li + ^{1}_{0}n \longrightarrow ^{3}_{1}H + ^{4}_{2}He$$

The fuel, deuterium, is abundant and can be obtained cheaply from sea water. Tritium, another isotope of hydrogen is also used.

The following considerations must be addressed concerning fusion reactors:

3. High Energy Requirement. Since each nucleus carries a positive charge, all nuclei repel one another with increasing strength as they are moved closer together. Consequently, for the nuclei to interact, they must have enough kinetic energy to overcome this repulsion.

The magnitude of repulsion increases with charge; therefore, only the nuclei of lowest possible charge can be used. Fusion with ordinary hydrogen, however, is very slow. Fusion reactions involving deuterium, or deuterium and tritium, are useful sources for the release of energy. Of these, the most rapid reaction is between deuterium and tritium. The thermonuclear approach, through the use of very high temperatures, appears to be very promising for controlled fusion. However, this requires temperatures of one billion (10^{9}) degrees Celsius.

Technical problems with nuclear fusion, such as the requirement of extremely high temperatures and their containment, continue to challenge nuclear scientists and engineers. For once begun, the containment of reaction with such high temperatures is extremely difficult.

In order that the reaction be carried on in an area where it does not come in contact with materials which can break down, "magnetic bottles" were designed to contain the reaction. **"Magnetic bottles"** make use of strong magnetic fields and confine the reaction to an area where it does not come in contact with any materials except the magnetic field itself.

Questions XI-1

1   When a uranium nucleus breaks up into fragments, which type of nuclear reaction occurs?
    (1) fusion   (2) fission   (3) replacement (4) redox

2   In the reaction: $X + {}_1^1 H \rightarrow {}_3^6 Li + {}_2^4 He$

    The nucleus represented by X is:

    (1) ${}_3^9 Li$ (2) ${}_5^{10} B$ (3) ${}_4^9 Be$ (4) ${}_6^{10} C$

3   In the reaction: ${}_3^6 Li + {}_0^1 n \rightarrow {}_2^4 He + {}_1^3 X$

    The species represented by X is:

    (1) ${}_1^2 H$ (2) ${}_1^3 H$ (3) ${}_2^3 He$ (4) ${}_3^4 He$

4   Which particle is electrically neutral?
    (1) proton   (2) positron   (3) neutron   (4) electron

5   Which equation represents artificial transmutation?

    (1) $H_2O \rightarrow H^+ + OH^-$ (3) ${}_{92}^{239} U \rightarrow {}_{90}^{234} Th + {}_2^4 He$

    (2) $UF_6 + 6Na \rightarrow 6NaF + U$ (4) ${}_{13}^{27} Al + {}_2^4 He \rightarrow {}_{15}^{30} P + {}_0^1 n$

6   Given the nuclear reaction:

    $${}_{92}^{235} U + {}_0^1 n \rightarrow {}_{56}^{141} Ba + {}_{36}^{92} Kr + 3{}_0^1 n + energy$$

    This equation can best be described as:
    (1) fission   (3) natural decay
    (2) fusion   (4) endothermic

7   The equation: ${}_1^2 H + {}_1^2 H \rightarrow {}_2^4 He$ represents:

    (1) alpha decay   (2) beta decay   (3) fission   (4) fusion

    $\therefore {}_{13}^{27} Al + {}_2^4 He \rightarrow {}_{15}^{30} P + X$

    represented by X is:
    n   (3) an electron
    article   (4) an alpha particle

9   What is represented by X in the equation below?

$$X = {}^{14}_{7}N + {}^{0}_{-1}e$$

(1) ${}^{14}_{6}C$ (2) ${}^{13}_{6}C$ (3) ${}^{12}_{6}C$ (4) ${}^{11}_{6}C$

10 In the reaction: ${}^{238}_{92}U \rightarrow X + {}^{4}_{2}He$

The particle represented by X is:

(1) ${}^{234}_{90}Th$ (2) ${}^{234}_{92}U$ (3) ${}^{236}_{93}Np$ (4) ${}^{242}_{94}Pu$

11 In the equation: ${}^{31}_{15}P \rightarrow {}^{31}_{16}S + X$

The particle represented by X is:

(1) ${}^{4}_{2}He$ (2) ${}^{1}_{0}n$ (3) ${}^{1}_{1}H$ (4) ${}^{0}_{-1}e$

C. Radioactive Wastes

Environmental ecosystem which are finely balanced are threatened by the presence of radioactive wastes. The nuclear energy industry has addressed this problem.

However, grave problems still plague the safe disposal of such products as radioactive strontium - 90, which has a half-life of 28.1 years. Even after 6 half-life periods, 168.6 years, enough will remain to cause great harm if it is not contained. The problem is that fission products from nuclear reactors are intensely radioactive and cannot be discarded. They must be stored for a long time or disposed of in special ways. Solid and liquid wastes, such as strontium-90 and cesium-137, are encased in special containers for permanent storage underground or in isolated areas.

Low level radioactive wastes may be diluted and released directly into the environment. Gaseous radioactive wastes such as radon-222, krypton-85, and nitrogen-16, are stored at safe levels for decay and then dispersed into the air.

D. Uses Of Radioisotopes

1. Based on chemical reactivity. Since radioisotopes are chemically similar to stable isotopes of the same element, they can be used as **tracers** to follow the course of a reaction without seriously altering the chemical conditions. Many organic reaction mechanisms are studied by the use of carbon-14 as a tracer.

2. Based on radioactivity. Radioisotopes are used in medical diagnos therapy, food preservation, and as a means of measuring the physical di sions of many industrial products.

Isotopes with very short half-lives and which will be quickly eliminated from the body are used for diagnostic injections. Technetium-99 is used for pinpointing brain tumors. Iodine-131 is used for diagnosing thyroid disorders. Radium and cobalt -60 are used in cancer therapy.

Radiation kills bacteria, yeasts, molds, and insect eggs in foods, permitting the food to be stored for a much longer time.

3. Based on half-life. Radioisotopes give a fairly consistent method of dating some geologic events. The ratio of uranium-238 to lead-206 in a mineral can be used to determine the age of the mineral.

Carbon -14 has a half-life of approximately 5,568 years. Part of the CO_2 in the air is carbon -14. Plants take in CO_2 from the air, consequently fixing a quantity in its protoplasm. A gram of carbon from a living plant will radiate about 15 beta particles per minute. When the plant dies, it no longer takes in carbon -14. Scientist determine the age of old trees by comparing the radiation of beta particles of the old materials with that of present plants. In this manner, they can very accurately determine their age.

Questions XI-2

1   In a nuclear reactor, the purpose of the moderator is to:
    (1) absorb neutrons   (3) produce neutrons
    (2) split neutrons   (4) slow down neutrons
2   In a nuclear reactor, the radioisotope U-235 serves as a:
    (1) shield   (3) neutron absorber
    (2) coolant   (4) fissionable material
3   Which particle cannot be accelerated by the electric or magnetic fields in a particle accelerator?
    (1) neutron   (3) alpha particle
    (2) proton   (4) beta particle
4   During which process would the ratio of uranium -238 to lead -206 be used?
    (1) diagnosing thyroid disorder   (3) detecting brain tumors
    (2) dating geologic formations   (4) treating cancer patients
5   Which substance may be used as both the coolant and moderator in a reactor?
    (1) boron   (2) cadmium   (3) heavy water (4) solid graphite
6   An isotope of which element may be used as a fuel in a fission reaction?
    (1) hydrogen   (2) carbon   (3) lithium   (4) plutonium
7   Which is a gaseous radioactive waste produced during some fission reactions?
    (1) nitrogen - 16   (3) uranium - 235
    ᵗ.ᵐ - 235   (4) plutonium - 239
                    ᵈotope technetium-99 can be used for
                        the age of a sample
                    ; medical disorders
                    fission reactions
                    , speeds of neutrons

Self-Help Questions XI

Unit XI - Nuclear Chemistry
I. Artificial radioactivity, mass defect, fission reactions,
and fission reactors. fusion reactions.

1 Given the reaction: $^{27}_{13}Al + ^{4}_{2}He \rightarrow X + ^{1}_{0}n$

   When the equation is correctly balanced, the nucleus represented by X is

   (1) $^{30}_{13}Al$ (2) $^{30}_{14}Si$ (3) $^{30}_{15}P$ (4) $^{30}_{16}S$

2 Which substance may be used as fuel in the nuclear reactor?
   (1) Ra-226   (3) Fr-220
   (2) U-235   (4) Pb-204

3 Which pair of nuclei can undergo a fusion reaction?
   (1) potassium-40 and cadmium-113
   (2) zinc-64 and calcium-44
   (3) uranium-238 and lead-208
   (4) hydrogen-2 and hydrogen-3

4 Which is the symbol for the deuterium isotope of hydrogen?

   (1) $^{1}_{1}H$ (2) $^{2}_{1}H$ (3) $^{3}_{1}H$ (4) $^{4}_{2}H$

5 In the nuclear reaction $^{9}_{4}Be + X \rightarrow ^{6}_{3}Li + ^{4}_{2}He$, the particle
   represented by X is

   (1) $^{0}_{-1}e$ (2) $^{0}_{+1}e$ (3) $^{2}_{1}H$ (4) $^{1}_{1}H$

6 Cadmium and boron are commonly used in a nuclear reactor as
   (1) external shielding   (3) control rods
   (2) internal shielding   (4) moderators

7 Which process occurs in a controlled fusion reaction?
   (1) Light nuclei collide to produce heavier nuclei.
   (2) Heavy nuclei collide to produce lighter nuclei.
   (3) Neutron bombardment splits light nuclei.
   (4) Neutron bombardment splits heavy nuclei.

8 Given the equation: $^{14}_{7}N + ^{4}_{2}He \rightarrow X + ^{17}_{8}O$

   When the equation is correctly balanced, the particle represented by the
   X will be

   (1) $^{0}_{-1}e$ (2) $^{1}_{0}n$ (3) $^{1}_{1}H$ (4) $^{2}_{1}H$

9 The atoms of some elements can be made radioactive by
   (1) placing them in a magnetic field
   (2) bombarding them with high-energy particles
   (3) separating them into their isotopes
   (4) heating them to a very high temperature

10 In the reaction $2H + 3H \rightarrow 4He + X$, what is the particle represented by X?
       1 1 2
    (1) a neutron   (3) an alpha particle
    (2) a positron   (4) a beta particle

11 Which substance is used in control rods of a nuclear reactor?
    (1) boron   (3) carbon dioxide
    (2) helium   (4) beryllium

12 Which substance may serve as both a moderator and coolant in some nuclear reactors?
    (1) carbon dioxide   (3) graphite
    (2) boron   (4) heavy water

13 Which particle can *not* be accelerated by the electric or the magnetic field in a particle accelerator?
    (1) electron   (3) helium nucleus
    (2) neutron   (4) hydrogen nucleus

II. Radioactive wastes; use of radioisotopes based on reactivity and half -life.

14 Which is a gaseous radioactive waste produced during some fission reactions?
    (1) nitrogen-16   (3) uranium-235
    (2) thorium-232   (4) plutonium-239

15 The radioactive isotope carbon-14 can be used for
    (1) determining the age of an organic sample
    (2) determining medical disorders
    (3) controlling fission reactions
    (4) controlling speeds of neutrons

16 Which radioisotope is used for diagnosing thyroid disorders?
    (1) cobalt-60   (3) lead-206
    (2) uranium-238   (4) iodine-131

17 Radiated food can be safely stored for a longer time because radiation
    (1) prevents air oxidation   (3) kills bacteria
    (2) prevents air reduction   (4) causes bacteria to mutate

18 Which two characteristics do radioisotopes have that are useful in medical diagnosis?
    (1) long half-lives and slow elimination from the body
    (2) short half-lives and slow elimination from the body
    (3) long half-lives and quick elimination from the body
    (4) short half-lives and quick elimination from the body

19 Iodine-131 is used for diagnosing thyroid disorders because it is absorbed by the thyroid gland and
    ⁀ a very short half-life   (3) emits alpha radiation
         ¹ong half-life   (4) emits gamma radiation

            ₊ing procedure to determine the age of a mineral
            ₊neral's remaining amounts of isotope ^{238}U and isotope
         (2) ^{206}Bi (3) ^{214}Pb (4) ^{214}Bi

I. Math Skills

Chemistry describes both chemical and physical phenomena and relationships in the world around us. Since mathematics is the most precise way of expressing these relationships, it is used extensively in the high school chemistry course. Students should have a basic knowledge of algebra and general problem solving.

A chemical equation differs from a pure mathematical equation because the chemical formula is based on measurement. No measurement can be "exactly" correct. Each measurement consists of three parts:

1st   There is a number reading, read off the measuring device.

2nd   There is a unit.

3rd   There is a statement of the measurement's accuracy. This accuracy is determined by the number of significant digits.

A measurement taken with a meter stick might be read as, 5.17 cm. The last digit is an estimate to the nearest tenth of a scale division and is considered significant. This measurement has three significant digits. This reading is the same as, 0.0517 meters. The zero (0) simply locates the decimal point. The reading still has the same accuracy and three significant digits.

When measurements combine arithmetically, the accuracy of the answer depends on both the accuracy of the measurements and the way in which they are combined.

A. Rules For Significant Figures

Students should apply the following rules for determining the number of significant digits.

- **Initial zeros** are never significant. Since they are only used to locate a decimal point, the number 0.0203 has three significant figures.

- **Final zeros** are ambiguous. The student cannot tell if they are significant, except in the context of the measurement process. Therefore, numbers should be written in scientific notation to eliminate ambiguity. The number 2500 could have two, three, or four significant figures.

- **Final zeros**, after the decimal point, are only written if they are significant. Therefore, 2.50×10^2 has three significant digits.

- In **adding** or **subtracting** measured quantities, first round off each quantity to the number of decimal places in the measurement having the least number of decimal places. Now, add or subtract the rounded-off quantities.

- In **multiplying** or **dividing** two measured numbers, the product or quotient should have the same number of significant digits as the least significant quantity.

- **Round off** final digits to the least accurate figure in the problem.

B. Scientific Notation

To simplify working with very large and very small numbers, scientists usually write numbers in scientific notation.

1. **The general procedure for writing any number** in scientific notation is listed below:

   1) Move the decimal point of the given number to the right or left until there is only one digit to the left of the decimal point.

   2) Count the number of places you moved the decimal point. Raise ten to the power of this number. Moving the decimal point to the left is positive, while moving the decimal point to the right is negative.

2. **To add or subtract numbers** in scientific notation, for example:

$$2.34 \times 10^3 + 1.33 \times 10^5$$

   1) Make sure the **powers** of ten are equal:

$$.0234 \times 10^5 + 1.33 \times 10^5$$

   2) Add or subtract the numbers **before** the power of ten:

$$\mathbf{0234} + 1.33 = 1.3535$$

   3) Retain the **same power** of ten in the answer :

$$1.35 \times 10^5$$

3. **To multiply numbers** in scientific notation:

   1) Multiply the two numbers **preceding** the power of ten:

$$\mathbf{2.3} \times 10^4 \cdot \mathbf{1.2} \times 10^3$$

   2) Multiply the **powers** of ten by adding exponents:

$$2.3 \times 10^4 \cdot 1.2 \times 10^3$$

   3) The answer includes both steps (1) and (2):

$$\mathbf{2.76 \times 10^7}$$

4. **To divide numbers** expressed in scientific notation:

$$4.9 \times 10^5 / 2.3 \times 10^2$$

1) Divide the number **preceding** the power of ten in the numerator, by the number **preceding** the power of ten in the denominator:

$$\mathbf{4.9 / 2.3} = 2.13$$

2) Subtract the **exponent** of the power of ten in the numerator. This gives the exponent of the power of ten in the answer:

$$10^5 / 10^2 = 10^3$$

3) The answer includes both steps (1) and (2):

$$\mathbf{2.13 \times 10^3}$$

C. Order Of Magnitude

Orders of magnitude are useful when it is desired to make a quick estimate of the measurements of some quantity. The power of ten that is the nearest approximation of a measurement is defined as the order of magnitude of that measurement.

D. Dimensional Analysis (Factor - Label Method)

Measurements in chemistry always have units associated with them. Addition and subtraction are only possible if the quantities have the same units. It is possible to multiply or divide quantities with different units and come out with still another kind of unit. In these cases, the units should be treated as algebraic quantities. For example, 2.2 lbs/kg x 0.5 kg = 1.1 lbs

In addition, mathematical expressions can be simplified by manipulating units. Units can be cancelled out if they appear in both the numerator and denominator or combined into other units.

Problem:

Sulfuric acid is used in automobile batteries. It has a density of 1.2 g/cm^3. What is the mass (in grams) of 400 mL (4.0 x 10^2 mL) of this acid?

Solution:

In liquid measure, we know that 1 cm^3 = 1 mL. Therefore, we can set up the following equation:

$$\text{x number of grams} = \left(4.0 \times 10^2 \, \cancel{mL} \right) \left(\frac{1 \, \cancel{cm^2}}{1 \, \cancel{mL}} \right) \left(\frac{1.2 \, g}{\cancel{cm^2}} \right)$$

$$\text{number of grams} = 4.8 \times 10^2 \, g \quad \text{(answer)}$$

II. Laboratory Activities

The skills and activities in this unit should be developed and reinforced throughout the year by your laboratory experiences.

Note: It is most important that you should be familiar with general laboratory safety procedures.

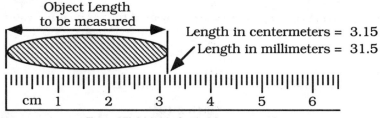

Object Length to be measured

Length in centermeters = 3.15
Length in millimeters = 31.5

Figure XII-1 Metric Scale Measurement

A. Measurement

You should be able to:

1. Use common measuring devices: balance, graduated cylinder, thermometer, buret, etc.

2. Interpolate to $\frac{1}{10}$th of the smallest scale division of a measuring device.

3. Round off numbers according to the correct number of **significant figures**. The accuracy of a measurement or calculated result can be indicated by the use of significant figures.

4. More precise measurements of the volume of acids and bases can be made by using a buret.

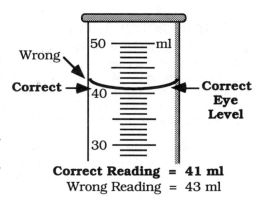

Correct Reading = 41 ml
Wrong Reading = 43 ml

Figure XII-2
Meniscus Reading
on a Graduated Cylinder

5. When adding or subtracting, the answer should be rounded off to contain the **least accurately known figure** as the final one. (see example at the right)

Add		Subtract	
32.6			
431.33		531.46	
6144.212		- 86.3	
6608.14	= **6608.1**	445.16	= **445.2**

6. When multiplying or dividing, the answer should be rounded off to contain only as many significant figures as are contained in the **least accurate number.** Examples:

Multiply	Divide
1.336	$\dfrac{5.1}{2.13} = 2.39 = \mathbf{2.4}$
x 4.2	
5.712 = **5.7**	

When the number dropped is 5 or more, the preceding number is increased by one. For example, 2.4179 taken to three significant figures becomes 2.42.

7. Determine percent error whenever applicable by using the following equation:

$$\text{Percent Error} = \frac{\overbrace{\left(\begin{array}{c}\text{Experimental}\\ \text{or}\\ \text{Observed Value}\end{array}\right) - \left(\begin{array}{c}\text{Theoretical}\\ \text{or}\\ \text{Accepted Value}\end{array}\right)}^{\textit{Actual Error}}}{\textit{Theoretical or Accepted Value}} \times 100\%$$

8. Observed value is the experimentally measured value or the value calculated from the experimental results. The **"accepted value"** is the most probable value taken from generally accepted references. The following is an example of how to report percentage error.

You are told to weigh a specific mass of $BaCl_2 \cdot 2H_2O$ and to find the percent by mass of water in the crystal. After heating the mass of hydrated crystal to constant mass in a crucible, you obtain the following results:

1. mass of the hydrated crystal   10.0 g
2. mass of the product after heating   8.8 g
3. mass of water   1.2 g
4. experimental percentage of water   12.0 %
5. theoretical percentage of water   14.0 %
6. actual error (step 4 - step 5)   -2.0 %

Substituting in the above equation, the percent error can be found:

$$\text{Percent Error} = \frac{\text{(actual error) -2\%}}{\text{(theoretical percent) 14\%}} = -0.143 \times 100\% = -14.3\%$$

B. Laboratory Skills
You should acquire at least the following basic laboratory skills:

1. The proper adjustment and use of the burner or heat source
In using the bunsen burner you should keep in mind the following:

1. Make sure all connections are secure and do not allow for gas leakage.

2. The gas valve is in the "off" position when the valve handle is at a right angle to the outlet. It is in the "on" position when the valve handle is parallel to the outlet.

3. Know how to adjust the gas supply with the burner gas adjustment device. In this manner, you can adjust the amount of gas supplied directly from the burner and not the gas valve.

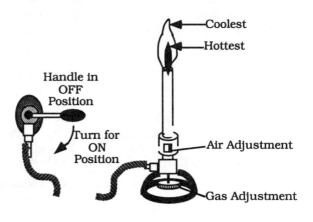

Figure XII-3
Laboratory Gas Burner

4. Know which part of the burner will adjust the amount of air intake so that the gas-air mixture will be at its optimum efficiency. If the air supply is too low, you will get a tall yellow flame which will blacken the piece of equipment you are using for heating purposes. This is due to the incomplete combustion of the gas.

5. Once the air-gas mixture is adjusted, the flame will produce two cones. It is the tip of the inner cone which is the hottest part of the flame.

6. If too much air is present in the line, a "flutter" or "roaring" noise will be heard.

2. How to cut, firepolish, and bend glass tubing.
In the laboratory we use two types of glass. "Hard" glass softens at about 800°C and is used to make pyrex test tubes. "Soft" glass is the glass that we use for glass tubing, and it melts at about 600°C.

1. **Cutting glass tubing.** (a) Using a flat surface on which to lay the glass and a triangular file, make a deep scratch at the desired point on the tubing. Holding the tubing so that the cut is away from the body, grasp the glass so that the thumbs are behind the scratch. (b) Using the thumbs as a fulcrum,

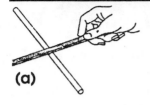
(a)

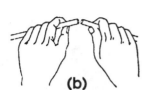
(b)

gently pull the ends of the glass towards the body. The glass should break easily at the scratch mark.

2. Firepolishing glass tubing. Glass tubing, when cut, will have sharp, cutting edges. Holding the tubing so that one end is just above the hot, inner blue cone of the bunsen burner flame, rotate it. When the flame gives off a yellow color, and the end appears smooth, remove it from the flame. If the tubing is left in the flame too long, the end will begin to close. Place the glass on a fireproof surface and allow it to cool.

Caution - Normally glass looks the same if hot or cold. Make sure that is has cooled down before touching it.

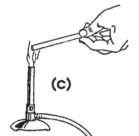
(c)

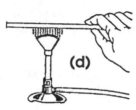
(d)

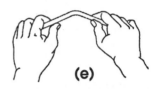
(e)

Figure XII-4
Cutting and Bending Glass

3. Bending glass tubing. (c) Before bending a piece of glass tubing make sure that the ends are firepolished. (d) With a "wing top" to spread the flame of the burner, take the piece of glass in both hands and rotate it above the spread-out flame until a yellow color appears, and the glass turns soft and pliable. (e) Remove it from the flame and bend it to form the desired angle.

3. The accepted use of a funnel and filter paper to separate a mixture.

When filtering a solid from a liquid, a good grade of filter paper should be used. Fold the filter paper many times to obtain a fluted filter. A fluted filter will more efficiently displace the air in the vessel into which the liquid is filtered, and the filtering process will proceed faster. The illustration below is the most commonly used method of folding filter paper.

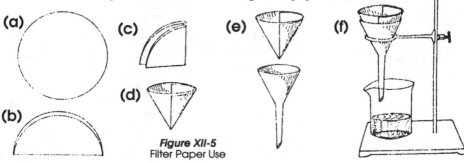

Figure XII-5
Filter Paper Use

4. The correct technique of removing solid chemicals from containers.

Before removing any chemical from its stock bottle, whether a solid or liquid, **read the label** for the following:

• Information concerning its identification, its purity, and most importantly, the caution information concerning the chemical inside.

• Never pour the solid chemical onto the balance pan. Instead, weigh a piece of weighing paper, then pour the chemical on the paper.

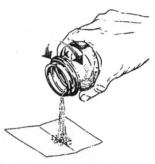

• Never return any excess to the stock bottle. Instead, discard it in a safe place, designated by your instructor.

• When pouring a chemical onto a piece of weighing paper which is on a balance, revolve the stock bottle in a forward and reverse manner. When close to the desired weight, tap the outer lip of the bottle lightly in order to have more control of the quantity of chemical poured.

Figure XII-6
Solid Chemical Pouring

5. The safe method of pouring liquids.

After reading the stock bottle for identification, purity, and caution information, grasp the stock bottle by the label and pour with the unlabeled side next to the container into which you are pouring. In this way, if there is excess dripping from the bottle, it will not drip onto the label.

• In transferring liquids into wide mouth bottles, pour the liquid onto a clean stirring rod so that the liquid clings to the stirring rod. This will minimize any splattering.

• When pouring into a test tube or other small mouthed container, pour along the inner edge in order to avoid splattering.

• When making a dilute acid solution, always add the acid to the water.

• When using a reagent bottle, grasp the flanged top with the back of the forefinger and middle fingers, holding it there while pouring, as shown in the illustration below.

Note: Do not set the stopper on a table; instead put it back directly onto the reagent bottle when you are finished pouring.

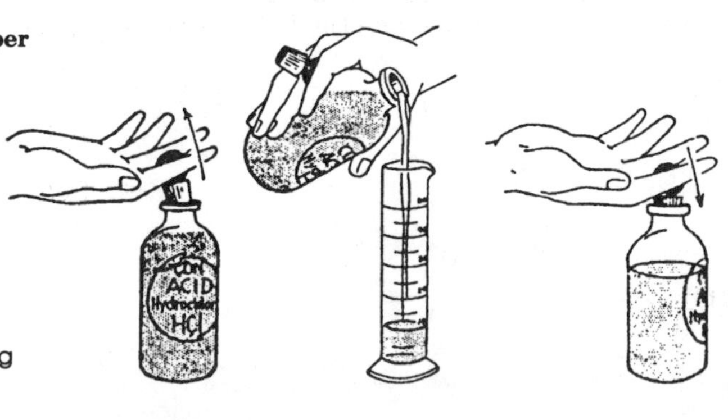

Figure XII-7
Liquid Chemical Pouring

6. Safely heating materials in test tubes, beakers, and flasks.

To help prevent accidents, follow some common sense rules when heating materials in the laboratory. They include:

• Use some protective device for the eyes.

• Never heat stoppered liquids or solids.

• Never heat a test tube containing any substance (including water) while it is pointed toward you or anyone else.

• Always use a holder while heating any substance.

• Place beakers, flasks, and evaporating dishes on a wire gauze to spread the heat when using a ring stand.

• Use a water bath to evaporate liquids that will catch fire easily.

• To heat a substance with direct heat, use a crucible and support the crucible on a clay triangle.

• The formation of large bubbles in boiling, due to local superheating and referred to as "bumping," can sometimes be prevented by the addition of a few glass beads or boiling chips

7. The evaporation of a liquid or the drying of a solid.

• **Use a fume hood** when evaporating hazardous solvents or materials.

• In situations where the temperature must be controlled with a high degree of precision, the use of drying ovens, heating mantle, or electric heater will be necessary.

Note: In the presence of flammable materials, do not use an open flame.

• **In drying a solid,** heat gently so that you do not lose some of your sample as the result of superheating in spots with the resultant popping out of some of the material being heated.

8. Identification of common laboratory apparatus.

Mortar / Pestle

Evaporating Dish

Crucible

Funnel

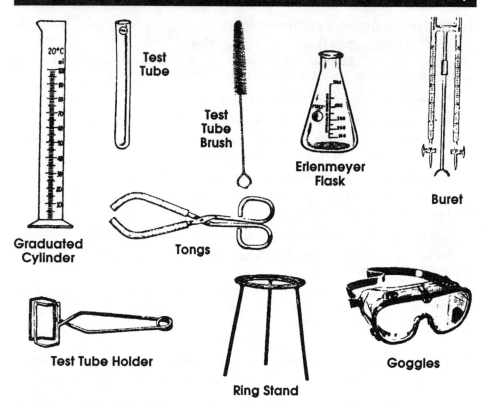

Questions XII-1

1 The process of filtration is performed in the laboratory to
   (1) form precipitates
   (2) remove water from solutions
   (3) separate dissolved particles from the solvent
   (4) separate insoluble substances in an aqueous mixture

2 The volume of an acid required to neutralize exactly 15.00 milliliters
   (mL) of a base could be measured most precisely if it were added to the
   base solution from a
   (1) 100. mL graduate   (3) 50. mL buret
   (2) 125 mL Erlenmeyer flask   (4) 50. mL beaker

3 Using the rules for significant figures, the sum of 0.027 gram and
   0.0023 gram should be expressed as
   (1) 0.029 gram   (3) 0.03 gram
   (2) 0.0293 gram   (4) 0.030 gram

4 The diagram at the right represents a portion of a buret.
   What is the reading of the meniscus?
   (1) 39.2 mL
   (2) 39.5 mL
   (3) 40.7 mL
   (4) 40.9 mL

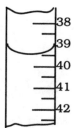

5   A bunsen burner flame is sooty black and mixed with an orange-yellow
    color. Which is the probable reason for this condition?
    (1) No oxygen is mixing with the oxygen.
    (2) No gas is mixing with the oxygen.
    (3) Insufficient oxygen is mixing with the gas.
    (4) Insufficient gas is mixing with the oxygen.
6   In an experiment the gram atomic mass of magnesium was determined to
    be 24.7. Compared to the accepted value 24.3, the percent error for this
    determination was
    (1) 0.400   (2) 1.65   (3) 2.47   (4) 98.4
7   Which diagram to the right represents a pipette?

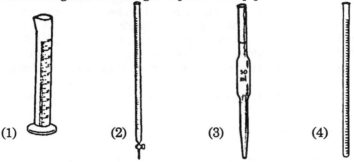

    (1)   (2)   (3)   (4)

8   Which piece of laboratory apparatus would most likely be used to
    evaporate a 1-milliliter sample of a solution to dryness?
    (1) volumetric flask   (3) pipette
    (2) buret   (4) watch glass
9   A student determined the melting point of a substance to be 55.2°C. If
    the accepted value is 50.1°C, the percent error is
    (1) 5.10   (2) 9.24   (3) 10.2   (4) 120
10 A cube has a volume of 8.0 cm^3 and a mass of 216 grams. The density of
    the cube, in grams per cubic centimeter, is best expressed as
    (1) 27   (2) 270   (3) 0.37   (4) 0.370
11 In an experiment, a student found that the percent of oxygen in a sample
    of $KClO_3$ was 42.3%. If the accepted value is 39.3%, the experimental
    percent error is

    (1) $\dfrac{42.3}{39.3}$ x 100%   (3) $\dfrac{3.0}{42.3}$ x 100%

    (2) $\dfrac{39.3}{42.3}$ x 100%   (4) $\dfrac{3.0}{39.3}$ x 100%

C. Laboratory Activities

You should do a variety of activities in the laboratory, including:

1. Carry out an activity involving phase change and interpret a simple
heating or cooling curve based on this activity. Ice or paradichlorobenzene or
naphthalene may be used to determine some portions of a heating and/or
cooling curve.

The illustration at the right represents the heating and cooling curves of paradichlorobenzene (mp. 54°C) superimposed on the same graph.

This allows a vivid demonstration of how the temperature remains constant during a phase change.

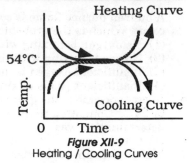

Heating Curve

54°C

Temp.

Cooling Curve

0　　　Time

Figure XII-9
Heating / Cooling Curves

2. Measure heats of simple reactions in calories. A simple activity in calorimetry may be done using water as a calorimetric liquid and a burning candle to determine the heat of combustion of the candle. First take the mass of the candle and the water, along with the temperature of the water before heating takes place. We can determine the amount of heat rise of the water, using the standard apparatus. Then, by reweighing the used candle, determine the amount of calories required to combust a quantity of the candle. Use the formula:

$$\text{calories} = \text{mass of } H_2O \ \times \ \Delta T$$

Then, divide the formula answer by the weight of the wax used up to obtain the answer in calories per gram of wax.

3. Identify endothermic and exothermic processes. It is possible to measure the temperature changes associated with certain chemical reactions or solution processes. The dissolution of ammonium chloride is endothermic, and that of sodium hydroxide is exothermic. By taking the mass of the water and the temperature change, we can determine the heat of dissolution of a specific mass of salt.

4. Do a solubility experiment and draw a solubility curve. Salts such as sodium nitrate, potassium nitrate, and potassium chlorate show large increases in solubility as the temperature increases.

One experiment that can be used in the study of solubility curves is the following. In this experiment the instructor assigns each "team" of students a specific temperature at which a saturated solution of a salt is allowed to remain for 5 minutes. The temperatures vary with each "team." After successfully keeping the saturated solution of the salt at their assigned temperature for 5 minutes, they must quickly filter a small amount into a preweighed evaporating dish. They then weigh the dish with the solution in it and evaporate the solution to constant weight. The dish and the dried salt is reweighed, and the mass of water evaporated is calculated. They then extrapolate the proportion of the mass of the salt to water to a proportion giving the mass of salt per 100 g of water and enter their result into a data table. Using values from other student "teams" to complete their data table, they can plot each value on a graph using grams of solute in 100 g of water as one vector, and temperature as the other vector. The graphs should look like those in Table *D* in the Reference Tables for Chemistry.

5. Identify differences between inorganic and organic substances. Differences in solubility, melting point, stability, and electrical conductivity may be qualitatively demonstrated using sodium chloride and sucrose.

These differences are due, in the most part, to types of bonds in organic and inorganic compounds in solvents. Organic compounds dissolve in nonpolar solvents and are generally insoluble in polar compounds except organic acids and alcohol. Because of the weak van der Waals forces between the molecules, the molecules dissociate easily, giving them low melting points. Also as a result, they do not conduct electricity. Whereas, inorganic compounds usually have ionic bonds that will dissolve in polar solvents but not in nonpolar solvents. They have high melting points, making them very stable, and will conduct electricity in solution.

There are exceptions to these general observations. For instance, although plastics are organic substances, some conduct electricity and have high melting points.

6. Identify metallic ions using a flame test. The absorption of energy by an electron causes it to jump to a higher energy level. At this point the atom is said to be in an excited state. When the electron returns to a ground state level it emits a photon of energy. If the frequency of the photon corresponds to those found in the visible spectrum, a specific colored light is seen. The identification of an element is made possible through the use of these spectral lines of color.

Normally the atoms of most elements require large amounts of energy to bring them to an excited state. However, some metallic ions can be excited by the flame of a bunsen burner. We can test for presence of some metallic elements by dipping a clean wire loop of platinum into solutions and then placing it into a bunsen burner flame. The color it exhibits can be used to identify the metallic ion present in the solution. Solutions of the salts of lithium (red), sodium (yellow), potassium (violet), calcium (yellowish-red), strontium (scarlet red), and barium (yellowish-green), may be used.

7. Find the percent by mass of water in a hydrated crystal. A hydrated crystal such as $CuSO_4 \cdot 5H_2O$ or $BaCl_2 \cdot 2H_2O$ is placed into a crucible, and the crucible and crystal are weighed. The crystal is then heated to **constant mass.** The mass which is "lost" during the heating is considered to be the water of hydration. The experimental result is used to find the percent of water of hydration in the original crystal and is found by using the following relationship:

$$\text{percent of water of hydration} = \frac{\text{mass of water}}{\text{mass of hydrated crystal}} \times 100\%$$

The end product after heating is a dry, anhydrous salt. Water is now added to the product in order to observe its transformation back into a hydrated crystal.

8. Determine the volume of a mole of gas in a reaction. The hydrogen gas generated from the reaction of approximately 0.05g magnesium, with excess dilute HCl, may be collected in a 50 mL buret of an eudiometer tube. Having adjusted the level of water left inside the buret with that in the trough outside of the buret, the water temperature is taken. Knowing the temperature of the water, we derive the partial pressure of the water from Table *O* in the Reference Tables of Chemistry, so that we can obtain the pressure of the dry hydrogen gas. Assuming that the temperature of the gas is the same as that of the water, volume of the hydrogen, corrected to STP, can be obtained by using the combined gas law. From the balanced equation we can determine the moles of hydrogen produced from the moles of magnesium used. The following proportion is then used:

$$\frac{\text{Volume Of Hydrogen Gas Produced}}{\text{Moles Of Hydrogen Used}} = \frac{\text{Experimental Volume Of Hydrogen}}{\text{1 Mole}}$$

The answer is subtracted from 22.4 liters to find the actual error, and the percentage error is calculated.

9. Acid-Base titration. Acid-base titration is a form of volumetric analysis by which we can determine the unknown concentration of an acid (or base) solution by neutralizing it with a known concentration of base (or acid).

Recall that one reason for a reaction to "go to an end or to completion" is that water appears as a product. Also recall that acids increase the amount of hydrogen ions (H^+) as they ionize in water, and most bases increase the amount of hydroxide ions (OH^-) as they ionize in water. So that, when an acid solution is neutralized by a basic solution, the net ionic equation is:

$$H^+ \ + \ OH^- \longrightarrow H_2O$$

When excess hydrogen ions are neutralized by hydroxide ions, the point of neutralization is reached ("equivalent point"). This point may be determined by use of an indicator whose end point (the pH at which it has a color change) is close to the pH at the equivalent point. Dilute solutions of sodium hydroxide or potassium hydroxide may be titrated against dilute hydrochloric or acetic acids. If burets are not available, calibrated medicine droppers or pipettes can be used. An example of an acid-base titration experiment is to find the unknown concentration of acetic acid in a solution of acetic acid.

After thoroughly cleaning the equipment, a few drops of phenolphthalein is added to the acetic acid sample. A standardized solution of sodium hydroxide is titrated into the acetic acid solution until the mixed solution just starts to turn pink for 30 seconds. At this point the volumes of both the acid and the base of known concentration are both noted, and the figures are keyed into the following formula:

$$\begin{array}{cccc}\text{Normality} & \text{Volume} & \text{Normality} & \text{Volume} \\ \text{of the Acid} & \text{x} \quad \text{of the Acid} & = \quad \text{of the Base} & \text{x} \quad \text{of the Base}\end{array}$$

$$N_a \ x \ V_a \ = \ N_b \ x \ V_b$$

The concentration of the acetic acid is calculated.

D. Laboratory Reports

You should be able to write concise reports of laboratory experiences. The student should be able to:

1. Organize information in a logical manner.
2. Put appropriate data in tables and graphs whenever possible.
3. Make a list of observations.
4. Draw conclusions based on observations.

Laboratory reports may seem tedious and redundant, but they are the essence of your personality. They convey in your words a concise form of communication which reflects the way that you work. If your laboratory report is sloppy, with misspelled words, and shows poor organization, it is an indication of your manner, and you will be judged by these criteria.

The most important discovery is **not** a contribution to science until it has been communicated to others.

In short, like any other form of communication, the laboratory report should include the following:

1. The content is **factual, objective,** and **easily read.**
2. The correct word is used. Knowledge of language is evident, and **spelling is flawless.**
3. All observations and conclusions are supported by **evidence.**
4. The scientific apparatus is **accurately described,** and the working details of the experiment are **complete.**
5. Self devised **graphs** and their **interpretation.**

Note: On ignition of a substance in a chemical reaction, the mass of the substance in grams divided by the mass of the residue in grams is called the **experimental mass ratio.**

Questions XII-2

1   As a result of dissolving a salt in water, a student found that the temperature of the water increased. From this observation alone, the student should conclude that the dissolving of the salt
    (1) produced an acid solution   (3) was endothermic
    (2) produced a basic solution   (4) was exothermic

2   How many milliliters of 0.4 M HCl are required to completely neutralize 200 milliliters of 0.16 M potassium hydroxide?
    (1) 500   (2) 200   (3) 80   (4) 30

3   During a titration, a student used 50. milliliters of 0.1 M acid. How many moles of acid, expressed to proper significance, were used?
    (1) 0.005   (3) 0.00500
    (2) 0.0050   (4) 0.005000

4   A student obtained the following data (at the right) in determining the solubility of $NaNO_3$:

Temp. (°C)	Solubility (g $NaNO_3$/100 g H_2O)
0	73
10	80
20	88
30	97
40	105
50	115
60	124
70	134
80	145

Which set of coordinates would graphically present the data in the table at the right most clearly?

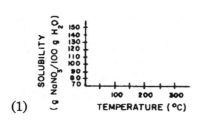

(1)

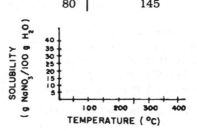

(3)

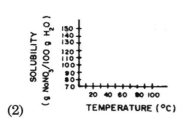

(2)

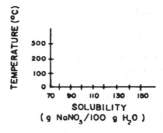

(4)

Base your answers to questions 5 and 6 on the table below which shows the data collected during the heating of a 5.0 gram sample of a hydrated salt.

Mass of Salt (g)	Heating Time (minutes)
5.0	0.0
4.1	5.0
3.1	10.
3.0	15.
3.0	30.
3.0	60.

5   After 60. minutes, how many grams of water appear to remain in the salt?
(1) 0.00
(2) 20
(3) 19
(4) 0.90

6   What is the percent of water in the original sample?
(1) 82. %   (2) 60. %   (3) 30. %   (4) 40. %

7   Given the following titration data:

         Volume of base (KOH) = 40.0 mL
         Molarity of base = 0.20 M
         Volume of acid (HCl) added = 20.0 mL

The concentration of HCl required for the acid to neutralize the base is
(1) 10. M   (2) 0.20 M   (3) 0.10 M   (4) 0.40 M

8   Which graph shown below could represent the uniform cooling of a substance, starting with the gaseous phase and ending with the solid phase?

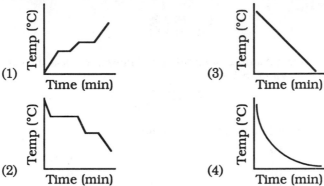

(1)   (3)

(2)   (4)

Self-Help Questions XII

Unit XII - Laboratory Activities
I. Measurement, including percent error, and laboratory skills

1   A student has to measure the diameter
of a test tube in order to calculate the tube's
volume. Based on the diagram, the tube's
diameter is closest to

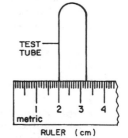

(1) 1.25 cm
(2) 2.32 cm
(3) 3.25 cm
(4) 12.5 cm

2   Which procedure is the safest to follow when using an open flame to heat
the contents of a test tube that contains a flammable mixture?
(1) Cork the test tube and then heat it gently near the bottom only.
(2) Heat the open test tube gently near the bottom only.
(3) Cork the test tube and place it in a beaker of water, then heat the
water in the beaker.
(4) Place the open test tube in a beaker of water, then heat the water in
the beaker.

3   Which milligram quantity contains a total of four significant figures?
(1) 0.3010 mg   (3) 3100. mg
(2) 3010. mg   (4) 30001 mg

4   The volume of acid used to neutralize 15.00 milliliters of a base solution
can be measured most precisely by the use of a
(1) beaker   (3) graduated cylinder
(2) buret   (4) volumetric flask

5   In the laboratory, a student determined the percent by mass of water in
    a hydrated salt to be 17.3 percent. If the accepted value is 14.8 percent,
    the percent error is
    (1) 2.50 %   (3) 16.9 %
    (2) 5.92 %   (4) 27.1 %

6   Expressed to the correct number of significant figures, what is the correct
    sum of (3.04 g + 4.134 g + 6.1 g)?
    (1) 13 g   (3) 13.27 g
    (2) 13.3 g   (4) 13.274 g

7   Which diagram represents a graduated cylinder?

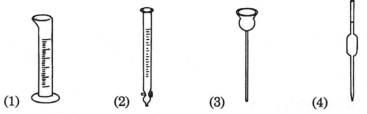

    (1)   (2)   (3)   (4)

8   Which procedure represents the safest technique to use for diluting a
    concentrated acid?
    (1) Add the acid to the water quickly.
    (2) Add the water to the acid quickly.
    (3) Add the acid slowly to the water with steady stirring.
    (4) Add the water slowly to the acid with steady stirring.

9   Which measurement contains three significant figures?
    (1) 0.05 g   (3) 0.056 g
    (2) 0.050 g   (4) 0.0563 g

*II. Laboratory activities including phase change, measure heats of reaction,
exothermic and endothermic processes, solubility experiment,
and draw solubility curves, identify between organic and
inorganic substances, identify metallic ions from a flame test,
find the percentage of water in a hydrated crystal, determine
the volume of a mole of gas in a reaction, and a titration experiment.*

10 A 9.90-gram sample of a hydrated salt is heated to a constant mass of
    6.60 grams. What was the percent by mass of water contained in the
    sample?
    (1) 66.7   (2) 50.0   (3) 33.3   (4) 16.5

11 In an experiment, a student
    determined the normal boiling points
    of four unknown liquids. The collected
    data was organized into the table:
    Which unknown liquid has the *weakest*
    attractive forces between its molecules?
    (1) *A*   (3) *C*
    (2) *B*   (4) *D*

Unknown Liquids	Normal Boiling Points (°C)
A	10
B	33
C	78
D	100

12 A student collected data in an experiment in which the uniform cooling of a water sample was observed from 50°C to -32°C. Which graph most likely represents the results obtained by the student?

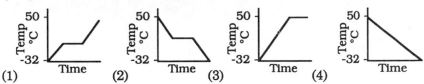

(1)   (2)   (3)   (4)

13 The graph at the right represents the decay or a radioactive isotope. What is the half-life of this isotope?

(1) 1 hour
(2) 2 hours
(3) 3 hours
(4) 6 hours

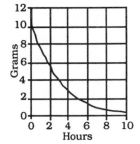

14 A student investigated samples of our different substances in the solid state. The table is a record of the behaviors observed (marked with an *X*) when each solid was tested.

Behavior Tested	Sample I	Sample II	Sample III	Sample IV
High Melting Point	X		X	
Low Melting Point		X		X
Soluble in Water	X			X
Insoluble in Water		X	X	
Decomposed Under High Heat		X		
Stable Under High Heat	X		X	X
Electrolyte	X			X
Nonelectrolyte		X	X	

Based on the tabulated results, which of the solids investigated had the characteristics most closely associated with those of an organic compound?
(1) Sample I   (2) Sample II   (3) Sample III   (4) Sample IV

15 The graph at the right represents four solubility curves. Which curve best represents the solubility of a gas in water?

(1) *A*
(2) *B*
(3) *C*
(4) *D*

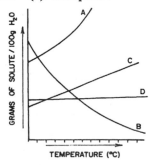

Laboratory Notes:

Chemistry Reference Tables

Including tables and explanations for:

Periodic Table of The Elements

Periodic Table of the Elements

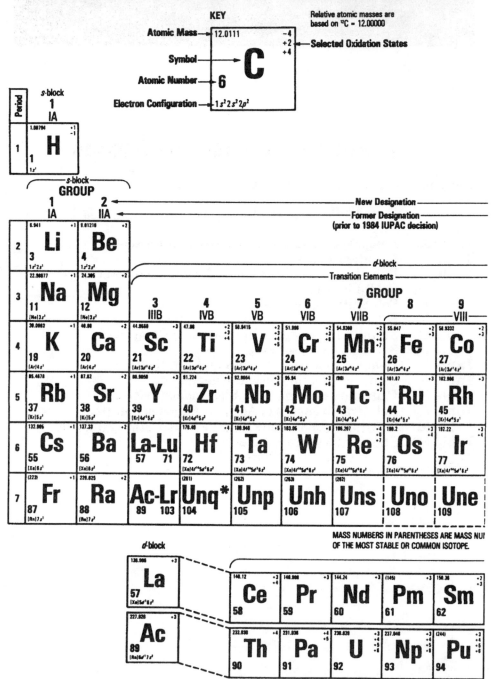

MASS NUMBERS IN PARENTHESES ARE MASS NU[...]
OF THE MOST STABLE OR COMMON ISOTOPE.

As of August 30, 1997, IUPAC (International Union of Pure and Applied Chemistry) adopted nomenclature for the short-lived Transfermium elements 101 through 109:

Element	Name	Symbol
101	Mendelevium	Md
102	Nobelium	No
103	Lawrencium	Lr
104	Rutherfordium	Rf
105	Dubnium	Db
106	Seaborgium	Sg
107	Bohrium	Bh
108	Hassium	Hs
109	Meitnerium	Mt

s-block

18
O

4.00260 — 0
He
2
$1s^2$

p-block
GROUP

13 IIIA	14 IVA	15 VA	16 VIA	17 VIIA	18 O
10.81 +3 **B** 5 $1s^2 2s^2 2p^1$	12.0111 −4,+2,+4 **C** 6 $1s^2 2s^2 2p^2$	14.0067 −3,−2,−1,+1,+2,+3,+4,+5 **N** 7 $1s^2 2s^2 2p^3$	15.9994 −2 **O** 8 $1s^2 2s^2 2p^4$	18.998403 −1 **F** 9 $1s^2 2s^2 2p^5$	20.179 0 **Ne** 10 $1s^2 2s^2 2p^6$
26.98154 +3 **Al** 13 $[Ne]3s^2 3p^1$	28.0855 −4,+2,+4 **Si** 14 $[Ne]3s^2 3p^2$	30.97376 −3,+3,+5 **P** 15 $[Ne]3s^2 3p^3$	32.06 −2,+4,+6 **S** 16 $[Ne]3s^2 3p^4$	35.453 −1,+1,+3,+5,+7 **Cl** 17 $[Ne]3s^2 3p^5$	39.948 0 **Ar** 18 $[Ne]3s^2 3p^6$

10	11 IB	12 IIB						
58.69 +2,+3 **Ni** 28 $[Ar]3d^8 4s^2$	63.546 +1,+2 **Cu** 29 $[Ar]3d^{10}4s^1$	65.39 +2 **Zn** 30 $[Ar]3d^{10}4s^2$	69.72 +3 **Ga** 31 $[Ar]3d^{10}4s^2 4p^1$	72.59 −4,+2,+4 **Ge** 32 $[Ar]3d^{10}4s^2 4p^2$	74.9216 −3,+3,+5 **As** 33 $[Ar]3d^{10}4s^2 4p^3$	78.96 −2,+4,+6 **Se** 34 $[Ar]3d^{10}4s^2 4p^4$	79.904 −1,+1,+5 **Br** 35 $[Ar]3d^{10}4s^2 4p^5$	83.80 0,+2 **Kr** 36 $[Ar]3d^{10}4s^2 4p^6$
106.42 +2,+4 **Pd** 46 $[Kr]4d^{10}5s^0$	107.868 +1 **Ag** 47 $[Kr]4d^{10}5s^1$	112.41 +2 **Cd** 48 $[Kr]4d^{10}5s^2$	114.82 +3 **In** 49 $[Kr]4d^{10}5s^2 5p^1$	118.71 +2,+4 **Sn** 50 $[Kr]4d^{10}5s^2 5p^2$	121.75 −3,+3,+5 **Sb** 51 $[Kr]4d^{10}5s^2 5p^3$	127.60 −2,+4,+6 **Te** 52 $[Kr]4d^{10}5s^2 5p^4$	126.905 −1,+1,+5,+7 **I** 53 $[Kr]4d^{10}5s^2 5p^5$	131.29 0,+2,+4,+6 **Xe** 54 $[Kr]4d^{10}5s^2 5p^6$
195.08 +2,+4 **Pt** 78 $[Xe]4f^{14}5d^9 6s^1$	196.967 +1,+3 **Au** 79 $[Xe]4f^{14}5d^{10}6s^1$	200.59 +1,+2 **Hg** 80 $[Xe]4f^{14}5d^{10}6s^2$	204.383 +1,+3 **Tl** 81 $[Xe]4f^{14}5d^{10}6s^2 6p^1$	207.2 +2,+4 **Pb** 82 $[Xe]4f^{14}5d^{10}6s^2 6p^2$	208.980 +3,+5 **Bi** 83 $[Xe]4f^{14}5d^{10}6s^2 6p^3$	(209) −2,+4 **Po** 84 $[Xe]4f^{14}5d^{10}6s^2 6p^4$	(210) **At** 85 $[Xe]4f^{14}5d^{10}6s^2 6p^5$	(222) 0 **Rn** 86 $[Xe]4f^{14}5d^{10}6s^2 6p^6$

* The systematic names and symbols for elements of atomic numbers greater than 103 will be used until the approval of trivial names by IUPAC.

f-block

151.96 +2,+3 **Eu** 63	157.25 +3 **Gd** 64	158.925 +3 **Tb** 65	162.50 +3 **Dy** 66	164.930 +3 **Ho** 67	167.26 +3 **Er** 68	168.934 +3 **Tm** 69	173.04 +2,+3 **Yb** 70	174.967 −3 **Lu** 71	Lanthanoid Series
(243) +3,+4,+5,+6 **Am** 95	(247) +3 **Cm** 96	(247) +3,+4 **Bk** 97	(251) +3 **Cf** 98	(252) +3 **Es** 99	(257) **Fm** 100	(258) **Md** 101	(259) **No** 102	(260) **Lr** 103	Actinoid Series

Table A

PHYSICAL CONSTANTS AND CONVERSION FACTORS

Name	Symbol	Value(s)	Units
Angstrom unit	Å	1×10^{-10} m	meter
Avogadro number	N_A	6.02×10^{23} per mol	
Charge of electron	e	1.60×10^{-19} C	coulomb
Electron volt	eV	1.60×10^{-19} J	joule
Speed of light	c	3.00×10^{8} m/s	meters/second
Planck's constant	h	6.63×10^{-34} J·s	joule-second
		1.58×10^{-37} kcal·s	kilocalorie-second
Universal gas constant	R	0.0821 L·atm/mol·K	liter-atmosphere/mole-kelvin
		1.98 cal/mol·K	calories/mole-kelvin
		8.31 J/mol·K	joules/mole-kelvin
Atomic mass unit	μ(amu)	1.66×10^{-24} g	gram
Volume standard, liter	L	1×10^{3} cm^3 = 1 dm^3	cubic centimeters, cubic decimeter
Standard pressure, atmosphere	atm	101.3 kPa	kilopascals
		760 mmHg	millimeters of mercury
		760 torr	torr
Heat equivalent, kilocalorie	kcal	4.18×10^{3} J	joules

Physical Constants for H$_2$O

Molal freezing point depression	1.86°C
Molal boiling point elevation	0.52°C
Heat of fusion	79.72 cal/g
Heat of vaporization	539.4 cal/g

Table B

STANDARD UNITS

Symbol	Name	Quantity	Selected Prefixes		
m	meter	length	Factor	Prefix	Symbol
kg	kilogram	mass			
Pa	pascal	pressure	10^6	mega	M
K	kelvin	thermodynamic temperature	10^3	kilo	k
mol	mole	amount of substance	10^{-1}	deci	d
J	joule	energy, work,	10^{-2}	centi	c
		quantity of heat	10^{-3}	milli	m
s	second	time	10^{-6}	micro	μ
C	coulomb	quantity of electricity	10^{-9}	nano	n
V	volt	electric potential, potential difference			
L	liter	volume			

How To Use The Reference Tables

Reference Table A
Physical Constants

Although Reference Table A is self explanatory, it is often overlooked. One of the most widely used constants is Avogadro's number (6.02×10^{23}), whose symbol is N_A. Another is the Molar freezing point depression constant for water (1.86°C) and the molar boiling point elevation of water (0.52°C). This means that a mole of particles in a liter of water will decrease the freezing point of water by 1.86° C and increase the boiling point by 0.52°C.

When you dissolve one mole of $CaCl_2$ in one liter of water, you will have one mole of calcium ions and two moles of chlorine ions in solution. Therefore, whereas the water itself had a freezing point of 0° C, the solution will have a freezing point of 3 x -1.86°C, or -5.58°C. Also, the boiling point of the water was 100°C, but the solution will have a boiling point of 3 x 0.52°C + 100°C, or 101.56°C.

Note: Add the following to your list of constants:

Pressure 1 atm = 101.3 kPa (kilopascals)

Reference Table B
Standard Units

Reference Table B is separated into two distinct parts:

I. Standard Units.
The symbol, name, and the quantity each represents is given. These units have been adopted in 1960 as the International System of Units. Also called Systeme International (SI) Units, they represent an extension of the metric system. Those shown here are:

  1. **Base Units** include the meter, kilogram, Kelvin, mole, and second.

  2. **Derived Units** include the joule, pascal, coulomb, and volt.

  3. **A Converted Unit** includes the liter (L) which has been considered a base unit for expressing volume. However, now the base SI unit which denotes volume is the cubic meter (m^3). The symbol for the liter, when expressed in SI units, is converted to cubic meters such as $10^{-3} m^3$.

II. Selected Prefixes.
There exist at least fourteen (14) prefixes which depict metric quantities. The prefixes shown here are a select few which are most commonly used.

Table **C**

DENSITY AND BOILING POINTS OF SOME COMMON GASES		Density grams/liter at STP*	Boiling Point (at 1 atm) K
Air	—	1.29	—
Ammonia	NH₃	0.771	240
Carbon dioxide	CO₂	1.98	195
Carbon monoxide	CO	1.25	82
Chlorine	Cl₂	3.21	238
Hydrogen	H₂	0.0899	20
Hydrogen chloride	HCl	1.64	188
Hydrogen sulfide	H₂S	1.54	212
Methane	CH₄	0.716	109
Nitrogen	N₂	1.25	77
Nitrogen (II) oxide	NO	1.34	121
Oxygen	O₂	1.43	90
Sulfur dioxide	SO₂	2.92	263
*STP is defined as 273 K and 1 atm			

Table **D**

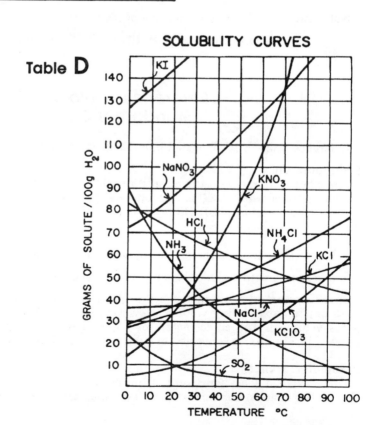

SOLUBILITY CURVES

Reference Table C
Density & Boiling Points of Some Common Gases

Table C is useful in four different ways:

1. It defines STP (standard temperature and pressure) as 273 K or 0°C and 1 atm or 760 torr.

2. It names and gives the formulas of the most common gases.

3. It gives the density of these common gases in grams/liter. Knowing that one mole of any gas occupies 22.4 liters of volume, it is easy to calculate the molecular mass of a gas from its density. You simply multiply both the numerator and denominator by 22.4. To find the density, given the molecular mass, you divide by 22.4. Then you can compare its density to those of the listed gases to identify the gas.

Example Problem:
An unknown gas has a molecular mass of 34 g. Identify the gas.

Solution: Since 34g is the weight of one mole, and we know that the molecular volume of any gas is 22.4 liters, simply divide by 22.4 to get its density.

Answer: The density of the unknown gas is 1.52 g/liter. When we examine Reference Table C, we find that the gas with a density nearest 1.52 g/liter is Hydrogen Sulfide.

4. The boiling point of the common gases is given in degrees Kelvin and at 1 atm. As you have studied in Unit III, the lower the boiling point, the weaker the bonds between the molecules. Also, the weaker the bonds between gas molecules, the higher their vapor pressure.

In a mixture of gases (the classical example is air), the table reveals the sequence by which this mixture can be separated. It is first liquified by compression or by lowering its temperature, then, as it is heated, hydrogen will boil off at 20 K, nitrogen at 77 K, and oxygen at 90 K.

Reference Table D
Solubility Curves

Three gases and seven salts are dissolved in 100 g of water at different temperatures. Their maximum solubility at that temperature is represented by a curved line. That is: any point on the solubility curve of each substance represents a saturated solution of that substance.

If the solubility was such that the dissolved amount was noted as a point above the curve of a particular substance, the resulting solution was termed to be supersaturated. If instead, the dissolved amount was noted as a point below the curve of the particular substance, it was considered unsaturated. Most salts become more soluble as the temperature rises. This prompted the saying, "solubility is favored by the endothermic reaction."

Ammonia gas (NH_3) represents a "typical" gas which always decreases in solubility as the temperature rises.

Table **E**

TABLE OF SOLUBILITIES IN WATER										
i — nearly insoluble ss — slightly soluble s — soluble d — decomposes n —not isolated	acetate	bromide	carbonate	chloride	chromate	hydroxide	iodide	nitrate	phosphate	sulfate	sulfide
Aluminum	ss	s	n	s	n	i	s	s	i	s	d
Ammonium	s	s	s	s	s	s	s	s	s	s	s
Barium	s	s	i	s	i	s	s	s	i	i	d
Calcium	s	s	i	s	s	ss	s	s	i	ss	d
Copper II	s	s	i	s	i	i	n	s	i	s	i
Iron II	s	s	i	s	n	i	s	s	i	s	i
Iron III	s	s	n	s	i	i	n	s	i	ss	d
Lead	s	ss	i	ss	i	i	ss	s	i	i	i
Magnesium	s	s	i	s	s	i	s	s	i	s	d
Mercury I	ss	i	i	i	ss	n	i	s	i	ss	i
Mercury II	s	ss	i	s	ss	i	i	s	i	d	i
Potassium	s	s	s	s	s	s	s	s	s	s	s
Silver	ss	i	i	i	ss	n	i	s	i	ss	i
Sodium	s	s	s	s	s	s	s	s	s	s	s
Zinc	s	s	i	s	s	i	s	s	i	s	i

Table **F**

SELECTED POLYATOMIC IONS			
Hg_2^{2+}	dimercury (I)	CrO_4^{2-}	chromate
NH_4^+	ammonium	$Cr_2O_7^{2-}$	dichromate
$C_2H_3O_2^-$ } CH_3COO^- } acetate		MnO_4^-	permanganate
		MnO_4^{2-}	manganate
CN^-	cyanide	NO_2^-	nitrite
CO_3^{2-}	carbonate	NO_3^-	nitrate
HCO_3^-	hydrogen carbonate	OH^-	hydroxide
		PO_4^{3-}	phosphate
$C_2O_4^{2-}$	oxalate	SCN^-	thiocyanate
ClO^-	hypochlorite	SO_3^{2-}	sulfite
ClO_2^-	chlorite	SO_4^{2-}	sulfate
ClO_3^-	chlorate	HSO_4^-	hydrogen sulfate
ClO_4^-	perchlorate	$S_2O_3^{2-}$	thiosulfate

Reference Table E
Solubilities in Water

An ionic or double replacement reaction of two or more ionic solutions produces a number of ions in solution. These ions are surrounded by water molecules which encapsulates each ion, thus separating the ions, not allowing them to come together again. This is called the **dielectric effect**.

Some ions in a solution of ions are capable of breaking this barrier, which allows them to combine with other ions and precipitate out of solution. This solubility of ions and the precipitation of others is given in Reference Table C.

If the product shows the symbols **i** or **ss**, a precipitate can be expected to form.

Reference Table F
Selected Polyatomic Ions

Polyatomic ions are usually made up of more than one element. Mercury (Hg^{2+}) is the only exception on the table. The intramolecular bonds between the elements which make up the ion are covalent. However, having formed an ion, the bond a polyatomic ion forms with another particle (forming a compound) will be ionic. Its reactions, like other ionic compounds, usually occur so that the polyatomic ion is transferred intact and stays together.

Example:

Sodium Acetate		Sodium Nitrate
+	$\longrightarrow$	+
Potassium Nitrate		Potassium Acetate

$$NaCH_3COO + KNO_3 \longrightarrow NaNO_3 + KCH_3COO$$

Table G

STANDARD ENERGIES OF FORMATION OF COMPOUNDS AT 1 atm AND 298 K		
Compound	*Heat (Enthalpy) of Formation* * kcal/mol ($\triangle H_f^0$)	*Free Energy of Formation* kcal/mol ($\triangle G_f^0$)
Aluminum oxide $Al_2O_3(s)$	−400.5	−378.2
Ammonia $NH_3(g)$	−11.0	−3.9
Barium sulfate $BaSO_4(s)$	−352.1	−325.6
Calcium hydroxide $Ca(OH)_2(s)$	−235.7	−214.8
Carbon dioxide $CO_2(g)$	−94.1	−94.3
Carbon monoxide $CO(g)$	−26.4	−32.8
Copper (II) sulfate $CuSO_4(s)$	−184.4	−158.2
Ethane $C_2H_6(g)$	−20.2	−7.9
Ethene (ethylene) $C_2H_4(g)$	12.5	16.3
Ethyne (acetylene) $C_2H_2(g)$	54.2	50.0
Hydrogen fluoride $HF(g)$	−64.8	−65.3
Hydrogen iodide $HI(g)$	6.3	0.4
Iodine chloride $ICl(g)$	4.3	−1.3
Lead (II) oxide $PbO(s)$	−51.5	−45.0
Magnesium oxide $MgO(s)$	−143.8	−136.1
Nitrogen (II) oxide $NO(g)$	21.6	20.7
Nitrogen (IV) oxide $NO_2(g)$	7.9	12.3
Potassium chloride $KCl(s)$	−104.4	−97.8
Sodium chloride $NaCl(s)$	−98.3	−91.8
Sulfur dioxide $SO_2(g)$	−70.9	−71.7
Water $H_2O(g)$	−57.8	−54.6
Water $H_2O(\ell)$	−68.3	−56.7

* Minus sign indicates an exothermic reaction.

Sample equations:

$$2Al(s) + \frac{3}{2}O_2(g) \rightarrow Al_2O_3(s) + 400.5 \text{ kcal}$$

$$2Al(s) + \frac{3}{2}O_2(g) \rightarrow Al_2O_3(s) \quad \triangle H = -400.5 \text{ kcal/mol}$$

Reference Table G
Standard Energies of Formation of Compounds at 1 atm and 298 K

As the sample equation at the bottom of this table shows, the compound Aluminum Oxide $Al_2O_3(s)$ is formed as a product directly by its two constituent elements. The other compounds listed in Reference Table G were also formed in the same manner. Therefore, these compounds represent the products of a direct composition reaction. The **Heat of Formation** as listed in the middle column represent the energy in Kcal absorbed or given off in the formation of **one mole** of these compounds. The greater the energy given off in the formation of a compound (indicated by a larger negative volume), the more stable the compound will be. This is true because the energy required to break the compound down will be greater. The **Free Energy of Formation** (ΔG) of each of the listed compounds represents the energy released or absorbed which can be used to do work for every **mole** of the compound formed. The entropy factor $-t\Delta S$ is the factor which results in the differences of the two columns.

The minus sign in the Free Energy of Formation column indicates ΔG is negative, and therefore, the reaction is spontaneous.

Note: In the Heat (Enthalpy) of Formation column only, the minus sign indicates an exothermic reaction.

Table H

SELECTED RADIOISOTOPES		
Nuclide	Half-Life	Decay Mode
^{198}Au	2.69 d	β^-
^{14}C	5730 y	β^-
^{60}Co	5.26 y	β^-
^{137}Cs	30.23 y	β^-
^{220}Fr	27.5 s	α
^{3}H	12.26 y	β^-
^{131}I	8.07 d	β^-
^{37}K	1.23 s	β^+
^{42}K	12.4 h	β^-
^{85}Kr	10.76 y	β^-
$^{85m}Kr*$	4.39 h	γ
^{16}N	7.2 s	β^-
^{32}P	14.3 d	β^-
^{239}Pu	2.44×10^4 y	α
^{226}Ra	1600 y	α
^{222}Rn	3.82 d	α
^{90}Sr	28.1 y	β^-
^{99}Tc	2.13×10^5 y	β^-
$^{99m}Tc*$	6.01 h	γ
^{232}Th	1.4×10^{10} y	α
^{233}U	1.62×10^5 y	α
^{235}U	7.1×10^8 y	α
^{238}U	4.51×10^9 y	α

y=years; d=days; h=hours; s=seconds
*m = meta stable or excited state of the same nucleus. Gamma decay from such a state is called an isomeric transition (IT).
Nuclear isomers are different energy states of the same nucleus, each having a different measurable lifetime.

Table I

HEATS OF REACTION AT 1 atm and 298 K	
Reaction	ΔH (kcal)
$CH_4(g) + 2O_2(g) \rightarrow CO_2(g) + 2H_2O(\ell)$	−212.8
$C_3H_8(g) + 5O_2(g) \rightarrow 3CO_2(g) + 4H_2O(\ell)$	−530.6
$CH_3OH(\ell) + \frac{3}{2}O_2(g) \rightarrow CO_2(g) + 2H_2O(\ell)$	−173.6
$C_6H_{12}O_6(s) + 6O_2(g) \rightarrow 6CO_2(g) + 6H_2O(\ell)$	−669.9
$CO(g) + \frac{1}{2}O_2(g) \rightarrow CO_2(g)$	−67.7
$C_8H_{18}(\ell) + \frac{25}{2}O_2(g) \rightarrow 8CO_2(g) + 9H_2O(\ell)$	−1302.7
$KNO_3(s) \xrightarrow{H_2O} K^+(aq) + NO_3^-(aq)$	+8.3
$NaOH(s) \xrightarrow{H_2O} Na^+(aq) + OH^-(aq)$	−10.6
$NH_4Cl(s) \xrightarrow{H_2O} NH_4^+(aq) + Cl^-(aq)$	+3.5
$NH_4NO_3(s) \xrightarrow{H_2O} NH_4^+(aq) + NO_3^-(aq)$	+6.1
$NaCl(s) \xrightarrow{H_2O} Na^+(aq) + Cl^-(aq)$	+0.9
$KClO_3(s) \xrightarrow{H_2O} K^+(aq) + ClO_3^-(aq)$	+9.9
$LiBr(s) \xrightarrow{H_2O} Li^+(aq) + Br^-(aq)$	−11.7
$H^+(aq) + OH^-(aq) \rightarrow H_2O(\ell)$	−13.8

Table J

SYMBOLS USED IN NUCLEAR CHEMISTRY		
alpha particle	^4_2He	α
beta particle (electron)	$^0_{-1}e$	β^-
gamma radiation		γ
neutron	^1_0n	n
proton	^1_1H	p
deuteron	^2_1H	
triton	^3_1H	
positron	$^0_{+1}e$	β^+

Reference Table H
Selected Radioisotopes

The time required for one-half of the nuclei of the listed radioactive isotopes in Reference Table H to disintegrate is noted in the middle column. The emitted particle, which results from this disintegration, is listed in the right column. When the half-life factor is required in solving a problem, the assumption is made that you will proceed to this table in order to find the half-life of the required radioisotope.

Reference Table I
Heats of Reaction at 1 atm and 298 K

The selected reactions in Reference Table I at 1 atm and 298 K are given in order to demonstrate example of exothermic (ΔH is negative) and endothermic (ΔH is positive) reactions. These examples are especially noted in the dissociation of two salts in water:

1. $NH_4Cl(s)$ requires energy to go into solution. The energy is provided by
   a) the bond energy given off when $NH_4Cl(s)$ combines with water, and
   b) the container and/or the atmosphere surrounding the container.
2. $NaOH(s)$ gives off energy when dissociating in water. This results in a water temperature rise and the warming of the container.

Reference Table J
Symbols Used in Nuclear Chemistry

There are two sets of symbols representing each of the subatomic particles noted in Reference Table J. The only item which has one symbol is gamma radiation which does not have any mass; therefore, it is not considered a particle. The electron is also called a beta particle, hence, the symbol B^-. In a nuclear equation, these particles are used as either reactants or products and can be identified by using Reference Table J.

Example Problem:
In the reaction: $^{27}_{13}Al + ^{4}_{2}He \rightarrow ^{30}_{15}P + X$, the particle represented by X is

(1) a neutron,
(2) a beta particle,

(3) an electron
(4) an alpha particle

Solution:
- First, by balancing the mass numbers (superscripts) on either side of the equation, you obtain a mass of 1 for X.
- Second, by balancing the charges (subscripts) on both sides of the equation, you determine the charge to be 0.
- Third, by looking at Table J, you find the particle to be a neutron.

*Note: The symbols in the middle column include **a superscript which represents the mass of the particle,** and the **subscript represents the charge on the particle.** This combination is different than a superscript and subscript designation of number used in isotopes.*

Table K

IONIZATION ENERGIES AND ELECTRONEGATIVITIES								

1								18
313 ← First Ionization Energy (kcal/mol of atoms) H **2.2** ← Electronegativity*								**567** He

2		13	14	15	16	17	

2		13	14	15	16	17	18
125 Li 1.0	215 Be 1.5	191 B 2.0	260 C 2.6	336 N 3.1	314 O 3.5	402 F 4.0	497 Ne
119 Na 0.9	176 Mg 1.2	138 Al 1.5	188 Si 1.9	242 P 2.2	239 S 2.6	300 Cl 3.2	363 Ar
100 K 0.8	141 Ca 1.0	138 Ga 1.6	182 Ge 1.9	226 As 2.0	225 Se 2.5	273 Br 2.9	323 Kr
96 Rb 0.8	131 Sr 1.0	133 In 1.7	169 Sn 1.8	199 Sb 2.1	208 Te 2.3	241 I 2.7	280 Xe
90 Cs 0.7	120 Ba 0.9	141 Tl 1.8	171 Pb 1.8	168 Bi 1.9	194 Po 2.0	At 2.2	248 Rn
Fr 0.7	122 Ra 0.9	* Arbitrary scale based on fluorine = 4.0					

Reference Table K
Ionization Energies and Electronegativities

The elements represented in Table K are those listed in the **A Groups** in the *older* Periodic Table of Elements. They are listed in **Groups 1 and 2** and in **Groups 13 through 18** in the *newer* Periodic Table of Elements. The first ionization energy and electronegativity is given for each of the elements listed.

The first ionization energy is defined as the energy required to remove the outer most electron of an atom. The energy is given in kcal per mole of atoms. The lower the ionization energy, the greater the tendency to form positive ions. In a group, as the atomic number increases and the atom becomes larger by adding on successive energy levels, the **screening effect** increases. The **screening effect** is caused by the electrons in the inner energy levels taking on extra nuclear charge leaving a disproportionate, smaller charge for the outer most electrons.

Therefore, as you move down a group the **screening effect** becomes more significant, and the ionization energy decreases. Since metals are defined as elements that have a tendency to give off electrons, the ionization energy decreases, and the metallic properties increase as you move down the group of elements. Moving from left to right on the table, the nuclear charge becomes greater without the addition of newer energy levels. This causes the outer most, or valance electrons, to be held more firmly. Therefore, the ionization energy increases as you go across a period.

Electronegativity is an arbitrary scale which measures the ability of an atom to attract the electron pair which it is sharing with another atom. Fluorine is assigned a value of 4.0 which is the highest value of all of the elements. The difference in electronegativity value between two atoms determines the percent of ionic character of the bond in a molecule or compound. The greater the difference, the greater the ionic character of the bond of the molecule or compound. Generally, an electronegativity value difference of 1.7 or more between two bonded atoms will result in an ionic bond. An electronegativity value difference of less than 1.7 generally denotes the formation of a covalent bond. *Note: There are exceptions to this rule.*

Table L

RELATIVE STRENGTHS OF ACIDS IN AQUEOUS SOLUTION AT 1 atm AND 298 K	
Conjugate Pairs *ACID BASE*	K_a
$HI = H^+ + I^-$	very large
$HBr = H^+ + Br^-$	very large
$HCl = H^+ + Cl^-$	very large
$HNO_3 = H^+ + NO_3^-$	very large
$H_2SO_4 = H^+ + HSO_4^-$	large
$H_2O + SO_2 = H^+ + HSO_3^-$	1.5×10^{-2}
$HSO_4^- = H^+ + SO_4^{2-}$	1.2×10^{-2}
$H_3PO_4 = H^+ + H_2PO_4^-$	7.5×10^{-3}
$Fe(H_2O)_6^{3+} = H^+ + Fe(H_2O)_5(OH)^{2+}$	8.9×10^{-4}
$HNO_2 = H^+ + NO_2^-$	4.6×10^{-4}
$HF = H^+ + F^-$	3.5×10^{-4}
$Cr(H_2O)_6^{3+} = H^+ + Cr(H_2O)_5(OH)^{2+}$	1.0×10^{-4}
$CH_3COOH = H^+ + CH_3COO^-$	1.8×10^{-5}
$Al(H_2O)_6^{3+} = H^+ + Al(H_2O)_5(OH)^{2+}$	1.1×10^{-5}
$H_2O + CO_2 = H^+ + HCO_3^-$	4.3×10^{-7}
$HSO_3^- = H^+ + SO_3^{2-}$	1.1×10^{-7}
$H_2S = H^+ + HS^-$	9.5×10^{-8}
$H_2PO_4^- = H^+ + HPO_4^{2-}$	6.2×10^{-8}
$NH_4^+ = H^+ + NH_3$	5.7×10^{-10}
$HCO_3^- = H^+ + CO_3^{2-}$	5.6×10^{-11}
$HPO_4^{2-} = H^+ + PO_4^{3-}$	2.2×10^{-13}
$HS^- = H^+ + S^{2-}$	1.3×10^{-14}
$H_2O = H^+ + OH^-$	1.0×10^{-14}
$OH^- = H^+ + O^{2-}$	$< 10^{-36}$
$NH_3 = H^+ + NH_2^-$	very small

Table M

CONSTANTS FOR VARIOUS EQUILIBRIA AT 1 atm AND 298 K	
$H_2O(\ell) = H^+(aq) + OH^-(aq)$	$K_w = 1.0 \times 10^{-14}$
$H_2O(\ell) + H_2O(\ell) = H_3O^+(aq) + OH^-(aq)$	$K_w = 1.0 \times 10^{-14}$
$CH_3COO^-(aq) + H_2O(\ell) = CH_3COOH(aq) + OH^-(aq)$	$K_b = 5.6 \times 10^{-10}$
$Na^+F^-(aq) + H_2O(\ell) = Na^+(OH)^- + HF(aq)$	$K_b = 1.5 \times 10^{-11}$
$NH_3(aq) + H_2O(\ell) = NH_4^+(aq) + OH^-(aq)$	$K_b = 1.8 \times 10^{-5}$
$CO_3^{2-}(aq) + H_2O(\ell) = HCO_3^-(aq) + OH^-(aq)$	$K_b = 1.8 \times 10^{-4}$
$Ag(NH_3)_2^+(aq) = Ag^+(aq) + 2NH_3(aq)$	$K_{eq} = 8.9 \times 10^{-8}$
$N_2(g) + 3H_2(g) = 2NH_3(g)$	$K_{eq} = 6.7 \times 10^5$
$H_2(g) + I_2(g) = 2HI(g)$	$K_{eq} = 3.5 \times 10^{-1}$

Compound	K_{sp}	Compound	K_{sp}
AgBr	5.0×10^{-13}	Li_2CO_3	2.5×10^{-2}
AgCl	1.8×10^{-10}	$PbCl_2$	1.6×10^{-5}
Ag_2CrO_4	1.1×10^{-12}	$PbCO_3$	7.4×10^{-14}
AgI	8.3×10^{-17}	$PbCrO_4$	2.8×10^{-13}
$BaSO_4$	1.1×10^{-10}	PbI_2	7.1×10^{-9}
$CaSO_4$	9.1×10^{-6}	$ZnCO_3$	1.4×10^{-11}

Reference Table L
Relative Strengths of Acids
in Aqueous Solution at 1 atm and 298 K

Reference Table L provides the following information:

1. It demonstrates the manner in which an acid ionizes to form a hydrogen ion (H^+) and its conjugate base at equilibrium.

2. It lists the acids in sequence of their strength. That is: the acids become sequentially weaker, so that HI is the strongest, and NH_3 is the weakest acid. Also, the weakest conjugate base is I^-, and as you move down the list, the bases become stronger. NH_2^- is the strongest conjugate base.

3. It lists the equilibrium constant (K_a) for each acid. Note: the value decreases as you move from top to bottom. Also, the first four acids ionize so completely, that the equilibrium constant is too high to be significantly measured for all practical purposes.

However, the sequence of their strengths is the same. That is: HI is the strongest, followed by HBr, HCl, and HNO_3.

4. Amphoprotic (also called amphoteric) substances, which can act as either an acid or a base, appear on both sides of the reactions. HSO_4^- and $H_2PO_4^-$ are examples of amphoprotic substances. One manner of preparing acids is to react such substances with strong acids which donate their proton to form the conjugate acid.

Example: $$H_2PO_4^- + HCl = H_3PO_4 + Cl^-$$

If instead, you react it with a strong base, it will donate its proton and act as an acid:

$$H_2PO_4^- + NH_2^- = NH_3 + HPO_4^-$$

Reference Table M
Constants for Various Equilibria at 1 atm and 298 K

The purpose of Reference Table M is:

1. To emphasize the significance of the equilibrium constant in understanding the amount of dissociation which slightly soluble substances undergo. That is: Since you arrive at the equilibrium constant by dividing the product of concentration of the products in moles per liter by the product of the concentrations of the reactants in moles per liter, the greater the concentration of the products, the greater the equilibrium constant.

2. To test you on you knowledge of scientific notation, by asking you to identify the most or least soluble of any substances in the table.

3. To identify the solubility product constant of various salts.

Note: The solubility constant for water (K_w) is 1×10^{-14} whether it is written:
$$H_2O(l) = H^+(aq) + OH^-(aq) \quad \text{or} \quad 2H_2O(l) = H_3O^+ + OH^-(aq)$$

Table N

STANDARD ELECTRODE POTENTIALS	
Ionic Concentrations 1 M Water At 298 K, 1 atm	
Half-Reaction	E^0 (volts)
$F_2(g) + 2e^- \rightarrow 2F^-$	+2.87
$8H^+ + MnO_4^- + 5e^- \rightarrow Mn^{2+} + 4H_2O$	+1.51
$Au^{3+} + 3e^- \rightarrow Au(s)$	+1.50
$Cl_2(g) + 2e^- \rightarrow 2Cl^-$	+1.36
$14H^+ + Cr_2O_7^{2-} + 6e^- \rightarrow 2Cr^{3+} + 7H_2O$	+1.23
$4H^+ + O_2(g) + 4e^- \rightarrow 2H_2O$	+1.23
$4H^+ + MnO_2(s) + 2e^- \rightarrow Mn^{2+} + 2H_2O$	+1.22
$Br_2(\ell) + 2e^- \rightarrow 2Br^-$	+1.09
$Hg^{2+} + 2e^- \rightarrow Hg(\ell)$	+0.85
$Ag^+ + e^- \rightarrow Ag(s)$	+0.80
$Hg_2^{2+} + 2e^- \rightarrow 2Hg(\ell)$	+0.80
$Fe^{3+} + e^- \rightarrow Fe^{2+}$	+0.77
$I_2(s) + 2e^- \rightarrow 2I^-$	+0.54
$Cu^+ + e^- \rightarrow Cu(s)$	+0.52
$Cu^{2+} + 2e^- \rightarrow Cu(s)$	+0.34
$4H^+ + SO_4^{2-} + 2e^- \rightarrow SO_2(aq) + 2H_2O$	+0.17
$Sn^{4+} + 2e^- \rightarrow Sn^{2+}$	+0.15
$2H^+ + 2e^- \rightarrow H_2(g)$	0.00
$Pb^{2+} + 2e^- \rightarrow Pb(s)$	-0.13
$Sn^{2+} + 2e^- \rightarrow Sn(s)$	-0.14
$Ni^{2+} + 2e^- \rightarrow Ni(s)$	-0.26
$Co^{2+} + 2e^- \rightarrow Co(s)$	-0.28
$Fe^{2+} + 2e^- \rightarrow Fe(s)$	-0.45
$Cr^{3+} + 3e^- \rightarrow Cr(s)$	-0.74
$Zn^{2+} + 2e^- \rightarrow Zn(s)$	-0.76
$2H_2O + 2e^- \rightarrow 2OH^- + H_2(g)$	-0.83
$Mn^{2+} + 2e^- \rightarrow Mn(s)$	-1.19
$Al^{3+} + 3e^- \rightarrow Al(s)$	-1.66
$Mg^{2+} + 2e^- \rightarrow Mg(s)$	-2.37
$Na^+ + e^- \rightarrow Na(s)$	-2.71
$Ca^{2+} + 2e^- \rightarrow Ca(s)$	-2.87
$Sr^{2+} + 2e^- \rightarrow Sr(s)$	-2.89
$Ba^{2+} + 2e^- \rightarrow Ba(s)$	-2.91
$Cs^+ + e^- \rightarrow Cs(s)$	-2.92
$K^+ + e^- \rightarrow K(s)$	-2.93
$Rb^+ + e^- \rightarrow Rb(s)$	-2.98
$Li^+ + e^- \rightarrow Li(s)$	-3.04

Table O

VAPOR PRESSURE OF WATER			
°C	torr (mmHg)	°C	torr (mmHg)
0	4.6	26	25.2
5	6.5	27	26.7
10	9.2	28	28.3
15	12.8	29	30.0
16	13.6	30	31.8
17	14.5	40	55.3
18	15.5	50	92.5
19	16.5	60	149.4
20	17.5	70	233.7
21	18.7	80	355.1
22	19.8	90	525.8
23	21.1	100	760.0
24	22.4	105	906.1
25	23.8	110	1074.6

Reference Table N
Standard Electrode Potentials

Using hydrogen as an arbitrary standard, Reference Table N demonstrates a number of **reduction half-cell reactions**. In each instance in the forward reaction, electrons are being added to a substance. The reactions take place in water at 298 K and 1 atm in a 1M ionic concentration. The difference is measured in volts.

In the electrochemical cell, a reduction half-cell reaction must be accompanied by an oxidation half-cell reaction. Therefore, you cannot directly measure the half-cell reduction potential without comparing it to another half-cell reaction. So, you use hydrogen as the standard, and assign it a E^0 value of 0.00. Having done this, you can now compare it to the other reactions on the table.

All of the reactions which show a positive E^0 value will accept electrons from hydrogen. In fact, any substance in the table will reduce those above it and will oxidize those below it. In order for an electrochemical cell to exist, it must be spontaneous, and it becomes spontaneous only when the algebraic sum of the E^0 values is positive. Since this table represents a list of reduction potentials only and since the other half-cell reaction is an oxidation reaction, in order to evaluate it correctly, you must reverse the arrow in the reaction and change the value of the sign in front of the E^0 value. That is: if hydrogen and aluminum were compared, the aluminum being below the hydrogen would be oxidized in order to reduce the hydrogen.

Therefore, reversing the arrow of the aluminum half-cell would give you an E^0 value of +1.66. Since the E^0 value of hydrogen is 0.00, the sum of the two half-cells will be +1.66, indicating that the reaction is spontaneous.

Reference Table O
Vapor Pressure of Water

When a gas is collected over water, Reference Table O is used to determine the vapor pressure of the water at a specific temperature. By subtracting the vapor pressure of the water from the air pressure, you are capable of obtaining the pressure of the gas which you are collecting. Knowing the pressure of the gas, you can use Boyle's Law to determine the gas volume.

Note: Even at 0°C (freezing), the vapor pressure of water is 4.6 torr.

Table P

RADII OF ATOMS

KEY

Symbol →	F
Covalent Radius, Å →	0.64
Atomic Radius in Metals, Å →	(–)
Van der Waals Radius, Å →	1.35

A dash (–) indicates data are not available.

Element	Covalent Radius (Å)	Atomic Radius in Metals (Å)	Van der Waals Radius (Å)
H	0.37	(–)	1.2
He	(–)	(–)	1.22
Li	1.23	1.52	(–)
Be	0.89	1.13	(–)
B	0.88	(–)	2.08
C	0.77	(–)	1.85
N	0.70	(–)	1.54
O	0.66	(–)	1.40
F	0.64	(–)	1.35
Ne	(–)	(–)	1.60
Na	1.57	1.54	2.31
Mg	1.36	1.60	(–)
Al	1.25	1.43	(–)
Si	1.17	(–)	2.0
P	1.10	(–)	1.90
S	1.04	(–)	1.85
Cl	0.99	(–)	1.81
Ar	(–)	(–)	1.91
K	2.03	2.27	2.31
Ca	1.74	1.97	(–)
Sc	1.44	1.61	(–)
Ti	1.32	1.45	(–)
V	1.22	1.32	(–)
Cr	1.17	1.25	(–)
Mn	1.17	1.24	(–)
Fe	1.17	1.24	(–)
Co	1.16	1.25	(–)
Ni	1.15	1.25	(–)
Cu	1.17	1.28	(–)
Zn	1.25	1.33	(–)
Ga	1.25	1.22	(–)
Ge	1.22	1.23	(–)
As	1.21	(–)	2.0
Se	1.17	(–)	2.0
Br	1.14	(–)	1.95
Kr	(–)	(–)	1.98
Rb	2.16	2.48	2.44
Sr	1.92	2.15	(–)
Y	1.62	1.81	(–)
Zr	1.45	1.60	(–)
Nb	1.34	1.43	(–)
Mo	1.29	1.36	(–)
Tc	(–)	1.36	(–)
Ru	1.24	1.33	(–)
Rh	1.25	1.35	(–)
Pd	1.28	1.38	(–)
Ag	1.34	1.44	(–)
Cd	1.41	1.49	(–)
In	1.50	1.63	(–)
Sn	1.40	1.41	(–)
Sb	1.41	(–)	2.2
Te	1.37	(–)	2.20
I	1.33	(–)	2.15
Xe	(–)	(–)	2.09
Cs	2.35	2.65	2.62
Ba	1.98	2.17	(–)
La–Lu			
Hf	1.44	1.56	(–)
Ta	1.34	1.43	(–)
W	1.30	1.37	(–)
Re	1.28	1.37	(–)
Os	1.26	1.34	(–)
Ir	1.26	1.36	(–)
Pt	1.29	1.38	(–)
Au	1.34	1.44	(–)
Hg	1.44	1.60	(–)
Tl	1.55	1.70	(–)
Pb	1.54	1.75	(–)
Bi	1.52	1.55	(–)
Po	1.53	1.67	(–)
At	(–)	(–)	(–)
Rn	(–)	(–)	2.14
Fr	(–)	2.7	(–)
Ra	(–)	2.20	(–)
Ac–Lr			

Lanthanide Series

Element	Covalent Radius (Å)	Atomic Radius in Metals (Å)	Van der Waals Radius (Å)
La	1.69	1.88	(–)
Ce	1.65	1.83	(–)
Pr	1.65	1.83	(–)
Nd	1.64	1.82	(–)
Pm	(–)	1.81	(–)
Sm	1.66	1.80	(–)
Eu	1.85	2.04	(–)
Gd	1.61	1.80	(–)
Tb	1.59	1.78	(–)
Dy	1.59	1.77	(–)
Ho	1.58	1.77	(–)
Er	1.57	1.76	(–)
Tm	1.56	1.75	(–)
Yb	1.70	1.94	(–)
Lu	1.56	1.73	(–)

Actinide Series

Element	Covalent Radius (Å)	Atomic Radius in Metals (Å)	Van der Waals Radius (Å)
Ac	(–)	1.88	(–)
Th	(–)	1.80	(–)
Pa	(–)	1.61	(–)
U	(–)	1.39	(–)
Np	(–)	1.31	(–)
Pu	(–)	1.51	(–)
Am	(–)	1.84	(–)
Cm	(–)	(–)	(–)
Bk	(–)	(–)	(–)
Cf	(–)	(–)	(–)
Es	(–)	(–)	(–)
Fm	(–)	(–)	(–)
Md	(–)	(–)	(–)
No	(–)	(–)	(–)
Lr	(–)	(–)	(–)

Reference Table P
Radii of Atoms

The **atomic radius** has been defined as the distance of closest approach that an atom has to another atom. At this distance and under specified conditions, the mutual repulsion of the electron clouds are in **equilibrium** with the mutual attraction of the electrons to the nuclear charge of the other atom.

The measurement will be taken as follows:

1. **For identical atoms** — one half the internuclear distance.

2. **For unlike atoms** — the internuclear distance is the sum of the individual radii.

Atomic Radii are classified into two basic groups:

1. **Radii concerned with mutually bonded electrons** which include:

   a) **Covalent radius.** In this bond, the bonding electrons are generally localized between the bonding atoms. *For example: Carbon, 1st line, value: 0.77* (For further discussion, refer to the text.)

   b) **Metallic radius.** The bonding electrons in this bond are highly delocalized. *For example: Cobalt, 2nd line, value: 1.25* (For further discussion, refer to the text.)

2. **Radii of mutually unbonded atoms** which, although unbonded, are held together by an attraction due to the equilibrium approach of the atoms involved. They are classified as follows:

   a) **Van der Waals Radius** — radii for atoms which are neutral in charge. *For example: Neon, 3rd line, value: 1.60* (For further discussion, refer to the text.)

   b) **Ionic Radius in a crystal** — radii for atoms which are charged. *For example: Cobalt $^{2+}$, not on Table P, value: 0.74* (For further discussion, see the text.)

Table P gives *only* the first three radii (explained above):

1) The **Covalent Radius**

2) The **Atomic Radius in Metals**

3) The **Van der Walls Radius**

Radii of Atoms
Key for Boron

Symbol	B
Covalent Radius, Å	0.88
Atomic Radius in Metals, Å	0.83
Van der Waals Radius, Å	2.08

Reference Table Notes:

Absolute zero: (12) -273 degrees Celsius.

Accelerators: (186) electric and magnetic fields used to accelerate charge particles.

Accepted value: (199) most probable value taken from generally accepted references.

Acetic acid: (166) CH_3COOH or ethanoic acid.

Acetylene: (159) 1st member of the Alkyne series with a molecular formula of C_2H_2.

Acid-base reactions: (122) neutralization reactions in which acids and bases react with one another.

Acids: (119, 120, 166) any species that can donate a proton to another substance.

Activated complex: (101) intermediate species formed when reactants form products in a reaction.

Activation energy: (99, 101) minimum energy required to initiate a reaction.

Addition polymerization: (169) a reaction in which monomers of unsaturated compounds join together.

Addition reactions: (163) addition of one or more atoms to an unsaturated organic molecule at the double or triple bond. This results in the saturation of the compound.

Alcohols: (165, 166) hydrocarbons in which one of the hydrogen atoms is replaced by one or more hydroxyl groups (O-H).

Aldehydes: (167) hydrocarbons in which an end carbon contains a functional group (CHO).

Alkadienes: (159) homologous series of hydrocarbons containing two double bonds.

Alkali metals: (67) common name for Group 1 metals.

Alkaline earth metals: (68) common name for Group 2 metals.

Alkanes: (157, 158) homologous series of hydrocarbons having the general formula $C_n H_{2n+2}$.

Alkenes: (158) see ethylene series.

Alkyl radicals: (160) an alkane which has lost a hydrogen atom, resulting in an open bond structure.

Alkynes: (159) homologous series of hydrocarbons having the general formula $C_n H_{2n-2}$.

Allotropic forms: (68-69) chemical element in two or more forms with differing physical properties; carbon exists as graphite and diamond.

Alpha decay: (38) result which occurs when a nucleus disintegrates with the emission of alpha particles.

Alpha emitter: (38) atom that emits an alpha particle.

Alpha particles: (38, 185) consists of two neutron and two protons and can be considered Helium nuclei.

Amphiprotic (amphoteric) substances: (121) substance which can act either as an acid or as a base.

Angstrom unit: (64) 1 Å = 1 x 10-10 meter; measurement of atomic and ionic radii.

Anode: (141, 144) pole where oxidation takes place in a chemical cell.

Apparatus, laboratory: (203-204) common laboratory apparatus pictured.

Applied research: (177) see research.

Arrhenius' theory: (120, 121) aqueous solutions, acids yield hydrogen ions, bases yield hydroxide ions.

Artificial radioactivity: (185) radioactive isotopes which result from artificial transmutation.

Artificial transmutation: (185) bombarding nuclei with high energy particles resulting in transmutations into new, usually stable, nuclei.

Atom: (29) the smallest elemental particle which retains all of the characteristics of that element.

Atomic energy: (186) see nuclear energy.

Atomic Mass Unit (amu): (29, 30, 81) unit of measure which is standardized as $\frac{1}{12}$ of the mass of a carbon atom.

Atomic number: (30) indicates the number of protons in the nucleus.

Atomic orbital model: (33) the average region of most probable electron location defined by quantum mechanics so that no two electrons will have the same energy level.

Atomic radius: (63, 64) (covalent) one-half of the measured internuclear distance between atoms in the solid phase.

Atomic structure: (29, 32) a nuclear model of an atom using quantum mechanical theory to account for probable energy levels of its electrons.

Avogadro's hypothesis: (21) equal volumes of gases, under the same conditions of temperature and pressure, contain equal numbers of particles.

Avogadro's number: (21, 79) 6.02×10^{23}; particles in a mole of matter.

Balancing an equation: **(58)** number of atoms for each element must be the same on both sides of the equation.

Balancing simple redox reactions: (148)

Bases: (57, 119, 121) any species (molecule or ion) which can accept a proton.

Battery: (180) spontaneous chemical cells in which spontaneous redox reactions are used to provide a source of electrical energy.

Benzene series: (160) cyclic homologous series with a general formula of C_nH_{2n-6} called the aromatic hydrocarbons.

Beta decay: (38) reaction which results in neutron disintegration and the emission of a high speed electron called a beta particle.

Beta emitter: (38) atom that emits a beta particle.

Binary acid: (56) acid in which there are only two elements in the formula, hydrogen being one of them.

Binary compounds: (8, 56) compounds made up of just two elements.

Binding energy: (186) energy released when the fusion of two particles occurs. It is a measure of the stability of the atom.

Blocks, Periodic Table: (66-73) division of elements in the Periodic Table.

Bohr model: (32) shows structure of the atom, named after Niels Bohr.

Boiling point: (22, 90) the temperature at which the vapor pressure of a liquid equals the pressure on the liquid.

Conceptual definitions: (119) definitions that attempt to answer the why and how statements of acids and bases.

Condensation: (168) reaction in which water is a product.

Condensation polymerization: (168) bonding of monomers into polymers by a dehydration reaction.

Conjugate acid-base pair: (125) in an acid-base reaction, an acid transfers a proton to become a conjugate base, this acid and the newly formed base form an acid-base pair.

Constants for various equilibria: (108, 230, 231) see Reference Table M.

Contact process: (178) process used in the manufacture of sulfuric acid.

Control rods: (188) see nuclear control rods.

Coolants: (188) see nuclear coolants.

Coordinate covalent bonds: (47) occurs when the two shared electrons forming a covalent bond are both donated by one of the atoms.

Corrosion: (179) gradual attack on a metal by its surroundings which results in the metal returning to its ionic form.

Covalent atomic radius: (64) effective distance from the center of the nucleus to outer valence shell of that atom in a typical or covalent bond.

Covalent bonds: (47, 52) bond in which the atoms share electrons in an overlap manner.

Cracking: (182) a process by which large molecules are broken down to smaller molecules.

Crystalline structures: (23, 110) see crystals.

Crystals: (23) solids in which atoms are arranged in a regular geometric pattern called the crystal lattice structure.

Cyclic hydrocarbons: (160) hydrocarbons arranged in a ring structure.

Daniell cells: (140) see chemical cells.

Decomposition reactions: (137) category of redox reactions.

Delta: (10) Greek letter sign (Δ) used in reactions to indicate change.

Density: (23) mass of unit volume of a substance.

Density & boiling points of common gases: (220, 221) see table C.

Deuterium: (30) isotope of hydrogen with one proton and one neutron.

Dihydroxy alcohols: (167) compounds containing two hydroxyl (O-H) groups.

Dilute solution: (87) solution in which a large amount of solvent is required to dissolve a small amount of solute.

Dimensional analysis: (197) measurements treated as algebraic quantities.

Dipoles: (50, 267) results from the asymmetric distribution of electrical charges in a molecule, and is polar in nature.

Directional nature - covalent bonds: (47) geometric structure in bonding.

Dissociation constant: (119) equilibrium constant for weak electrolytes in aqueous solution.

Double bond: (157) bond between carbon atoms by the sharing of two pairs of electrons.

Dynamic: (104) term that implies motion.

Dynamic equilibrium: (104) a state of balance in which the reaction rates are for opposing reactions are equal.

Effective collisions: (101) see rate of reaction.

Electrochemical cells (140, 142, 144) see chemical cells.

Electrochemistry: (139, 141) study of the relationship between electricity and chemistry.

Electrodes: (141, 144) conductor used to attract cations to the cathode and anions to the anode.

Electrolysis: (144) application of an electrolytic cell.

Electrolytes: (91, 119) substance whose water solution conducts an electrical current.

Electrolytic cell: (141, 143, 144) used in electrolysis reactions.

Electron configuration: (35) distribution of the electrons among the various orbitals of an atom.

Electron density formulas: (52-53) of H_2O, CO_2, and carbon tetrafluoride.

Electron-dot diagrams: (36, 52) diagram showing the valence electrons of an atom by using the elements letter representation and the number of valence electrons in dots.

Electronegativity: (46, 68, 70) measure of the ability of an atom to attract the electrons that form a bond between it and another atom.

Electrons: (29, 31) all of the atoms negative charges and particles located outside of the nucleus in orbitals.

Electroplating: (145) when the surface of an item has been coated usually by the process of electrolysis.

Element: (7, 73) substance that cannot be decomposed into two or more substances by means of chemical change.

Emanations: (38, 39) result from radioactivity in which particles are emitted from the nucleus.

Empirical formula: (55, 82, 89) representation of the simplest ratio in which elements combine to form a compound.

"Empty space" concept: (29) Rutherford's gold foil experiment: atom has a very small nucleus with electrons orbiting around it in a large space area.

Endothermic reaction: (10, 13, 46, 100, 111, 206) reaction that absorbs heat from its surroundings.

End point: (105) point in titration at which an indicator shows that the equivalence point has been reached.

Energy, conservation law: (9) energy may be converted from one form to another but is never created nor destroyed.

Energy forms: (9) include mechanical, heat, electrical, radiant, chemical, and nuclear energy.

Energy, measurement of: (10) chemical reactions measured in units of heat called calories or joules.

Energy and chemical bonds: (45) potential energy associated with changes in chemical bonds (chemical energy).

Energy (nuclear): (186) reactions that involve a great amount of energy.

Energy and spontaneous reactions: (110) energy changes in reactions.

Energy levels in atoms: (33) the shells in which an atom's electrons orbit.

Enthalpy: (99) represents the total energy of a particle or species.

Entropy: (13, 110) randomness or disorder of regularity in a system.

Enzymes: (163) catalysts secreted by living organisms.

Equations, balanced: (58) equations that are equal in reactants and products and fulfill the Law of the Conservation of Atoms.

Equations, chemical: (58) examples of chemical equations and balancing.

Equations, stoichiometry: (81) complete single reactions to consumption.

Equations, nuclear: (186) atomic numbers and masses in equations.

Equilibrium: (99, 104, 108, 111, 144, 177) when any reaction takes place under fixed conditions so that both forward and reverse reactions occur at the same rate; chemical cell equilibrium is reached when voltage is measured at 0.

Equilibrium constants (K_{eq}): (108) ratio between the concentration of products divided by the concentration of reactants.

Equilibrium shift in the Haber reaction: (178) when ammonia is drawn out of the reaction container, the concentration of the ammonia is decreased causing an equilibrium shift toward the making of more ammonia in order to offset this decreased concentration.

Equivalence point: (122) point of neutralization.

Error: (199) see percent error.

Esterification: (163) reaction of acid with alcohol to give an ester and water.

Esters: (164) covalent compounds with pleasant odors made by reacting organic acids with alcohols.

Ethanoic acid: (166) common name for acetic acid whose molecular formula is CH_3COOH.

Ethanol: (166) ethyl alcohol C_2H_5OH (primary)

Ethers: (168) two primary alcohols that are treated with a dehydrating agent, water is removed from the molecules and the two alcohol chains are joined by an oxygen bridge.

Ethylene series (Alkene): (159) series of unsaturated hydrocarbons having one double bond with the general formula of C_nH_{2n}.

Evaporation: (22) when a liquid substance changes to gas.

Evaporation of a liquid in a laboratory activity: (203) drying of a solid.

Excited state: (32) occurs when atoms absorb energy and electrons shift to a higher energy level.

Exothermic reaction: (10, 13, 46, 99, 101, 110, 206) reaction that releases heat into its surroundings.

Experimental mass ratio: (209) mass of the substance in grams divided by the mass of the residue in grams.

External shield: (188) see nuclear reactors, made of high density concrete, acts as a radiation containment vessel.

Factor – label method: (197) see dimensional analysis.

Families - periodic table: (63) elements with similar properties that fall into the same vertical column of the periodic table.

Fats: (164) esters made up of glycerine and of higher fatty acids.

Fatty acids: (164) long-chain carboxylic acids.

Fermentation: (163) chemical change in an organic compound.

Filtration: (201) process used to separate suspended matter in a liquid usually through filter paper.

Final zero: (195) see math skills, rules for significant figures.

Firepolishing in a laboratory activity: (201) cutting and bending glass.

Fission products: (187) products formed after a fission reaction.

Fission reaction: (187) splitting of heavier nuclei into lighter ones.

Flame tests: (207) test used to identify metallic ions.

Fluorine: (70) element with highest electronegativity in G. 17 of Per. Table.

Formaldehyde: (167) common name for methanal.

Formic acid: (166) HCOOH or methanoic acid.

Formulas: (54, 55) statements that use chemical symbols to represent the composition of a substance.

Formulas (structural): (166) chemical formula, indicates arrangement of atoms in a molecule by using connecting lines from one atom to another.

Fractional distillation: (181, 182) utilizing the differences in the boiling points of two or more liquids in order to separate them from one another.

Free energy: (111) tendency for a change to occur spontaneously.

Free energy of formation: (111) the energy released or absorbed which can be used to do work for every mole of the compound formed.

Freezing point: (12) temperature at which a liquid becomes a solid.

Freezing point depression: (91) lowering of the freezing point of a solvent.

Fuel rods in fission reactors: (187) control rods inside of the reactor used to control or adjust the number of neutrons available.

Functional groups: (165) particular arrangement of a few atoms which gives characteristic properties to an organic molecule.

Funnel and filter paper use in a laboratory activity: (201) illustrated.

Fusion reaction: (189) when two light nuclei fuse into a heavier nucleus at high temperature and pressure, usually an element of more stable configuration is formed.

Fusion reactor: (189) where fusion reactions can be completed at a controlled rate, wherein two atoms are fused into one element.

Gamma **radiation: (39)** extremely high frequency electromagnetic waves which do not have mass or charge.

Gases and gaseous state of matter: (13, 15, 63, 80) in this phase all molecules in a species have vibrating, rotating, and translating motions.

Gases behavior: (19) manner in which certain gases behave in reactions and takes into account certain variables which include temperature, pressure, and volume.

Gases, kinetic theory of: (19) combined theories of Boyle, Charles, Graham, and Dalton based on the assumption that gases are made of separate individual particles in continuous motion.

Glass tubing: (200) glass tubes used in experiments.

Glycols: (167) see alcohols; compounds with two –OH groups.

Graham's law: (94) relationship between mass and velocity. Theory - "Under the same conditions of temperature and pressure, gases diffuse at a rate inversely proportionate to the square roots of their molecular masses," therefore, H_2 diffuses faster than O_2.

Gram atom: (31) see gram atomic mass.

Gram atomic mass: (31, 79) quantity of an element that has a mass in grams equal to its atomic mass in atomic mass units.

Gram molecular mass: (79, 90) quantity of substance that has a mass in grams equal to its molecular mass in atomic mass units.

Gram formula mass: (79) sum of gram atoms represented in a formula.

Gram molecular volume: (80) volume occupied by the molecular mass of a gas or vapor, usually 22.4 liters at standard conditions.

Ground state atoms: (32) Atoms whose electrons are not in an excited state; lowest energy state of an atom.

Groups, periodic table: (63, 66, 67) vertical combination of elements that have similar characteristics or properties in the periodic table.

Groups, functional: (165) particular arrangement of a few atoms which gives characteristic properties to an organic molecule.

H aber process: (177) a process in which nitrogen and hydrogen are combined to make ammonia.

Hafnium series: (72, 73) transition series (atomic numbers 72-80) found on the periodic table.

Half-cell: (140, 180) each vessel in which one half-reaction is taking place during a redox reaction.

Half-cell potential: (141) also see Reference Table N.

Half-life: (39, 192) Time required for a radioactive isotope nuclei to disintegrate to one half its mass.

Half-reactions: (139, 142) reaction either representing a loss of electrons or the gain of electrons.

Half-cell electrode potential: (141) comparison of the driving force of a half-reaction with that of the hydrogen standard to establish a scale of voltages.

Half-reaction of redox Equations: (139)

Halides: (70) compounds formed when elements of Group 17 react with hydrogen.

Halogenation: (163) replacement of the hydrogen atoms in saturated hydrocarbons by an active halogen atom.

Halogen derivatives: (163) products of halogenation.

Halogen group: (70) elements in Group 17 of the periodic table.

Heat energy: (10) energy associated with the temperature of a body or system of bodies.

Heat of formation: (100) amount of heat energy either absorbed or released in the formation of one mole of the products of a reaction.

Heat of fusion: (23, 93) amount of energy required to convert one gram of any solid substance to a liquid at its melting point.

Heat of reaction: (99, 206) heat energy released or absorbed in the formation of the products.

Heat of vaporization: (22, 93) energy needed to vaporize a unit mass of a liquid at a constant temperature.

Heating materials safely: (203) accident prevention during heating.

Heats of reaction: (226, 227) see Reference Table I.

Heterogeneous mixture: (8) mixture samples that are not uniform in composition throughout.

Homogeneous mixture: (8) see solutions.

Homogeneous substance: (7) substances that contain properties and compositions that are the same throughout.

Homologous series: (157) classification of related structures and properties into groups, in which each member of the group differs from its neighbor by a definite increment.

Hydrated crystal: (207) a crystal in which water molecules are attached by coordinate covalent bonds to a central metallic ion or ionic compound.

Hydration: (52, 85) solute particle dissociation when water is the solvent.

Hydration of the ions: (52, 207) see hydrated crystal.

Hydro–: (56) see binary acids; example: HCl - hydrochloric acid.

Hydrocarbons: (157) compounds containing only carbon and hydrogen.

Hydrogenation: (163) addition of hydrogen to an unsaturated substance.

Hydrogen bond: (51) bond formed between a hydrogen atom in one molecule and a highly electronegative atom in another molecule.

Hydrolysis: (124) a process by which some salts react with water to form solutions that are acidic or basic.

Hydronium ion: (120) results when a water molecule forms a coordinate covalent bond with a proton (or hydrogen ion).

Hydroxyl group: (165) -OH group that replace hydrogen atoms of hydrocarbons.

Ide: (56, 57) "-ide" ending of nonmetallic ions when named or written.

Ideal gas model: (20, 269) used in the study of the behavior of gases.

Immiscible solution: (87) a mixture of two liquids which do not dissolve in one another.

Indicators: (120) substances (usually weak acids) which have difference colors in acid and base solutions.

Industrial applications: (177) equilibrium, reaction rates, redox, petroleum.

Inert gases: (71) see noble gases, common name for Group 18 elements.

Initial zero: (195) see math skills, rules for significant figures.

Inorganic substances and how they differ from organic compounds: (207)

Internal shield: (188) see nuclear reactors, made of a steel lining, protects walls of reactor from radiation damage.

Ionic bonds: (46) occurs when the atoms of two or more different elements combine, resulting in the transfer of one or more electrons from one element to another, so that when the bond is broken, ions are formed.

Ionic compounds: (46) hetero-nuclear compound containing ionic bonds.

Ionic radius: (64) radii of ions in solid state.

Ionic solids: (47) solid compound containing ionic bonds.

Ionization constant: (126, 127) K_a — K_w — K_{eq}.

Ionization energy: (36, 68) energy required to remove an electron from an atom.

Ionization energies & electronegativities: (228) see Reference Table K.

Ions: (46) charged particles formed by the transfer of electrons.

Isomers: (157) compounds which have the same molecular formula but different structural formulas.

Isotopes: (30) atoms of one element with the same number of protons but a different number of neutrons.

IUPAC: (6, 65, 158, 165) International Union of Pure and Allied Chemists.

Joules: (11) the basic SI unit of energy.

Kelvin: (12, 17) absolute temperature scale; temperature scale with fixed points at 273 K at ice water equilibrium and 1 atmosphere of pressure, and 373 K at the steam water equilibrium and 1 atmosphere of pressure.

Ketones: (168) organic compounds which contain a carbonyl (CO) functional group and hydrocarbon side groups.

Kernel: (**36, 64, 67**) refers to the atom exclusive of the valence electrons.
Kilocalorie (kcal.): (**10**) one thousand calories.
Kinetic energy (K.E.): (**94**) energy of motion or reaction.
Kinetics, chemical: (**99**) branch of chemistry concerned with the rates of chemical reactions, and the mechanisms by which these reactions occur.
Kinetic theory of gases: (**19, 267**) also see "ideal gas" model.

Laboratory activities: (**195, 205**) measurements, skills, safety, & reports.

Laboratory apparatus: (**203-204**) identification of common lab equipment.
Laboratory reports: (**209**)
Laboratory skills: (**200**)
Law of chemical equilibrium: (**108**) see chemical equilibrium.
Law of the conservation of energy: (**9**) energy may be converted from one form to another but is never created or destroyed.
Lead-acid battery: (**181**) see batteries.
LeChatelier's principle: (**106, 177**) see chemical equilibrium.
Light: (**32**) see spectrum; visible radiant energy.
Limiting reactants: (**269**) control in industrial chemical applications.
Liquids: (**21, 63, 105**) phase of matter: has a definite volume but an indefinite shape.
Liquid state of matter: (**13**) matter which has attained both vibrational and rotational movements.
Litmus: (**120**) pH indicator, blue in basic solution and red in acid solution.

Magnesium oxide: (**47**) example of an ionic solid.

Magnetic bottles: (**190**) designed to confine nuclear fusion through the use of strong magnetic fields.
Mass defect: (**186**) energy equal to the binding energy when the fusion of two or more particles occur.
Mass number: (**30**) number which indicates the total number of protons and neutrons in the nucleus of an atom.
Mass problems: (**83**) see mass and volume problems.
Mass-volume problems: (**83**) problems involving equations.
Math skills: (**195**) significant figures, scientific notation, order of magnitude.
Mathematics of chemistry: (**79, 88**)
Matter: (**7**) anything which occupies space and has mass.
Measurement: (**198**) laboratory devices to determine size, weight, etc.
Melting point: (**23**) temperature at which a solid and liquid are in equilibrium.
Metal ion: (**56, 207**) see Stock system; examples: FeO - iron (II) oxide and Fe_2O_3 - iron (III) oxide.
Metallic bonding: (**48**) bonding between the atoms of metals.
Metalloids (semimetals): (**65, 68**) atoms which have properties of both metals and nonmetals.
Metals: (**56, 64, 65, 136, 179**) atoms which tend to lose electrons when combining with other atoms.
Methane: (**158**) first member of the alkanes whose formula is CH_4.
Methanoic acid: (**166**) first member of organic acids whose formula is HCOOH.

Miscible: (8) property enabling two or more liquids to dissolve when brought together.

Miscible solution: (87) solution of liquid solutes - soluble in liquid solvents.

Mixtures: (8) varying amount of two or more distinct substances which differ in properties and composition.

Moderators: (188) see nuclear moderators, materials that have the ability to slow down neutrons quickly.

Molality: (90) expression of solution concentration which indicates the number of moles of solute in 1 kilogram of solvent.

Molar volume: (21, 80) the volume of one mole of particles in gases, a volume of 22.4 liters.

Molarity: (85, 123) expression of solution concentration which indicates the number of moles of solute in one liter of solution.

Mole: (21, 58, 79, 85, 88, 208) name for Avogardo's number of particles.

Molecular attraction: (50, 52) see dipoles and bonding.

Molecular formula: (55, 82) formula of a covalently bonded particle; molecular ratio (mole ratio) determines partial pressure of gases.

Molecular shapes: (266) shape is based on bond pairs and lone pairs.

Molecular solids: (48) solid composed of molecules.

Molecule, defined: (47, 48, 58) two definitions: 1) A discrete particle formed by covalently bonded atoms. 2) The smallest particle of an element or compound capable of independent existence.

Mole volume of a gas: (80) 22.4 liters of any gas at S.T.P.

Monatomic gases: (70) Group 18 elements in the Periodic Table.

Monatomic ion: (48, 70) a single atom with a charge.

Monohydroxy alcohols: (166) alcohols which have 1 hydroxyl (-O-H) group.

Monomer: (164) repeating chemical units which make up a polymer.

Naming and writing of chemical compound formulas: (55)

Naming elements with atomic numbers greater than 100: (73)

Natural radioactivity: (38) the spontaneous disintegration of the nucleus of an atom with the emission of subatomic particles, usually occurs in nature in elements with an atomic number higher than 82.

Negative oxidation state: (70) see oxidation state.

Negligible bond rearrangements: (101) often rapid reactions at room temperature, such as reactions of ionic substances in aqueous solutions.

Networks solids: (48) solid consisting of covalently bonded atoms linked into a network which extends throughout the sample with an absence of simple discrete particles.

Neutralization: (120, 122, 124) in acid-base reactions, occurs when equal quantities of an acid and base are mixed; the resulting products are a salt and water.

Neutrons: (29) subatomic particle with a mass of approximately one atomic mass unit and a unit charge of zero.

Nickel oxide-cadmium battery: (181) see batteries.

Noble gases: (71) common name for Group 18 (formally Group O) elements.

Nonelectrolytes: (91, 155) substances which when dissolved in solution form molecules which are not charged and will not conduct electricity.

Nonmetals: (56, 64, 65, 69) atoms which tend to gain electrons when in combination with other atoms.

Nonpolar covalent bond: (47) the equal sharing of electrons between atoms of the same element.

Normality: (123, 208) manner of expressing concentration which is concerned with the gram equivalent weight of H^+ ions supplied or accommodated.

Nuclear chemistry: (185) radioactive element study.

Nuclear control rods: (188) boron and cadmium rods which absorb neutrons in a fission reactor.

Nuclear coolants: (188) water, heavy water, air, helium, carbon dioxide, molten sodium, and molten lithium are used to keep the temperatures generated in fission reactions at reasonable levels within the reactor and to carry heat to heat exchanges and turbines.

Nuclear energy: (186) energy released when mass is converted to energy.

Nuclear fission: (187) the "splitting" of heavier nuclei into lighter ones with the release of energy.

Nuclear fuels: (187, 189) fuels used in nuclear reactors and include: uranium - 233, uranium - 235, and plutonium - 239.

Nuclear fusion: (189) the fusion of two lighter nuclei into a heavier nucleus at high temperatures and pressures.

Nuclear moderators: (188) materials which slow down the speed of neutrons; they include: water, heavy water, beryllium, and graphite.

Nuclear shielding: (188) internal and external shields used in reactors.

Nucleons: (29) any particle located in the nucleus.

Nucleus: (30) core of an atom in which are located the protons and neutrons.

Observed value: (199) experimentally measured value, value calculated from experimental results.

Oil of vitriol: (178) common name for sulfuric acid.

Olefin series: (159) common name for the Alkene series of hydrocarbons.

Open system: (106) product removal from the reaction "go to completion."

Operational definition: (119, 120) properties and reactions of substances based on experimental observation which includes a set of conditions.

Orbital notation: (35) depiction of orbital electron configuration.

Orbitals: (34) the average region of the most probable electron location.

Order of magnitude: (197) see math skills.

Organic acids: (166) acids contain an carboxyl (-COOH) functional group.

Organic chemistry: (155) chemistry of the compounds of carbon.

Organic compounds: (155) discussion of compounds of carbon.

Organic reactions: (156, 163, 207) substitution, addition, fermentation, and esterification.

Oxidation: (164) combustion or burning with oxygen.

Oxidation, defined: (135) loss or an apparent loss of electrons.

Oxidation number: (55, 135) see oxidation state.

Oxidation state: (72, 135) number (or state) of an atom, which represents the charge which that atom has, or appears to have, when electrons are counted according to certain arbitrary rules.

Oxidizing agents: (135, 140) species in a redox reaction which is reduced.

Paraffin series: (158) common name for the Alkane series.

Partial pressure of gases: (19) pressure exerted by each gas in a mixture.

Particle accelerator: (186) instrument used to accelerate charged particles by using electric and magnetic fields.

Percent composition: (81, 89)

Percent error: (199) see measurement in laboratory activities.

Periodic table: (55, 63, 66, 216-217)

Periodic table key: (66) how to read and use the periodic table.

Periods of the periodic table: (63) horizontal rows of the periodic table, also called "rows" or "series."

Petroleum: (181) includes fractional distillation and cracking.

pH: (127, 128) negative log, base 10, of the hydrogen ion concentration.

Phase change: (13, 14, 205) see phases of matter.

Phase equilibrium: (104, 105) reversible phase changes in a closed system.

Phases of matter: (13) an expression which refers to the state in which matter exists; they include: gases, liquids, and solids.

Phenolphthalein: (120) pH indicator, pink in basic solution and colorless in acid solution.

Photons: (32) also called quantum, is a unit of electromagnetic radiation.

Physical constants: (218, 219) see Reference Table A.

Polar covalent bonds: (47, 50) unequal sharing of electrons.

Polyatomic ion: (48, 49, 222-223) see Reference Table F, compound of two or more covalently bonded atoms with a charge.

Polyatomic ion endings: (57) "-ate" and "-ite."

Polymers: (168) long chain molecules made up of repeating smaller molecules called monomers. An example is polyethylene.

Polymerization: (164) the formation of large molecules from smaller ones.

Potential energy: (100, 101) energy of position.

Pouring liquids: (202) see laboratory safety skills.

Pressure: (106, 177) see industrial applications in chemistry.

Primary alcohols: (166) alcohols in which the hydroxyl group is attached to an end carbon atom.

Principal energy levels: (32, 33) see atomic structure models.

Principal quantum number: (33) see atomic orbital model.

Protium: (30) an isotope of hydrogen which contains one proton in its nucleus with a mass of 1 amu.

Proton acceptor: (121) ion or molecule which accepts a proton from another species, and is classified as a Brönsted-Lowry base.

Proton donor: (121) ion or molecule which donates a proton to another species is a proton donor, and is classified as a Brönsted-Lowry acid.

Protons: (29) subatomic particle, found in the nucleus, with a unit positive charge of one and a mass of one atomic mass unit.

Pure research: (177) see research.

Quantum, defined: (32) distinct, discrete amount of energy.

Quantum number: (33-34) numerical designation of the rules of quantum mechanics which refer to the probable location and energy level of the electron.

Radii of atoms: (234, 235) see Reference Table P, the distance of closest approach that an atom has to another.

Radioactive dating: (192) dating based on the radioactive decay of elements, also see half-life.

Radioactive isotopes: (185) see radioisotopes.

Radioactive tracers: (191) radioactive isotopes such as Carbon-14 used in tracing the course of a reaction without altering the chemical conditions.

Radioactive wastes: (191) discussion of nuclear energy industrial wastes.

Radioactivity: (38, 39) spontaneous disintegration of the nucleus of an atom with the emission of particles and/or radiant energy.

Radioisotopes: (185, 191) radioactive isotopes of an element.

Radioisotopes: (226, 227) see Reference Table H.

Radius, covalent, of atom: (64) one-half the measured internuclear distance, in the solid phase.

Rare earth series: (73) Lanthanide series of elements. They are difficult to extract from the "earth" or oxide salt, hence the term "rare earth."

Rare gases: (71) Group 18 (formally group O) elements.

Reaction completion: (105) removal of a product causing the reactants to exhaust themselves.

Reaction rates: (101, 102) measure, in terms of the number of moles of reactant consumed (or moles of products formed) per unit volume in a unit of time.

Reactors, nuclear: (187) see fission reactors.

Reactors, fission: (187) reactors in which controlled fission reactions take place.

Redox: (135, 141, 148, 178) abbreviation for reduction-oxidation reactions.

Redox reactions, defined: (137) reactions in which atoms and ions compete for electrons, resulting in the reduction or oxidation of these particles. This process is referred to as redox.

Reducing agents: (135) agent which supplies electrons in a redox reaction.

Reduction, defined: (135, 179) gain or apparent gain of electrons.

Reduction of metals: (179) the reduction of the positively charged metallic ions in ores to metallic atoms.

Reference tables in chemistry: (215-235) Reference Tables A – P.

Refining of metallic elements: (179) "Reduction of Metals."

Removing chemicals from containers: (202) see laboratory safety skills.

Research, applied: (177) research directed toward obtaining the maximum yield of products with maximum economic efficiency in industrial processes.

Research, pure: (177) research directed toward the acquiring of knowledge without the consideration of the immediate practical application of that knowledge.

"Roasting" of metallic ores: (179) reaction of a metallic ore with oxygen in order to form the oxide of the ore.

Rotating: (13) spinning or revolving, electron motion; as in gas and liquid.

Rutherford: (30) gold foil experiments which indicated the atom to be mostly empty space; nucleus very small compared to the size of atom.

Salt bridge: **(140)** bridge, containing a solution of a salt which allows the migration of ions in a Daniell electrochemical cell.

Salts: (124) ionic compound containing positive ions other than hydrogen ions and negative ions other than hydroxide ions.

Saponification: (164) hydrolysis of fats by bases.

Stock system: (56) naming of compounds of metals with more than one oxidation state, Roman numerals denote the oxidation state of the metal.

Stoichiometry: (81) study of quantitative relationships implied by chemical formulas and chemical equations.

STP: (19) see standard temperature and pressure.

Strengths of acids in aqueous solution: (230-231) see Reference Table L.

Structural formula: (156) formula showing the manner in which atoms bond, with bond structures shown.

Subatomic particles: (29) particles smaller than an atom.

Sublevels: (33, 72) second quantum numbers.

Sublimation: (23) change from the solid phase to a gaseous phase without passing through an apparent liquid phase.

Substance, defined: (7) homogeneous matter having identical properties and composition.

Substitution reactions: (163) replacement of one kind of atom or group with another kind of atom or group.

Super–cooled liquids: (23) materials that behave as highly viscous liquids with crystalline structures.

Supersaturated solution: (87) solution in which more solute is dissolved than can be dissolved under normal conditions.

Surface area: (102) reactive area of a substance, when increased, the rate of reaction increases.

Symbol: (54) see formula writing, a representation of one atom or one mole of atoms of an element.

Symbols used in nuclear chemistry: (226-227) see Reference Table J.

Table of Standard Electrode Potentials: (67, 232) see Reference Tables in Chemistry.

Temperature, defined (thermometry): (12) measure of the average kinetic energy in a system.

Temperature and pressure relationship: (18) at constant volume, in a direct proportion, if temperature increases, pressure increases.

Tertiary Alcohols: (167) hydroxyl group in the alcohol is attached to a carbon atom which is in turn attached to three other alkyl radicals.

Ternary acids: (57) acids which combine hydrogen with a polyatomic ion and contain three elements in their formulas.

Ternary compounds: (8) compounds with three elements.

Tetrahedron bonds: (156) four equidistant bonds whose vertices form a tetrahedron.

Thermometer: (12) instrument used to measure temperature.

Titration: (122) process of metering a standard solution into a solution of unknown concentration.

Toluene: (160) second member of the benzene series of carbohydrates.

Tracers, radioactive: (191) radioisotopes used to follow the course of a reaction without seriously altering any chemical conditions which exist.

Transition elements: (72) elements found in Groups 3 through 12 of the periodic table.

Translating: (13) changing or transforming, as in gases.

Transmutation, defined: (38) when one element is changed to another as a result of changes in the nucleus.

Transmutation, artificial: (185) when one element is bombarded with subatomic particles forming newer elements.

Trihydroxy alcohols: (167) compounds containing three hydroxyl groups (O-H) in their formulas.

Triple bond: (69, 157) bond formed between carbon atoms by the sharing of three pairs of electrons.

Tritium: (30) an isotope of hydrogen in which the nucleus contains one proton and two neutrons.

Unsaturated compounds: **(157)** organic compounds with one or more double or triple bonds between carbon atoms.

Unsaturated solution: (87) solution in which less solute is dissolved than is capable of being dissolved under normal conditions.

Valence electrons: **(36)** electrons found in the outermost principal energy level of an atom.

Van der Waal's forces: (51, 64, 70, 157) half the internuclear distance or radius of closest approach of an atom with another atom with which it forms no bond.

Vapor pressure: (21, 22) pressure exerted by the vapor escaping from a liquid or solid in equilibrium.

Vapor pressure of water: (232, 233) see Reference Table O.

Vibrating: (13) shaking of particles as in solids, liquids, and gases.

Volume-volume problems: (84) problems concerning reacting gases at the same temperature and pressure. Wherein, the volume ratio is equal to the mole ratio.

Wastes, radioactive: **(191)** radioactive material no longer be usable.

Water, ionization constant of: (127) 1×10^{-14}.

Water of hydration: (207) also called "water of crystallization". The water chemically combined with another substance in a hydrate.

Zymase: **(163)** catalyst used in the fermentation of carbohydrates to alcohol and carbon dioxide.

Self-Help Questions/Answers

Col. 1 - **Question Number**;   Col. 2 - **Answer**;   Col. 3 - **Page Reference**

Unit I			Unit II			Unit III			Unit IV		
1.	3	8	1.	2	30	1.	1	46	1.	1	70
2.	3	7	2.	3	30	2.	1	45	2.	2	65
3.	1	7	3.	2	31	3.	1	46	3.	4	65
4.	2	8	4.	3	30	4.	1	46	4.	2	64
5.	1	8	5.	3	31	5.	3	46	5.	1	64
6.	4	7	6.	2	30	6.	2	48	6.	1	63
7.	2	7	7.	1	38	7.	1	46	7.	2	70
8.	3	10	8.	2	30	8.	2	46	8.	2	63
9.	4	10	9.	3	31	9.	1	46	9.	3	65
10.	4	22	10.	2	30	10.	3	48	10.	1	64
11.	3	10	11.	2	30	11.	2	48	11.	1	65
12.	4	12	12.	3	30	12.	3	47	12.	3	65
13.	4	12	13.	3	30	13.	1	47	13.	2	64
14.	3	12	14.	3	30	14.	4	47	14.	1	69
15.	2	12	15.	4	34	15.	2	47	15.	3	67
16.	2	14	16.	1	32	16.	1	48	16.	1	67
17.	1	14	17.	2	34	17.	2	48	17.	3	70
18.	3	14	18.	3	32	18.	4	46	18.	2	68
19.	2	13	19.	4	34	19.	3	48	19.	4	68
20.	4	15	20.	3	32	20.	2	47	20.	3	70
21.	4	19	21.	2	34	21.	1	51	21.	4	68
22.	3	17	22.	3	34	22.	2	49	22.	1	67
23.	4	17	23.	2	34	23.	3	51	23.	3	67
24.	2	18	24.	2	36	24.	2	52	24.	3	70
25.	1	20	25.	2	36	25.	1	51	25.	4	70
26.	1	20	26.	1	36	26.	4	47	26.	4	68
27.	4	20	27.	3	32	27.	1	51	27.	4	69
28.	1	12	28.	4	36	28.	2	50	28.	3	68
29.	1	21	29.	4	36	29.	4	51	29.	1	70
30.	3	21	30.	2	36	30.	4	51	30.	3	70
31.	1	21	31.	4	51	31.	4	51	31.	4	67
32.	1	21	32.	3	32	32.	4	56	32.	4	70
33.	3	20	33.	1	30	33.	3	54	33.	3	68
34.	4	22	34.	4	39	34.	4	55	34.	2	68
35.	2	21	35.	4	39	35.	4	56	35.	4	70
36.	2	21	36.	4	38	36.	1	56	36.	2	68
37.	1	21	37.	3	39	37.	4	54	37.	4	68
38.	4	22	38.	2	39	38.	4	56	38.	4	70
39.	4	21	39.	3	39	39.	1	56	39.	3	73
40.	3	23	40.	1	38	40.	2	58	40.	4	73
41.	3	23	41.	1	38	41.	4	58	41.	3	72
42.	1	23	42.	3	38				42.	1	72
									43.	2	72
									44.	3	72
									45.	1	73

Self-Help Questions/Answers

Col. 1 - **Question Number**;   Col. 2 - **Answer**;   Col. 3 - **Page Reference**

Unit V			Unit VI			Unit VII			Unit VIII		
1.	4	79	1.	3	99	1.	3	121	1.	3	136
2.	3	80	2.	1	100	2.	4	120	2.	1	135
3.	1	80	3.	2	102	3.	4	119	3.	4	135
4.	1	80	4.	2	102	4.	4	121	4.	2	139
5.	4	79	5.	2	101	5.	4	119	5.	2	140
6.	2	79	6.	1	101	6.	1	120	6.	4	140
7.	2	81	7.	2	100	7.	2	120	7.	3	135
8.	2	81	8.	1	101	8.	3	121	8.	4	139
9.	1	82	9.	2	100	9.	3	120	9.	1	140
10.	2	83	10.	1	101	10.	4	119	10.	1	136
11.	3	84	11.	1	100	11.	1	121	11.	2	136
12.	4	82	12.	4	107	12.	2	119	12.	3	140
13.	1	81	13.	2	109	13.	1	120	13.	1	139
14.	2	83	14.	4	108	14.	1	120	14.	2	140
15.	3	82	15.	2	106	15.	3	123	15.	3	136
16.	1	82	16.	2	107	16.	1	125	16.	4	136
17.	2	81	17.	1	106	17.	4	122	17.	3	139
18.	2	84	18.	3	106	18.	1	122	18.	2	140
19.	4	84	19.	2	106	19.	3	124	19.	1	140
20.	3	84	20.	2	105	20.	3	122	20.	1	139
21.	3	83	21.	3	107	21.	4	123	21.	1	141
22.	3	85	22.	4	106	22.	2	124	22.	2	141
23.	4	85	23.	4	105	23.	2	125	23.	3	141
24.	1	86	24.	4	104	24.	4	122	24.	1	141
25.	4	86	25.	3	110	25.	4	122	25.	2	141
26.	4	85	26.	4	111	26.	4	125	26.	3	140
27.	3	85	27.	1	112	27.	2	119	27.	4	142
28.	3	90	28.	1	111	28.	1	125	28.	2	140
29.	2	89	29.	4	112	29.	4	124	29.	3	141
30.	4	90	30.	2	112	30.	1	125	30.	3	145
31.	1	91	31.	3	112	31.	4	124	31.	1	140
32.	4	90	32.	3	111	32.	1	122	32.	1	140
33.	1	90	33.	1	112	33.	2	122	33.	4	142
34.	1	94	34.	2	110	34.	1	126	34.	4	145
35.	1	93				35.	2	127	35.	2	140
36.	4	94				36.	2	127	36.	4	141
37.	3	93				37.	3	127	37.	4	142
38.	3	92				38.	2	127	38.	3	140
39.	3	94				39.	1	127	39.	3	141
40.	3	94				40.	1	127	40.	3	142
41.	1	94				41.	1	127	41.	4	148
42.	4	92				42.	1	126	42.	1	148
						43.	3	127	43.	4	148
						44.	1	126	44.	2	148
						45.	1	125			

Self-Help Questions/Answers

Col. 1 - **Question Number**;   Col. 2 - **Answer**;   Col. 3 - **Page Reference**

Unit IX			Unit X			Unit XI			Unit XII		
1.	3	157	1.	1	177	1.	3	185	1.	1	198
2.	2	159	2.	3	179	2.	2	187	2.	4	203
3.	3	156	3.	2	177	3.	4	189	3.	1	199
4.	3	159	4.	4	179	4.	2	185	4.	2	198
5.	3	155	5.	2	178	5.	4	185	5.	3	199
6.	3	159	6.	1	179	6.	3	188	6.	2	195
7.	4	160	7.	2	179	7.	1	189	7.	1	204
8.	2	159	8.	4	179	8.	3	185	8.	3	202
9.	2	159	9.	1	179	9.	2	186	9.	4	195
10.	4	158	10.	3	179	10.	1	185	10.	3	207
11.	3	155	11.	2	179	11.	1	188	11.	1	207
12.	2	157	12.	3	177	12.	4	188	12.	2	205
13.	1	159	13.	2	177	13.	2	185	13.	2	209
14.	1	160	14.	4	181	14.	1	191	14.	2	207
15.	4	158	15.	3	180	15.	1	192	15.	2	206
16.	3	163	16.	4	181	16.	4	192			
17.	1	168	17.	3	181	17.	3	192			
18.	2	163	18.	1	180	18.	4	191			
19.	3	163	19.	4	182	19.	1	192			
20.	1	163	20.	4	181	20.	1	192			
21.	1	163									
22.	4	163									
23.	4	165									
24.	2	167									
25.	2	166									
26.	3	168									
27.	4	168									
28.	3	168									
29.	4	167									
30.	2	166									
31.	1	167									
32.	1	166									
33.	3	167									
34.	2	168									
35.	2	167									
36.	4	168									
37.	1	168									
38.	3	168									

We expresse our appreciation to the College Board and the Educational Testing Service for providing their permission to use materials from The College Board Achievement Tests in Science.
Also, for students who need a more "in depth" explanation of the Achievement Test in Chemistry, we recommend purchasing the official publication of The College Board:
The College Board Achievement Tests (14 Tests in 13 Subjects) ISBN: 0-87447-162-1.

The Chemistry Achievement Test©

Excerpts from:
The College Board Achievement Tests in Science©

The Chemistry Achievement Tests consists of 85 multiple-choice questions. The test content is based on the assumption that you have had a one-year introductory course in chemistry at a level suitable for college preparation. The test covers the topics listed on the next page. Different aspects of each topic are stressed from year to year. Because high school courses differ, both in the percentage of time devoted to each major topic and in the specific subtopics covered, you may encounter questions on topics with which you have little or no familiarity. However, in any typical high school chemistry course, more topics are usually covered - and in more detail - than there are questions in the Chemistry Achievement Test. So, even if high school chemistry curriculums were the same, the questions in any particular test edition could only be a sample of all the questions that might be asked. The questions in all of the editions, however, test knowledge and abilities that might reasonably be expected of high school chemistry students intending to go to college.

No one instructional approach is better than another in helping you prepare for the test provided that you are able to recall and understand the major concepts of chemistry and to apply the chemical principles you have learned to solve specific scientific problems. In addition, you should be able to organize and interpret results obtained by observation and experimentation and to draw conclusions or make inferences from experimental data. Laboratory experience is a significant factor in developing reasoning and problem-solving skills. Although laboratory skills can be tested only in a limited way in a standardized test, reasonable laboratory experience is an asset in helping you prepare for the Chemistry Achievement Test.

The Chemistry Achievement Test assumes that your preparation in mathematics enables you to handle simple algebraic relationships, to understand the concepts of ratio and proportion, and to apply these concepts to word problems.

A periodic table indicating the atomic numbers and atomic weights of the elements is provided. You will not be allowed to use electronic calculators during the chemistry test. Numerical calculations are limited to simple arithmetic. In this test, the metric system of units is used.

Content of the Test

Topics Covered   Approximate Percentage of Test

I.   **Atomic Theory and Structure,** **14**
         (including periodic relationships)

II.   **Chemical Bonding and Molecular Structure** **10**

III.   **States of Matter and Kinetic Molecular Theory** **9**

IV.   **Solutions** **6**
         (inc. concentration units, solubility, colligative properties)

V.   **Acids and Bases** **9**

IV.   **Oxidation-reduction and Electrochemistry** **7**

VII. **Stoichiometry** **11**

VIII. **Reaction Rates** **2**

IX.   **Equilibrium** **5**

X.   **Thermodynamics,** **4**
         (including energy changes in chemical reactions, randomness,
         and criteria for spontaneity)

XI.   **Descriptive Chemistry:** **16**
         (physical and chemical properties of elements and their more
         familiar compounds, including simple examples from organic
         chemistry; periodic properties)

XII. **Laboratory:** **7**
         (equipment, procedures, observations, safety, calculations,
         and interpretation of results)

Note: Every edition contains approximately five questions on equation balancing and/or predicting products of chemical reactions. These are distributed among the various content categories.

Skills Specifications   Approximate Percentage of Test

Level I: **Essentially Recall:** **30**
         (remembering information and understanding facts)

Level II: **Essentially Application:** **55**
         (applying knowledge to unfamiliar and/or practical
         situations; solving mathematical problems)

Level III: Essentially Interpretation: **15**
         (inferring and deducing from available data and
         integrating information to form conclusions)

85 Questions: Time - 60 minutes

Questions Used in the Test Include:
        Classification Questions,
        Multiple-Completion Questions,
        Relationship Analysis Questions, and
        Five-choice Completion Questions

Note for N.Y.S. students: The following question types are NOT generally included on the New York State Regents Examination in Chemistry.

The **Relationship Analysis Question** consists of a specific statement or assertion followed by an explanation of the assertion. The question is answered by determining if the assertion and the explanation are each true statements and if so, whether the explanation (or reason) provided does in fact properly explain the statement given in the assertion.

This type of question tests your ability to identify proper cause - and - effect relationships. It probes whether you can assess the correctness of the original assertion and then evaluate the truth of the "reason" proposed to justify it. The analysis required by this type of question provides you with an opportunity to demonstrate developed reasoning skills and the scope of your understanding of a particular topic.

A **special type of five-choice completion question** is used in the Chemistry Achievement Test, to allow for the possibility of multiple correct answers. Unlike many quantitative problems that must by their nature have a unique solution, some questions have more than one correct response. For these questions, you must evaluate each response independently of the others in order to select the most appropriate combination. In questions of this type, several (usually three to five) statements labeled by Roman numerals are given with the question. One or more of these statements may correctly answer the question.

The statements are followed by five lettered choices (A to E) consisting of some combination of the Roman numerals that label the statements. You must select from among the five lettered choices the one combination of statements that best answers the question. In the test, questions of this type are intermixed among the more standard five-choice completion questions.

The five - choice completion question also tests problem - solving skills. With this type of question, you may be asked to convert the information given in a word problem into graphical forms or to select and apply the mathematical relationship necessary to solve the scientific problem. Alternatively, you may be asked to interpret experimental data, graphical stimulus, or mathematical expressions.

Chemistry Achievement Test Questions

The following questions have been selected (with permission from the Educational Testing Service) from a published Chemistry Achievement Test. These questions have been chosen because we feel that the questions are
   (1) "typical" of the style and format of the actual achievement test questions that you will be expected to answer correctly.
   (2) "typical" in respect to their difficulty and usage of graphs, illustrations, and diagrams.
   (3) "typical" as to the broad range of material that will be covered on the actual achievement test.
   (4) "typical" of the most often missed questions (generally, over 50%).

Achievement Test Questions For Practice

1   Which of the following best describes the shape of BI_3.
   (A) Linear   (C) Trigonal, planar   (E) Regular tetrahedral
   (B) Bent (V-shaped)   (D) Pyramidal

Questions 2-3

   (A) $3O_2(g) \leftrightarrow 2O_3(g)$

   (B) $OH^- + H_3O^+ \leftrightarrow 2H_2O$

   (C) $BaCl_2 \cdot 2H_2O(s) \leftrightarrow BaCl_2(s) + 2H_2O(g)$

   (D) $Ca^{2+} + CO_3^{2-} \leftrightarrow CaCO_3(s)$

   (E) $Fe + Cu^{2+} \leftrightarrow Fe^{2+} + Cu$

2   Represents an oxidation-reduction reaction

3   Involves the formation of an ionic precipitate from a solution

Questions 4-6

   (A) $3d$ Transition metals

   (B) Alkali metals

   (C) Halogens

   (D) Noble gases

   (E) Actinides

4   Are the most readily oxidized elements within a given period

5   Have the highest first ionization energies (potentials) of the elements in
   their respective rows of the periodic table

6   Are all radioactive elements

7   Which of the following has the lowest pH?
   (A) 0.010-molar HCl
   (B) 0.010-molar NaOH
   (C) 0.010-molar $Ba(OH)_2$
   (D) 0.010-molar H_2SO_4
   (E) 0.010-molar $C_{12}H_{22}O_{11}$ (cane sugar)

Questions 8-9

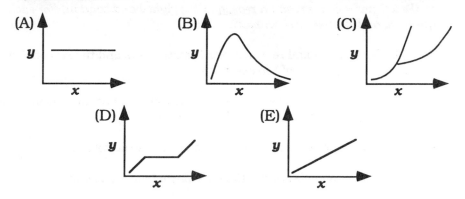

8 Shows how a plot of the pressure-volume product y varies with absolute temperature x for 1 mole of an ideal gas

9 Shows a plot of the fraction of molecules in a gas sample y versus kinetic energy x at a given temperature

Questions 10-12: For each of the questions below, ONE or MORE of the responses given are correct. Decide which of the responses is (are) correct and on the answer sheet mark

    A if 1, 2, and 3 are correct;
    B if only 1 and 2 are correct;
    C if only 2 and 3 are correct;
    D if only 1 is correct;
    E if only 3 is correct.

10 A piece of zinc may be distinguished from a piece of magnesium by
    (1) determining which conducts electricity
    (2) determining which releases H_2 gas from a 1-molar hydrochloric acid solution
    (3) measuring the density of each

11 Molecules that involve both s and p orbital electrons in their bonding include which of the following?
(Atomic numbers: H = 1, C = 6, N = 7, Cl = 17)
    (1) NH_3 (2) CCl_4 (3) HCl

12 ... $CaO(s) +$ ... H_3O^+ ... $Cl^- \rightarrow$
Addition of solid calcium oxide to a solution of hydrochloric acid results in the formation of
    (1) Ca^{2+} (2) H_2O (3) H_2

Questions 13-17: Each question below consists of an <u>assertion</u> (statement) in the left-hand column and a <u>reason</u> in the right-hand column. For each question, on the answer sheet mark

A   if both assertion and reason are true statements and the reason is a <u>correct explanation</u> of the assertion;

B   if both assertion and reason are true statements, but the reason is <u>NOT a correct explanation</u> of the assertion;

C   if the assertion is true, but the reason is a false statement;

D   if the assertion is false, but the reason is a true statement;

E   if both assertion and reason are false statements.

<u>Assertion</u>		<u>Reason</u>
13 The molecule CO_2 has a net dipole moment of zero	BECAUSE	the arrangement of atoms in the CO_2 molecule is linear and symmetrical, and the bond polarities within the molecule are canceled out.
14 At 25°C and 1 atmosphere pressure, H_2O is a liquid but H_2S is a gas	BECAUSE	the molecular weight of H_2O is less than that of H_2S.
15 The equation below is balanced $Fe^{2+} + NO_3^- + 4H^+ \leftrightarrow Fe^{3+} + NO + 2H_2O$	BECAUSE	all the atoms and charges in the equation shown for this reaction are conserved.
16 ^{50}Ti and ^{50}Cr are isotopes of each other	BECAUSE	atoms of ^{50}Ti and ^{50}Cr have the same mass number.

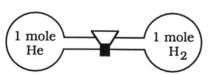

17 When the valve in the system shown above is opened and the system is allowed to reach equilibrium, most of the helium remains in the bulb on the left.	BECAUSE	when two bulbs containing gases are connected, the randomness of the system decreases.

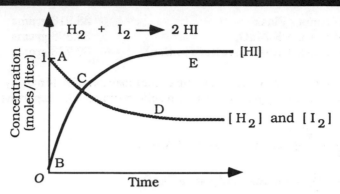

18 The graph above shows the variation in concentration of reactants and products as H_2 and I_2 react to form HI at a given temperature. Equilibrium is reach when the concentration of HI is
(A) A   (C) C   (E) E
(B) B   (D) D

19

Element	Electronegativity
P	2.1
H	2.2
Br	2.8
O	3.5

On the basis of the electronegativity values given above, in which of the following are the bonds most polar?
(A) PH_3 (C) HBr   (E) PBr_3
(B) H_2O (D) P_4

20

Liquid	Vapor Pressure (mm Hg), 25°C
A	20
B	35
C	56
D	112
E	224

In which of the liquids listed in the table above are the intermolecular forces of attraction the strongest?
(A) A   (C) C   (E) E
(B) B   (D) D

21 The range of oxidation states exhibited by most elements in group V of the periodic table (N, P, As, Sb, Bi) is
(A) -3 to +1   (C) -3 to +5   (E) +1 to +5
(B) -3 to +3   (D) -2 to +3

22 Mass of Empty Flask   38.913 grams
    Mass of Flask + K MnO$_4$ 39.898 grams
    Mass of Flask + Residue (after ignition)   39.773 grams

The data above were taken for the decomposition of K MnO$_4$ by ignition, which liberates oxygen and leaves a solid residue. The experimental mass ratio is

(A) $\dfrac{0.125}{0.985}$ gram residue/gram K MnO$_4$

(B) $\dfrac{0.125}{0.860}$ gram residue/gram K MnO$_4$

(C) $\dfrac{0.860}{0.985}$ gram residue/gram K MnO$_4$

(D) $\dfrac{0.985}{0.860}$ grams residue/gram K MnO$_4$

(E) $\dfrac{0.860}{0.125}$ grams residue/gram K MnO$_4$

23 A sample of 90.0 grams of glucose is dissolved in enough water to yield 200. milliliters of solution. What is the molar concentration of glucose (molecular weight 180.)?

(A) 0.500 *M*   (C) 2.00 *M*   (E) 10.0 *M*
(B) 1.00 *M*   (D) 2.50 *M*

24      $CO_2(g) + C(s) \rightarrow 2CO(g)$

A sample of 100 liters of carbon dioxide at 25°C and 1 atmosphere pressure is reduced by being passed over hot coke according to the equation above. When measured at 25°C and 1 atmosphere pressure, the volume of carbon monoxide formed is

(A) 22.4 liters   (C) 50 liters   (E) 200 liters
(B) 44.8 liters   (D) 100 liters

25      $\ldots S_2O_8{}^{2-} + \ldots I^- \rightarrow \ldots SO_4{}^{2-} + \ldots I_2$

What is the maximum yield of iodine that can be obtained when 1 mole of $Na_2S_2O_8$ reacts completely with excess iodide ion according to the equation above? (Equation above is not balanced.)

(A) 1 mole   (C) 4 moles   (E) 8 moles
(B) 2 moles   (D) 6 moles

26 The volume occupied by 0.50 mole of propane gas, C_3H_8, at a temperature of 27°C and a pressure of 2.0 atmospheres is best expressed by which of the following? (R = 0.082 liter-atm/mole°K)

(A) $\dfrac{0.50 \times 0.082 \times 27}{2}$ liters

(D) $\dfrac{0.50 \times 0.082 \times 300}{2 \times 760}$ liters

(B) $\dfrac{0.50 \times 0.082 \times 300}{2}$ liters

(E) $\dfrac{0.50 \times 0.082 \times 27}{2 \times 760}$ liters

(C) $\dfrac{0.50 \times 0.082 \times 273}{300}$ liters

27 In qualitative analysis the separation of Ag^+ ions from Cu^{2+} ions by the addition of HCl depends on the fact that

(A) Cu^{2+} forms an insoluble chloride and Ag^+ does not
(B) Ag^+ forms an insoluble chloride and Cu^{2+} does not
(C) Cu^{2+} forms a complex with HCl and Ag^+ does not
(D) Cu reacts with HCl and Ag does not
(E) Ag^+ is oxidized by HCl whereas Cu^{2+} is in its highest oxidation state

28 A battery jar contained a solution of copper sulfate. Two electrodes, one made of copper, the other a metal object to be copper plated, were placed in the jar and connected to a source of direct current. Which of the following statements concerning this system is correct?

(A) The object to be plated is the anode.
(B) Oxidation occurs at the anode.
(C) The sulfate ions migrate toward the cathode.
(D) The concentration of the copper sulfate solution increases as electrolysis proceeds.
(E) The copper electrode increases in mass.

	I		II		III		IV		V
29	$2s$	→	$2p$	→	$3d$	→	$3p$	→	$4s$	→	$2p$

The electronic transitions shown above are observed when lithium atoms are sprayed into a hot flame. The various transitions are numbered for identification. Which of these transitions would result in emission of electromagnetic radiation (light)?

(A) I and II only   (C) I and V only   (E) III and V only
(B) I and III only   (D) III and IV only

30 $HCl(g) + NH_3(g) \rightarrow NH_4Cl(s)$

If 3.0 moles of HCl gas and 5.0 moles of NH_3 gas, each measured at 20°C and 1.0 atmosphere pressure, are allowed to react completely according to the equation above, the final mixture will contain

(A) 3 moles of solid NH_4Cl only
(B) 5 moles of solid NH_4Cl only
(C) 3 moles of solid NH_4Cl + 2 moles of NH_3 gas
(D) 3 moles of solid NH_4Cl + 2 moles of HCl gas
(E) 2 moles of HCl gas, 4 moles of NH_3 gas, and 1 mole of solid NH_4Cl

31 In an electrolysis cell, the passage of 6.02×10^{23} electrons can produce

(A) 22.4 liters of H_2 gas (measured at standard conditions) from dilute H_2SO_4 solution
(B) 22.4 liters of O_2 gas (measured at standard conditions) from dilute H_2SO_4 solution
(C) 1 mole of Cl_2 gas from HCl solution
(D) 1 mole of metallic silver from $AgNO_3$ solution
(E) 1 mole of metallic copper from $CuSO_4$ solution

Achievement Test
Answers and Explanations

After completing the preceding questions, check your answers. For more information concerning the question, refer to the area of the text indicated or review the additional material provided with the answer.

1. **Answer C: Molecular Shapes.**
    The shape of a molecule depends on:
    1) The number of **bond pairs**.
    2) The number of unbonded pairs of electrons called **lone pairs**.
    3) The understanding that electron pairs (both bond pairs and lone pairs) try to move as far apart from each other as possible.

    The following table incorporates these three conditions and points out the resulting molecular shape:

Bond Pairs	Lone Pairs	Shape	Example
2	0	linear	CO_2
3	0	trigonal planar	BI_3
2	1	bent	NO_2
2	2	bent	H_2O
3	1	pyramidal	NH_3
4	0	tetrahedral	CH_4

2.   **Answer E:** Information is included in Unit VIII, page 137 of text.

3.   **Answer D:** Information is included in Unit VI, page 106 of text.

4.   **Answer B:** Information is included in Unit IV, page 67 of text.

5.   **Answer D:** Information is included on Table K, pages 228-229 of text.

6.   **Answer E:** Information is included in Unit II, page 38 of text.

7.   **Answer D:** Information is included in Unit VII, page 127 of text.

8.   **Answer E:** Information is included in Unit I, page 16 of text.

9.   **Answer B: Illustration of the Kinetic Theory of Gases.**
     As the temperature in a system of gases rises, the average kinetic
     energy of the molecules increases. This can be shown on an energy
     distribution curve at two temperatures (illustrated below).

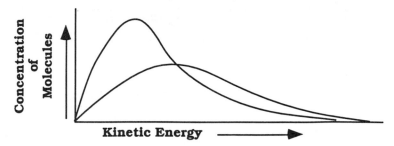

10. **Answer E:** Information is included in Unit I, page 23 of text.

11. **Answer A:** Information is included in Unit III, page 47 of text.

12. **Answer B:** Information is included in Unit VII, page 120 of text.

13. **Answer A: Dipole Moment.**
     The direction and charge of a polar bond can be measured as any
     other vector quantity. To go one step further, vector quantities can be
     added or subtracted from each other. A molecule may have one end with
     a slight positive charge equal to a negative charge on the other end. The
     intensity of the negative charge multiplied by its distance from the
     positive charge of equal units. One *debye* unit equals a negative charge
     of 0.208, the charge of one electron separated by 1×10^{-8} cm from an
     equal but opposite positive charge.
     The measurement of the dipole moment character of molecule allows
     us to predict if a molecule will be polar or nonpolar.
     For example, in the CO_2 molecule, the dipole vectors have the same
     magnitude but are opposite in direction. Therefore, the net dipole
     moment is zero, and the molecule is nonpolar. However, in the water
     molecule, the O—H polar bonds are bent and pointing in one direction.
     Therefore, we have a net positive value of 1.94 debye, and the molecule
     is polar.

14. **Answer B**: Information is included in Unit III, page 51 of text.

15. **Answer E**: Information is included in Unit VII, page 148 of text.

16. **Answer D**: Information is included in Unit II, page 30 of text.

17. **Answer E**: Information is included in Unit VI, page 104 of text.

18. **Answer E**: **Illustrations of Chemical Equilibrium Graphs**.
     The graphs below depict the concentrations of reactants and
     products as they react and reach a state of equilibrium. The three
     possible results are as follows:

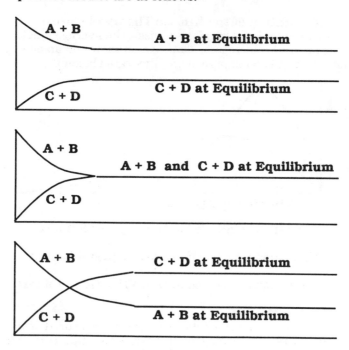

19. **Answer B**: Information is included in Unit III, page 46 of text.

20. **Answer A**: Information is included in Unit III, page 51 of text.

21. **Answer C**: Information is included in Unit IV, page 68 of text.

22. **Answer C**: Information is included in Unit XII, page 199 of text.

23. **Answer D**: Information is included in Unit V, page 85 of text.

24. **Answer E**: Information is included in Unit V, page 84 of text.

25. **Answer A**: Information is included in Unit III, page 58 of text.

26. **Answer B: Ideal Gas Equation**.
The variables which determine the state of a gaseous system are related in the following manner:

1) If **n** = the number of moles of a gas, and the pressure and temperature is kept constant, the volume (**V**) of a gas is directly related to the number of moles of the gas $V \propto n$.

2) From Boyle's Law: $V \propto \dfrac{1}{p}$ (**n** and **T** are constant)

From Charles' Law: $V \propto T$ (**n** and **P** are constant)

Therefore, $V \propto \dfrac{nT}{p}$

A proportionality can be made into an equality by including a constant into the expression. Let **R** = the gas constant with definitive units of

$$\frac{\textbf{atm x liters}}{\textbf{moles x K}}$$

The following equation can then be used to solve most gas problems:

$$\textbf{PV} \propto \textbf{nRT}$$

27. **Answer B: Solubility Rules**.

1) All salts of chlorides are soluble except salts of Ag^+, Hg_2^{2+}, and Pb^{2+}.
2) Almost all sodium, potassium, and ammonium are soluble.
3) All nitrates, chlorates, perchlorates, and acetates are soluble.
4) All sulfates are soluble except Ba^{2+}, Sr^{2+}, and Pb^{2+} salts. (Ca^{2+} and Hg_2^{2+} are moderately soluble sulfate salts.
5) All the rest are moderately or partially soluble.)

28. **Answer B**: Information is included in Unit VIII, page 145 of text.

29. **Answer E**: Information is included in Unit II, page 32 of text.

30. **Answer C**: Information is included in Unit V, page 83 of text.

Limiting Reactants: In industrial applications, low product yields and time limitations require the use of an excess amount of one of the reactants (usually the cheapest) in order that a reaction will go to completion quickly and more completely. The reactant which is not in excess and will be used up completely is called the limiting reactant. In the reaction: $HCl(g) + NH_3(g) \rightarrow NH_4Cl(s)$, the mole rations are 1:1::1. Theoretically, if 3 Moles of $HCl(g)$ are used, only 3 Moles of NH_3 are needed to make 3 Moles of $NH_4Cl(s)$. However, 5 Moles of $NH_3(g)$ were used. Therefore, the NH_3 is the reactant in excess and the limiting reactant is the $HCl(g)$. The excess NH_3 is found in the products as follows:

$$3\,HCl(g) \;+\; 5\,NH_3 \;\rightarrow\; 3\,NH_4Cl(s) \;+\; 2\,NH_3(g)$$

31. **Answer D**: Information is included in Unit VIII, page 145 of text.

Achievement Test Notes:

696   **Chemistry Practice Examination 1**

Part 1 - Answer all 56 questions in this part.

Directions (1-56): For *each* statement or question, select the word or expression that, of those given, best completes the statement or answers the question.

Page 271

1   What is the vapor pressure of a liquid at its normal boiling temperature?
     (1) 1 atm   (2) 2 atm   (3) 273 atm   (4) 760 atm

2   A sealed container has 1 mole of helium and 2 moles of nitrogen at 30°C. When the total pressure of a mixture is 600 torr, what is the partial pressure of the nitrogen?
     (1) 100 torr   (2) 200 torr   (3) 400 torr   (4) 600 torr

3   Solid X is placed in contact with solid Y. Heat will flow spontaneously from X to Y when
     (1) X is 20°C and Y is 20°C   (3) X is -25°C and Y is -10°C
     (2) X is 10°C and Y is 5°C   (4) X is 25°C and Y is 30°C

4   Which graph represents the relationship between volume and Kelvin temperature for an ideal gas at constant pressure?

  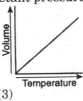

     (1)   (2)   (3)   (4)

5   An example of a binary compound is
     (1) potassium chloride   (3) potassium chlorate
     (2) ammonium chloride   (4) ammonium chlorate

6   Which kind of radiation will travel through an electric field on a pathway that remains unaffected by the field?
     (1) a proton   (3) an electron
     (2) a gamma ray   (4) an alpha particle

7   The major portion of an atom's mass consists of
     (1) electrons and protons   (3) neutrons and positrons
     (2) electrons and neutrons   (4) neutrons and protons

8   Which atom contains exactly 15 protons?
     (1) phosphorus-32   (3) oxygen-15
     (2) sulphur-32   (4) nitrogen-15

9   Element X has two isotopes. If 72.0% of the element has an isotopic mass of 84.9 atomic mass units, and 28.0% of the element has an isotopic mass of 87.0 atomic mass units, the average atomic mass of element X is numerically equal to

     (1) $(72.0 + 84.9) \times (28.0 + 87.0)$ (3) $\dfrac{(72.0 \times 84.9)}{100} + \dfrac{(28.0 \times 87.0)}{100}$

     (2) $(72.0 - 84.9) \times (28.0 + 87.0)$ (4) $(72.0 \times 84.9) + (28.0 \times 87.0)$

10   Given the equation: $^{14}_{6}C \rightarrow ^{14}_{7}N + X$
     Which particle is represented by the letter X?
     (1) an alpha particle   (3) a neutron
     (2) a beta particle   (4) a proton

11   The atom of which element in the ground state has 2 unpaired electrons
     in the $2p$ sublevel?
     (1) fluorine   (2) nitrogen   (3) beryllium   (4) carbon

12   Which atoms contain the same number of neutrons?

     (1) ^1_1H and ^3_2He (3) ^3_1H and ^3_2He

     (2) ^2_1H and ^4_2He (4) ^3_1H and ^4_2He

13   Which hydrocarbon formula is also an empirical formula?
     (1) CH_4 (2) C_2H_4 (3) C_3H_6 (4) C_4H_8

14   The potential energy possessed by a molecule is dependent upon
     (1) its composition, only
     (2) its structure, only
     (3) both its composition and its structure
     (4) neither its composition nor its structure

15   Which is a correctly balanced equation for a reaction between hydrogen
     gas and oxygen gas?

     (1) $H_2(g) + O_2(g) \rightarrow H_2O(l) +$ heat

     (2) $H_2(g) + O_2(g) \rightarrow 2H_2O(l) +$ heat

     (3) $2H_2(g) + 2O_2(g) \rightarrow H_2O(l) +$ heat

     (4) $2H_2(g) + O_2(g) \rightarrow 2H_2O(l) +$ heat

16   The atom of which element has an ionic radius smaller than its atomic
     radius?
     (1) N   (2) S   (3) Br   (4) Rb

17   Which molecule contains a polar covalent bond?

     (1) $\overset{x\,x}{\underset{x\,x}{\overset{x}{I}} \, \underset{\bullet\bullet}{\overset{\bullet\bullet}{I}}}$ (2) $H\overset{x}{\cdot}H$ (3) $H \overset{\bullet\bullet}{\underset{H}{\overset{x}{N}} \overset{x}{\cdot}} H$ (4) $\overset{\bullet\bullet}{\underset{\bullet\bullet}{N}} \overset{x}{\underset{x}{\overset{x}{N}}} \overset{x}{x}$

18   Which is the correct formula for nitrogen (I) oxide?
     (1) NO   (2) N_2O (3) NO_2 (4) N_2O_3

19   Which element in Group 15 has the strongest metallic character?
     (1) Bi   (2) As   (3) P   (4) N

20   Which halogens are gases at STP?
     (1) chlorine and fluorine   (3) iodine and fluorine
     (2) chlorine and bromine   (4) iodine and bromine

21   What is the total number of atoms represented
     in the formula $CuSO_4 \bullet 5H_2O$?

     (1) 8   (2) 13   (3) 21   (4) 27

22 When combining with nonmetallic atoms, metallic atoms generally will
    (1) lose electrons and form negative ions
    (2) lose electrons and form positive ions
    (3) gain electrons and form negative ions
    (4) gain electrons and form positive ions

23 Which set of elements contains a metalloid?
    (1) K, Mn, As, Ar   (3) Ba, Ag, Sn, Xe
    (2) Li, Mg, Ca, Kr   (4) Fr, F, O, Rn

24 Atoms of elements in a group on the Periodic Table have similar chemical
    properties. This similarity is most closely related to the atoms'
    (1) number of principal energy levels
    (2) number of valence electrons
    (3) atomic numbers
    (4) atomic masses

25 Which element in Period 2 of the Periodic Table is the most reactive
    nonmetal?
    (1) carbon   (2) nitrogen   (3) oxygen   (4) fluorine

26 What is the gram formula mass of $(NH_4)_3PO_4$?
    (1) 113 g   (2) 121 g   (3) 149 g   (4) 404 g

27 Given the reaction: $CH_4 + 2O_2 \rightarrow CO_2 + 2H_2O$ What amount of
    oxygen is needed to completely react with 1 mole of CH_4?
    (1) 2 moles   (2) 2 atoms   (3) 2 grams   (4) 2 molecules

28 Based on Reference Table E, which of the following saturated solutions
    would be the *least* concentrated?
    (1) sodium sulfate   (3) copper (II) sulfate
    (2) potassium sulfate   (4) barium sulfate

29 What is the total number of moles of H_2SO_4 needed to prepare 5.0 liters
    of a 2.0 M solution of H_2SO_4?
    (1) 2.5   (2) 5.0   (3) 10.   (4) 20.

30 Given the reaction: $Ca(s) + 2H_2O(l) \rightarrow Ca(OH)_2(aq) + H_2(g)$ When
    40.1 grams of Ca(s) reacts completely with the water, what is the total
    volume, at STP, of $H_2(g)$ produced?
    (1) 1.00 L   (2) 2.00 L   (3) 22.4 L   (4) 44.8 L

31 Which is the correct equilibrium expression for the reaction below?

$$4NH_3(g) + 7O_2(g) \leftrightarrow 4NO_2(g) + 6H_2O(g)$$

(1) $K = \dfrac{[NO_2][H_2O]}{[NH_3][O_2]}$ (3) $K = \dfrac{[NH_3][O_2]}{[NO_2][H_2O]}$

(2) $K = \dfrac{[NO_2]^4[H_2O]^6}{[NH_3]^4[O_2]^7}$ (4) $K = \dfrac{[NH_3]^4[O_2]^7}{[NO_2]^4[H_2O]^6}$

32 The potential energy diagram at the right shows the reaction

$$X + Y \leftrightarrow Z.$$

When a catalyst is added to the reaction, it will change the value of

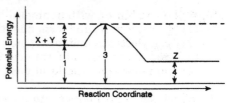

(1) 1 and 2   (3) 2 and 3
(2) 1 and 3   (4) 3 and 4

33 Which conditions will increase the rate of a chemical reaction?
(1) decreased temperature and decreased concentration of reactants
(2) decreased temperature and increased concentration of reactants
(3) increased temperature and decreased concentration of reactants
(4) increased temperature and increased concentration of reactants

34 A solution exhibiting equilibrium between the dissolved and undissolved solute must be
(1) saturated   (3) dilute
(2) unsaturated   (4) concentrated

35 Which 0.1 M solution has a pH greater than 7?
(1) $C_6H_{12}O_6$ (3) KCl
(2) CH_3COOH (4) KOH

36 What color is phenolphthalein in a basic solution?
(1) blue   (2) pink   (3) yellow   (4) colorless

37 According to Reference Table L, which of the following is the strongest Brönsted-Lowry acid?

(1) HS^- (2) H_2S (3) HNO_2 (4) HNO_3

38 When HCl(aq) is exactly neutralized by NaOH(aq), the hydrogen ion concentration in the resulting mixture is
(1) always less than the concentration of the hydroxide ions
(2) always greater than the concentration of the hydroxide ions
(3) always equal to the concentration of the hydroxide ions
(4) sometimes greater and sometimes less than the concentration of the hydroxide ions

39 If 20. milliliters of 4.0 M NaOH is exactly neutralized by 20. milliliters of HCl, the molarity of the HCl is
(1) 1.0 M   (2) 2.0 M   (3) 5.0 M   (4) 4.0 M

40 The value of the ionization constant of water, K_w, will change when there is a change in
(1) temperature   (3) hydrogen ion concentration
(2) pressure   (4) hydroxide ion concentration

41 Based on Reference Table L, which species is amphoteric?
(1) NH_2^- (2) NH_3 (3) I^- (4) HI

42 A redox reaction is a reaction in which
(1) only reduction occurs
(2) only oxidation occurs
(3) reduction and oxidation occur at the same time
(4) reduction occurs first and then oxidation occurs

43 Given the reaction: ___Mg + ___Cr $^{3+}$ → ___Mg $^{2+}$ + ___Cr   When the equation is correctly balanced using the smallest whole numbers, the sum of the coefficients will be
    (1) 10   (2) 7   (3) 5   (4) 4

44 Oxygen has an oxidation number of –2 in
    (1) O_2 (2) NO_2 (3) Na_2O_2 (4) OF_2

45 Given the statements:
    A   The salt bridge prevents electrical contact between the solutions of half-cells.
    B   The salt bridge prevents direct mixing of one half-cell solution with the other.
    C   The salt bridge allows electrons to migrate from one half-cell to the other.
    D   The salt bridge allows ions to migrate from one half-cell to the other.

    Which two statements explain the purpose of a salt bridge used as part of a chemical cell?
    (1) *A* and *C*   (2) *A* and *D*   (3) *C* and *D*   (4) *B* and *D*

46 When a substance is oxidized, it
    (1) loses protons   (3) acts as an oxidizing agent
    (2) gains protons   (4) acts as a reducing agent

47 In the reaction Cu + 2Ag $^{+}$ → Cu $^{2+}$ + 2Ag, the oxidizing agent is
    (1) Cu   (2) Cu^{2+} (3) Ag $^{+}$ (4) Ag

48 A compound that is classified as organic must contain the element
    (1) carbon   (2) nitrogen   (3) oxygen   (4) hydrogen

49 Which substance is a product of a fermentation reaction?
    (1) glucose   (2) zymase   (3) ethanol   (4) water

50 Which of the following hydrocarbons has the *lowest* normal boiling point?
    (1) ethane   (2) propane   (3) butane   (4) pentane

51 What type of reaction is $CH_3CH_3 + Cl_2 → CH_3CH_2Cl + HCl$?
    (1) an addition reaction   (3) a saponification reaction
    (2) a substitution reaction   (4) an esterification reaction

52 Which compound is a saturated hydrocarbon?
    (1) ethane   (2) ethene   (3) ethyne   (4) ethanol

Note that questions 53 through 56 have only three choices

53 As atoms of elements in group 16 are considered in order from top to bottom, the electronegativity of each successive element
    (1) decreases   (2) increases   (3) remains the same

54 As the pressure of a gas at 760 torr is changed to 380 torr at constant temperature, the volume of the gas
    (1) decreases   (2) increases   (3) remains the same

55 Given the change of phase: $CO_2(g) → CO_2(s)$
    As $CO_2(g)$ changes to $CO_2(s)$, the entropy of the system
    (1) decreases   (2) increases   (3) remains the same

56 In heterogeneous reactions, as the surface area of the reactants increases, the rate of the reaction
    (1) decreases   (2) increases   (3) remains the same

Part II

This part consists of twelve groups, each containing five questions. Each group tests a major area of the course. Choose seven of these twelve groups. Be sure that you answer all five questions in each group chosen. [35]

Group 1 – Matter and Energy
If you choose this group, be sure to answer questions 57–61.

57 What is the total number of calories of heat energy absorbed by 15 grams of water when it is heated from 30.°C to 40.°C?
    (1) 10.   (2) 15   (3) 25   (4) 150

58 The graph at the right represents the uniform cooling of a sample of a substance, starting with the substance as a gas above its boiling point.

Which segment of the curve represents a time when both the liquid and the solid phases are present?
    (1) *EF*   (2) *BC*   (3) *CD*   (4) *DE*

59 Which change of phase is exothermic?

    (1) $NaCl(s) \rightarrow NaCl(l)$ (3) $H_2O(l) \rightarrow H_2O(s)$

    (2) $CO_2(s) \rightarrow CO_2(g)$ (4) $H_2O(l) \rightarrow H_2O(g)$

60 According to the kinetic theory of gases, which assumption is correct?
    (1) Gas particles strongly attract each other.
    (2) Gas particles travel in curved paths.
    (3) The volume of gas particles prevents random motion.
    (4) Energy may be transferred between colliding particles.

61 A compound differs from a mixture in that a compound always has a
    (1) homogeneous composition
    (2) maximum of two components
    (3) minimum of three components
    (4) heterogeneous composition

Group 2 – Atomic Structure
If you choose this group, be sure to answer questions 62–66.

62 An ion with 5 protons, 6 neutrons, and a charge of 3+ has an atomic number of
    (1) 5   (2) 6   (3) 8   (4) 11

63 Electron X can change to a higher energy level or a lower energy level. Which statement is true of electron X?
    (1) Electron X emits energy when it changes to a higher energy level.
    (2) Electron X absorbs energy when it changes to a higher energy level.
    (3) Electron X absorbs energy when it changes to a lower energy level.
    (4) Electron X neither emits nor absorbs energy when it changes energy level.

64 What is the highest principal quantum number assigned to an electron in an atom of zinc in the ground state?
    (1) 1   (2) 2   (3) 5   (4) 4

65 The first ionization energy of an element is 176 kilocalories per mole of atoms. An atom of this element in the ground state has a total of how many valence electrons?
    (1) 1   (2) 2   (3) 3   (4) 4

66 What is the total number of occupied s orbitals in an atom of nickel in the ground state?
    (1) 1   (2) 2   (3) 3   (4) 4

Group 3 – Bonding
If you choose this group, be sure to answer questions 67–71.

67 What is the chemical formula for nickel (II) hypochlorite?
    (1) $NiCl_2$ (2) $Ni(ClO)_2$ (3) $NiClO_2$ (4) $Ni(ClO)_3$

68 Based on Reference Table G, which of the following compounds is most stable?
    (1) $CO(g)$ (2) $CO_2(g)$ (3) $NO(g)$ (4) $NO_2(g)$

69 The attractions that allow molecules of krypton to exist in the solid phase are due to
    (1) ionic bonds   (3) molecule-ion forces
    (2) covalent bonds   (4) van der Waals forces

70 Oxygen, nitrogen, and fluorine bond with hydrogen to form molecules. These molecules are attracted to each other by
    (1) ionic bonds   (3) electrovalent bonds
    (2) hydrogen bonds   (4) coordinate covalent bonds

71 An atom of which of the following elements has the greatest ability to attract electrons?
    (1) silicon   (2) sulfur   (3) nitrogen   (4) bromine

Group 4 – Periodic Table
If you choose this group, be sure to answer questions 72–76.

72 Which electron configuration represents the atom with the largest covalent radius?
    (1) $1s^1$ (2) $1s^2 2s^1$ (3) $1s^2 2s^2$ (4) $1s^2 2s^2 2p^1$

73 A solution of $Cu(NO_3)_2$ is colored because of the presence of the ion
    (1) Cu^{2+} (2) N^{5+} (3) O^{2-} (4) NO_3^{1-}

74 Which element is more reactive than strontium?
    (1) potassium   (2) calcium   (3) iron   (4) copper

75 At STP, which substance is the best conductor of electricity?
(1) nitrogen   (2) neon   (3) sulfur   (4) silver

76 The oxide of metal X has the formula XO. Which group in the Periodic Table contains metal X?
(1) Group 1   (2) Group 2   (3) Group 13   (4) Group 17

Group 5 – Mathematics of Chemistry
If you choose this group, be sure to answer questions 77–81.

77 Given the same condition of temperature and pressure, which noble gas will diffuse most rapidly?
(1) He   (2) Ne   (3) Ar   (4) Kr

78 What is the total number of molecules of hydrogen in 0.25 mole of hydrogen?
(1) 6.0×10^{23} (2) 4.5×10^{23} (3) 3.0×10^{23} (4) 1.5×10^{23}

79 The volume of a 1.00-mole sample of an ideal gas will decrease when the
(1) pressure decreases and the temperature decreases
(2) pressure decreases and the temperature increases
(3) pressure increases and the temperature decreases
(4) pressure increases and the temperature increases

80 A 0.100-molal aqueous solution of which compound has the *lowest* freezing point?
(1) $C_6H_{12}O_6$ (2) CH_3OH (3) $C_{12}H_{22}O_{11}$ (4) NaOH

81 What is the empirical formula of a compound that contains 85% Ag and 15% F by mass?
(1) AgF   (2) Ag_2F (3) AgF_2 (4) Ag_2F_2

Group 6 – Kinetics and Equilibrium
If you choose this group, be sure to answer questions 82–86.

82 Based on Reference Table M, which compound is less soluble in water than $PbCO_3$ at 298 K and 1 atmosphere?
(1) AgI   (2) AgCl   (3) $CaSO_4$ (4) $BaSO_4$

83 Given the equilibrium reaction at constant pressure:

$$2HBr(g) + 17.4 \text{ kcal} \leftrightarrow H_2(g) + Br_2(g)$$

When the temperature is increased, the equilibrium will shift to the
(1) right, and the concentration of HBr(g) will decrease
(2) right, and the concentration of HBr(g) will increase
(3) left, and the concentration of HBr(g) will decrease
(4) left, and the concentration of HBr(g) will increase

84 A system is said to be in the state of dynamic equilibrium when the
(1) concentration of products is greater than the concentration of reactants
(2) concentration of products is the same as the concentration of reactants
(3) rate at which products are formed is greater than the rate at which reactants are formed
(4) rate at which products are formed is the same as the rate at which reactants are formed

85 Which reaction will occur spontaneously? [Refer to Reference Table G.]

(1) $\frac{1}{2}N_2(g) + \frac{1}{2}O_2(g) \rightarrow NO(g)$ (3) $2C(s) + 3H_2(g) \rightarrow C_2H_6(g)$

(2) $\frac{1}{2}N_2(g) + O_2(g) \rightarrow NO_2(g)$ (4) $2C(s) + 2H_2(g) \rightarrow C_2H_4(g)$

86 Which potential energy diagram represents the reaction $A + B \rightarrow C +$ energy?

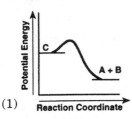

(1)

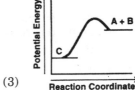

(3)

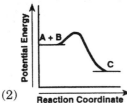

(2)

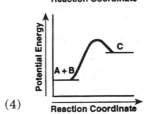

(4)

Group 7 – Acids and Bases
If you choose this group, be sure to answer questions 87–91.

87 Potassium chloride, KCl, is a salt derived from the neutralization of a
(1) weak acid and a weak base
(2) weak acid and a strong base
(3) strong acid and a weak base
(4) strong acid and a strong base

88 Given the recation: $HSO_4^- + H_2O \leftrightarrow H_3O^+ + SO_4^{2-}$
Which is a Brönsted-Lowry conjugate acid-base pair?

(1) HSO_4^- and H_3O^+ (3) H_2O and SO_4^{2-}

(2) HSO_4^- and SO_4^{2-} (4) H_2O and HSO_4^-

89 An aqueous solution that has a hydrogen ion concentration of 1.0×10^{-8} mole per liter has a pH of
(1) 6, which is basic   (3) 8, which is basic
(2) 6, which is acidic   (4) 8, which is acidic

90 The $[OH^-]$ of a solution is 1×10^{-6}. At 298 K and 1 atmosphere, the product $[H_3O^+][OH^-]$ is
(1) 1×10^{-2} (2) 1×10^{-6} (3) 1×10^{-8} (4) 1×10^{-14}

91 Given the recation: $KOH + HNO_3 \rightarrow KNO_3 + H_2O$
Which process is taking place?
(1) neutralization   (3) substitution
(2) esterification   (4) addition

Group 8 – Redox and Electrochemistry
If you choose this group, be sure to answer questions 92–96.

92 Given the unbalanced equation:

$$\_\_ MnO_2 + \_\_HCl \rightarrow \_\_MnCl_2 + \_ H_2O + \_\_Cl_2$$

When the equation is correctly balanced using smallest whole-number coefficients, the coefficient of HCl is
(1) 1   (2) 2   (3) 3   (4) 4

93 Based on Reference Table N, which half-cell has a lower electrode potential than the standard hydrogen half-cell?
(1) $Au^{3+} + 3e^- \rightarrow Au(s)$ (3) $Cu^+ + e^- \rightarrow Cu(s)$
(2) $Hg^{2+} + 2e^- \rightarrow Hg(l)$ (4) $Pb^{2+} + 2e^- \rightarrow Pb(s)$

94 According to Reference Table N, which reaction will take place spontaneously?
(1) $Ni^{2+} + Pb(s) \rightarrow Ni(s) + Pb^{2+}$
(2) $Au^{3+} + Al(s) \rightarrow Au(s) + Al^{3+}$
(3) $Sr^{2+} + Sn(s) \rightarrow Sr(s) + Sn^{2+}$
(4) $Fe^{2+} + Cu(s) \rightarrow Fe(s) + Cu^{2+}$

95 Given the recation: $Mg(s) + Zn^{2+}(aq) \rightarrow Mg^{2+}(aq) + Zn(s)$
What is the cell voltage (E^0) for the overall reaction?
(1) + 1.61 V   (3) + 3.13 V
(2) − 1.61 V   (4) − 3.13 V

96 The diagram at the right represents a chemical cell at 298 K.

When the switch is closed, electrons flow from

(1) $Al(s)$ to $Ni(s)$
(2) $Ni(s)$ to $Al(s)$
(3) $Al^{3+}(aq)$ to $Ni^{2+}(aq)$
(4) $Ni^{2+}(aq)$ to $Al^{3+}(aq)$

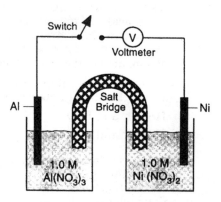

$$2Al(s) + 3Ni^{2+}(aq) \rightarrow 2Al^{3+} + 3Ni(s)$$

Group 9 – Organic Chemistry
If you choose this group, be sure to answer questions 97–101.

97 The compound C_4H_{10} belongsd to the series of hydrocarbons with the general formula
(1) C_nH_{2n} (2) C_nH_{2n+2} (3) C_nH_{2n-2} (4) C_nH_{2n-6}

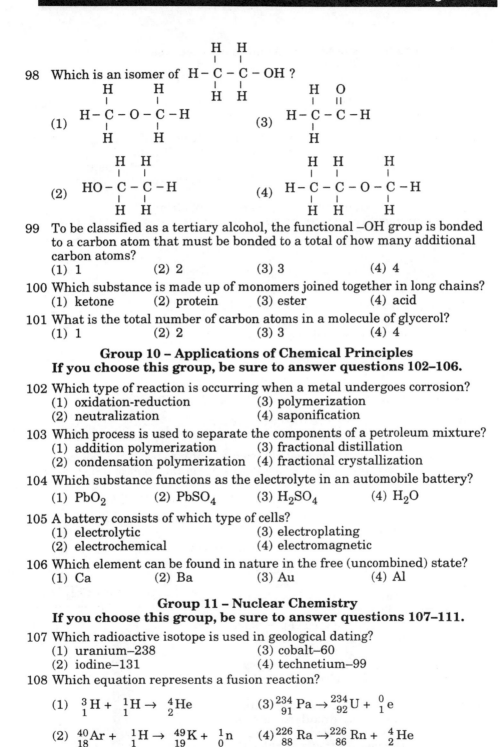

98 Which is an isomer of

$$H - \overset{\overset{\displaystyle H}{|}}{C} - \overset{\overset{\displaystyle H}{|}}{C} - OH \ ?$$

(1)
$$H - \overset{\overset{\displaystyle H}{|}}{\underset{\underset{\displaystyle H}{|}}{C}} - O - \overset{\overset{\displaystyle H}{|}}{\underset{\underset{\displaystyle H}{|}}{C}} - H$$

(3)
$$H - \overset{\overset{\displaystyle H}{|}}{\underset{\underset{\displaystyle H}{|}}{C}} - \overset{\overset{\displaystyle O}{||}}{C} - H$$

(2)
$$HO - \overset{\overset{\displaystyle H}{|}}{\underset{\underset{\displaystyle H}{|}}{C}} - \overset{\overset{\displaystyle H}{|}}{\underset{\underset{\displaystyle H}{|}}{C}} - H$$

(4)
$$H - \overset{\overset{\displaystyle H}{|}}{\underset{\underset{\displaystyle H}{|}}{C}} - \overset{\overset{\displaystyle H}{|}}{\underset{\underset{\displaystyle H}{|}}{C}} - O - \overset{\overset{\displaystyle H}{|}}{\underset{\underset{\displaystyle H}{|}}{C}} - H$$

99 To be classified as a tertiary alcohol, the functional –OH group is bonded to a carbon atom that must be bonded to a total of how many additional carbon atoms?
    (1) 1   (2) 2   (3) 3   (4) 4

100 Which substance is made up of monomers joined together in long chains?
    (1) ketone   (2) protein   (3) ester   (4) acid

101 What is the total number of carbon atoms in a molecule of glycerol?
    (1) 1   (2) 2   (3) 3   (4) 4

Group 10 – Applications of Chemical Principles
If you choose this group, be sure to answer questions 102–106.

102 Which type of reaction is occurring when a metal undergoes corrosion?
    (1) oxidation-reduction   (3) polymerization
    (2) neutralization   (4) saponification

103 Which process is used to separate the components of a petroleum mixture?
    (1) addition polymerization   (3) fractional distillation
    (2) condensation polymerization   (4) fractional crystallization

104 Which substance functions as the electrolyte in an automobile battery?
    (1) PbO_2 (2) $PbSO_4$ (3) H_2SO_4 (4) H_2O

105 A battery consists of which type of cells?
    (1) electrolytic   (3) electroplating
    (2) electrochemical   (4) electromagnetic

106 Which element can be found in nature in the free (uncombined) state?
    (1) Ca   (2) Ba   (3) Au   (4) Al

Group 11 – Nuclear Chemistry
If you choose this group, be sure to answer questions 107–111.

107 Which radioactive isotope is used in geological dating?
    (1) uranium–238   (3) cobalt–60
    (2) iodine–131   (4) technetium–99

108 Which equation represents a fusion reaction?

(1) $^{3}_{1}H + \ ^{1}_{1}H \rightarrow \ ^{4}_{2}He$

(3) $^{234}_{91}Pa \rightarrow \ ^{234}_{92}U + \ ^{0}_{1}e$

(2) $^{40}_{18}Ar + \ ^{1}_{1}H \rightarrow \ ^{49}_{19}K + \ ^{1}_{0}n$

(4) $^{226}_{88}Ra \rightarrow \ ^{226}_{86}Rn + \ ^{4}_{2}He$

109 Which substance is used as a coolant in a nuclear reactor?
   (1) neutrons   (3) hydrogen
   (2) plutonium   (4) heavy water

110 Which substance has chemical properties similar to those of radioactive ^{235}U?
   (1) ^{235}Pa (2) ^{233}Pa (3) ^{233}U (4) ^{206}Pb

111 Control rods in nuclear reactors are commonly made of boron and
    cadmium because these two elements have the ability to
   (1) absorb neutrons   (3) decrease the speed of neutrons
   (2) emit neutrons   (4) increase the speed of neutrons

Group 12 – Laboratory Activities
If you choose this group, be sure to answer questions 112–116.

Base your answers to questions 112 and
113 on the table at the right which
represents the production of 50 milliliters
of CO_2 in the reaction of HCl with
$NaHCO_3$. Five trials were performed under
different conditions as shown. (The same
mass of $NaHCO_3$ was used in each trial.)

Trial	Particle Size of NaHCO$_3$	Concentration of HCl	Temperature (°C) of HCl
A	small	1 M	20
B	large	1 M	20
C	large	1 M	40
D	small	2 M	40
E	large	2 M	40

112 Which two trials could be used to measure
    the effect of surface area?
   (1) trials A and B (3) trials A and D
   (2) trials A and C (4) trials B and D

113 Which trial would produce the fastest reaction?
   (1) trial A (2) trial B (3) trial C (4) trial D

114 A student determined the heat of fusion of water to be 88 calories per
    gram. If the accepted value is 80. calories per gram, what is the student's
    percent error?
   (1) 8.0%   (2) 10.%   (3) 11%   (4) 90.%

115 Given: (52.6 cm) (1.214 cm)   What is the product expressed to the
    correct number of significant figures?
   (1) 64 cm^2 (2) 63.9 cm^2 (3) 63.86 cm^2 (4) 63.8564 cm^2

116 The diagram at the right represents a
    metal bar and two centimeter rulers, A
    and B. Portion of the rulers have been
    enlarged to show detail.
    What is the greatest degree of precision
    to which the metal bar can be measured
    by ruler A and by ruler B?
   (1) to the nearest tenth by both rulers
   (2) to the nearest hundredth by both
       rulers
   (3) to the nearest tenth by ruler A and
       to the nearest hundredth by ruler B
   (4) to the nearest hundredth by ruler A
       and to the nearest tenth by ruler B

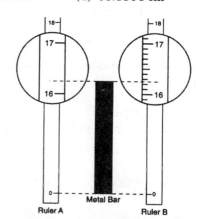

Ruler A   Ruler B

Chemistry Practice Exam 2
Part 1
Answer all 56 questions in this part.

Directions (1-56): For each statement or question, select the word or expression that, of those given, best completes the statement or answers the question.

1 Which Kelvin temperature is equal to −73°C?
(1) 100 K   (2) 173 K   (3) 200 K   (4) 346 K

2 A substance that is composed only of atoms having the same atomic number is classified as
(1) a compound
(2) an element
(3) a homogeneous mixture
(4) a heterogeneous mixture

3 At which temperature will water boil when the external pressure is 17.5 torr?
(1) 14.5°C   (2) 16.5°C   (3) 20°C   (4) 100°C

4 At which point do a liquid and a solid exist at equilibrium?
(1) sublimation point
(2) vaporization point
(3) boiling point
(4) melting point

5 When 7.00 moles of gas A and 3.00 moles of gas B are combined, the total pressure exerted by the gas mixture is 760. mmHg. What is the partial pressure exerted by gas A in this mixture?
(1) 76.0 mmHg
(2) 228 mmHg
(3) 532 mmHg
(4) 760. mmHg

6 Which radioactive emanations have a charge of 2+?
(1) alpha particles
(2) beta particles
(3) gamma rays
(4) neutrons

7 Which symbols represent atoms that are isotopes of each other?
(1) ^{14}C and ^{14}N
(2) ^{16}O and ^{18}O
(3) ^{131}I and ^{131}I
(4) ^{222}Rn and ^{222}Ra

8 Which orbital notation correctly represents the outermost principal energy level of a nitrogen atom in the ground state?

(1) s [↑↓] P [↑][↑][↑]

(2) s [↑↓] P [↑↓][↑][ ]

(3) s [↑↓] P [↑↓][↑][↑]

(4) s [↑↓] P [↑↓][↑↓][ ]

9 The atomic mass of an element is defined as the weighted average mass of that element's
(1) most abundant isotope
(2) least abundant isotope
(3) naturally occurring isotopes
(4) radioactive isotopes

10 When electrons in an atom in an excited state fall to lower energy levels, energy is
    (1) absorbed, only   (3) neither released nor absorbed
    (2) released, only   (4) both released and absorbed

11 A neutron has approximately the same mass as
    (1) an alpha particle   (3) an electron
    (2) a beta particle   (4) a proton

12 What is the formula for sodium oxalate?
    (1) $NaClO$ (2) Na_2O (3) $Na_2C_2O_4$ (4) $NaC_2H_3O_2$

13 Given the unbalanced equation: $Al + O_2 \rightarrow Al_2O_3$
When this equation is completely balanced using smallest whole numbers, what is the sum of the coefficients?
    (1) 9   (2) 7   (3) 5   (4) 4

14 One mole of which substance contains a total of 6.02×10^{23} atoms?
    (1) Li (2) NH_3 (3) O_2 (4) CO_2

15 Which formula represents a molecular substance?
    (1) CaO (2) CO (3) Li_2O (4) Al_2O_3

16 In a aqueous solution of an ionic salt, the oxygen atom of the water molecule is attracted to the
    (1) negative ion of the salt, due to oxygen's partial positive charge
    (2) negative ion of the salt, due to oxygen's partial negative charge
    (3) positive ion of the salt, due to oxygen's partial positive charge
    (4) positive ion of the salt, due to oxygen's partial negative charge

17 What is the empirical formula of the compound whose molecular formula is P_4O_{10}?
    (1) PO (2) PO_2 (3) P_2O_5 (4) P_8O_{20}

18 Which sequence of Group 18 elements demonstrates a gradual *decrease* in the strength of the van der Waals forces?
    (1) $Ar(l)$, $Kr(l)$, $Ne(l)$, $Xe(l)$ (3) $Ne(l)$, $Ar(l)$, $Kr(l)$, $Xe(l)$
    (2) $Kr(l)$, $Xe(l)$, $Ar(l)$, $Ne(l)$ (4) $Xe(l)$, $Kr(l)$, $Ar(l)$, $Ne(l)$

19 In the ground state, atoms of the elements in Group 15 of the Periodic Table all have the same number of
    (1) filled principal energy levels
    (2) occupied principal energy levels
    (3) neutrons in the nucleus
    (4) electrons in the valence shell

20 Which elements have the most similar chemical properties?
    (1) K and Na   (2) K and Cl   (3) K and Ca   (4) K and S

21 Which three groups of the Periodic Table contain the most elements classified as metalloids (semimetals)?
    (1) 1, 2, and 13   (3) 14, 15, and 16
    (2) 2, 13, and 14   (4) 16, 17, and 18

22   In which classification is an element placed if the outermost 3 sublevels of its atoms have a ground state electron configuration of $3p^63d^54s^2$?
(1) alkaline earth metals   (3) metalloids (semimetals)
(2) transition metals   (4) nonmetals

23   A diatomic element with a high first ionization energy would most likely be a
(1) nonmetal with a high electronegativity
(2) nonmetal with a low electronegativity
(3) metal with a high electronegativity
(4) metal with a low electronegativity

24   As the elements in Period 3 are considered from left to right, they tend to
(1) lose electrons more readily and increase in metallic character
(2) lose electrons more readily and increase in nonmetallic character
(3) gain electrons more readily and increase in metallic character
(4) gain electrons more readily and increase in nonmetallic character

25   An atom of an element has 28 innermost electrons and 7 outermost electrons. In which period of the Periodic Table is this element located?
(1) 5   (2) 2   (3) 3   (4) 4

26   Which solution is the most concentrated?
(1) 1 mole of solute dissolved in 1 liter of solution
(2) 2 moles of solute dissolved in 3 liters of solution
(3) 6 moles of solute dissolved in 4 liters of solution
(4) 4 moles of solute dissolved in 8 liters of solution

27   What is the gram formula mass of K_2CO_3?
(1) 138 g   (2) 106 g   (3) 99 g   (4) 67 g

28   What is the total number of atoms contained in 2.00 moles of nickel?
(1) 58.9   (2) 118   (3) 6.02×10^{23} (4) 1.20×10^{24}

29   Given the reaction at STP: $2KClO_3(s) \rightarrow 2KCl(s) + 3O_2(g)$
What is the total number of liters of $O_2(g)$ produced from the complete decomposition of 0.500 mole of $KClO_3(s)$?
(1) 11.2 L   (2) 16.8 L   (3) 44.8 L   (4) 67.2 L

30   What is the percent by mass of oxygen in magnesium oxide, MgO?
(1) 20%   (2) 40%   (3) 50%   (4) 60%

31   A solution in which the crystallizing rate of the solute equals the dissolving rate of the solute must be
(1) saturated   (3) concentrated
(2) unsaturated   (4) dilute

32   Which statement explains why the speed of some chemical
     reactions is increased when the surface area of the reactant is
     increased?
     (1) This change increases the density of the reactant particles.
     (2) This change increases the concentration of the reactant.
     (3) This change exposes more reactant particles to a possible
         collision.
     (4) This change alters the electrical conductivity of the reactant
         particles.

33   According to Reference Table G, which compound forms
     exothermically?
     (1) hydrogen fluoride   (3) ethene
     (2) hydrogen iodide   (4) ethyne

34   The potential energy diagram shown
     at the right represents the reaction
     $A + B \rightarrow AB$. Which statement
     correctly describes this reaction?
     (1) It is endothermic and energy is
         absorbed.
     (2) It is endothermic and energy is
         released.

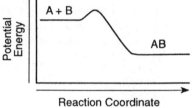

     (3) It is exothermic and energy is absorbed.
     (4) It is exothermic and energy is released.

35   Given the reaction at equilibrium: $N_2(g) + 3H_2(g) \leftrightarrow 2NH_3(g)$
     Increasing the concentration of $N_2(g)$ will increase the forward
     reaction rate due to
     (1) a decrease in the number of effective collisions
     (2) an increase in the number of effective collisions
     (3) a decrease in the activation energy
     (4) an increase in the activation energy

36   Based on Reference Table L, which of the following aqueous
     solutions is the best conductor of electricity?
     (1) 0.1 M HF   (3) 0.1 M H_2SO_4
     (2) 0.1 M H_2S (4) 0.1 M H_3PO_4

37   Which substance is classified as an Arrhenius base?
     (1) HCl   (2) NaOH   (3) $LiNO_3$ (4) $KHCO_3$

38   The conjugate acid of the HS- ion is
     (1) H^+ (2) S^{2-} (3) H_2O (4) H_2S

39   If 20. milliliters of 1.0 M HCl was used to completely neutralize
     40. milliliters of an NaOH solution, what was the molarity of the
     NaOH solution?
     (1) 0.50 M   (2) 2.0 M   (3) 1.5 M   (4) 4.0 M

40 According to Reference Table L, which species is amphoteric (amphiprotic)?
    (1) NCl   (2) HNO_2 (3) HSO_4^- (4) H_2SO_4

41 In the reaction $H_2O + CO_3^{2-} \leftrightarrow OH^- + HCO_3^-$, the two Brönsted-Lowry acids are
    (1) H_2O and OH^- (3) CO_3^{2-} and OH^-
    (2) H_2O and HCO_3^- (4) CO_3^{2-} and HCO_3^-

42 What happens to reducing agents in chemical reactions?
    (1) Reducing agents gain protons.
    (2) Reducing agents gain electrons.
    (3) Reducing agents are oxidized.
    (4) Reducing agents are reduced.

43 What is the oxidation number of carbon in $NaHCO_3$?
    (1) +6   (2) +2   (3) −4   (4) +4

44 Which statement correctly describes a redox reaction?
    (1) The oxidation half-reaction and the reduction half-reaction occur simultaneously.
    (2) The oxidation half-reaction occurs before the reduction half-reaction.
    (3) The oxidation half-reaction occurs after the reduction half-reaction.
    (4) The oxidation half-reaction occurs spontaneously but the reduction half-reaction does not.

45 Given the redox reaction: $Co(s) + PbCl_2(aq) \rightarrow CoCl_2(aq) + Pb(s)$
    Which statement correctly describes the oxidation and reduction that occur?
    (1) $Co(s)$ is oxidized and $Cl^-(aq)$ is reduced.
    (2) $Co(s)$ is oxidized and $Pb^{2+}(aq)$ is reduced.
    (3) $Co(s)$ is reduced and $Cl^-(aq)$ is oxidized.
    (4) $Co(s)$ is reduced and $Pb^{2+}(aq)$ oxidized.

46 Which half-reaction correctly represents reduction?
    (1) $Cr^{3+} + 3e^- \rightarrow Cr(s)$ (3) $Cr(s) \rightarrow Cr^{3+} + 3e^-$
    (2) $Cr^{3+} \rightarrow Cr(s) + 3e^-$ (4) $Cr(s) + 3e^- \rightarrow Cr^{3+}$

47 Which statement best describes how a salt bridge maintains electrical neutrality in the half-cells of an electrochemical cell?
    (1) It prevents the migration of electrons.
    (2) It permits the migration of ions.
    (3) It permits the two solutions to mix completely.
    (4) It prevents the reaction from occurring spontaneously.

48 What is the name of a compound that has the molecular formula C_6H_6?
    (1) butane   (2) butene   (3) benzene   (4) butyne

49   The fermentation of $C_6H_{12}O_6$ will produce CO_2 and
   (1) $C_3H_5(OH)_3$ (2) C_2H_5OH (3) $Ca(OH)_2$ (4) $Cr(OH)_3$

50   Which is the correct name for the substance at the
   right?
   (1) ethanol   (3) ethane
   (2) ethyne   (4) ethene

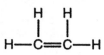

51   Which structural formula represents an organic acid?

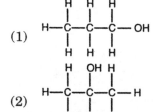

   (1)   (3)

   (2)   (4)

52   In a molecule of CH_4, the hydrogen atoms are spatially oriented
   toward the corners of a regular
   (1) pyramid   (2) tetrahedron (3) square   (4) rectangle

Note that questions 53 through 56 have only three choices.

53   Given the reaction at equilibrium: $N_2(g) + O_2(g) \leftrightarrow 2NO(g)$
   As the concentration of $N_2(g)$ increases, the concentration of $O_2(g)$
   will
   (1) decrease   (2) increase   (3) remain the same

54   As the temperature of a sample of a radioactive element decreases,
   the half-life of the element will
   (1) decrease   (2) increase   (3) remain the same

55   As ice cools from 273 K to 263 K, the average kinetic energy of its
   molecules will
   (1) decrease   (2) increase   (3) remain the same

56   As the hydrogen ion concentration of an aqueous solution increases,
   the hydroxide ion concentration of this solution will
   (1) decrease   (2) increase   (3) remain the same

Part II

This part consists of twelve groups, each containing five questions. Each group tests a major area of the course. Choose seven of these twelve groups. Be sure that you answer all five questions in each group chosen.

Group 1 - Matter and Energy
If you choose this group, be sure to answer questions 57–61.

57 The phase change represented by the equation $I_2(s) \rightarrow I_2(g)$ is called
   (1) sublimation   (3) melting
   (2) condensation   (4) boiling

58 The graph at the right represents the relationship between temperature and time as heat is added uniformly to a substance, starting when the substance is a solid below its melting point.

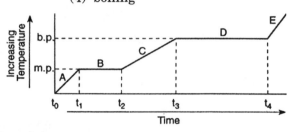

Which portions of the graph represent times when heat is absorbed and potential energy increases while kinetic remains constant?
   (1) *A* and *B*   (2) *B* and *D*   (3) *A* and *C*   (4) *C* and *D*

59 The heat of fusion is defined as the energy required at constant temperature to change 1 unit mass of a
   (1) gas to a liquid   (3) solid to a gas
   (2) gas to a solid   (4) solid to a liquid

60 Given the equation: $2Na + 2H_2O \rightarrow 2NaOH + H_2$
   Which substance in this equation is a binary compound?
   (1) Na   (2) H_2 (3) H_2O (4) NaOH

61 At STP, 1 liter of $O_2(g)$ and 1 liter of Ne(g) have the same
   (1) mass   (3) number of atoms
   (2) density   (4) number of molecules

Group 2 - Atomic Structure
If you choose this group, be sure to answer questions 62–66.

62 The diagram at the right represents radiation passing through an electric field. Which type of emanation is represented by the arrow labeled 2?
   (1) alpha particle
   (2) beta particle
   (3) positron
   (4) gamma radiation

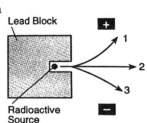

63 Which sample will decay *least* over a period of 30 days? [Refer to Reference Table *H*.]
(1) 10 g of Au-198   (3) 10 g of P-32
(2) 10 g of I-131   (4) 10 g of Rn-222

64 A particle has a mass of 1.0 atomic mass unit. What is the approximate mass of this particle in grams?
(1) 1.0 g   (3) 1.7 x 10^{-24} g
(2) 2.0 g   (4) 6.0 x 10^{-23} g

65 Which equation represents nuclear disintegration resulting in release of a beta particle?

(1) $^{220}_{87}Fr + ^{4}_{2}He \rightarrow ^{224}_{89}Ac$ (3) $^{32}_{15}P + ^{0}_{-1}e \rightarrow ^{32}_{14}Si$

(2) $^{239}_{94}Pu \rightarrow ^{235}_{92}U + ^{4}_{2}He$ (4) $^{198}_{79}Au \rightarrow ^{198}_{80}Hg + ^{0}_{-1}e$

66 Which electron configuration represents a potassium atom in the excited state?
(1) $1s^22s^22p^63s^23p^3$ (3) $1s^22s^22p^63s^23p^64s^1$
(2) $1s^22s^22p^63s^13p^4$ (4) $1s^22s^22p^63s^23p^54s^2$

Group 3 – Bonding
If you choose this group, be sure to answer questions 67–71.

67 Which type of attraction is directly involved when KCl dissolves in water?
(1) molecule–molecule   (3) molecule–ion
(2) molecule–atom   (4) ion–ion

68 In which compound have electrons been transferred to the oxygen atom?
(1) CO_2 (2) NO_2 (3) N_2O (4) Na_2O

69 A strontium atom differs from a strontium ion in that the atom has a greater
(1) number of electrons   (3) atomic number
(2) number of protons   (4) mass number

70 Which substance is an example of a network solid?
(1) nitrogen dioxide   (3) carbon dioxide
(2) sulfur dioxide   (4) silicon dioxide

71 Which combination of atoms can form a polar covalent bond?
(1) H and H   (2) H and Br   (3) N and N   (4) Na and Br

Group 4 – Periodic Table
If you choose this group, be sure to answer questions 72–76.

72 Which element has the highest first ionization energy?
(1) sodium   (2) aluminum   (3) calcium   (4) phosphorus

73 Which compound forms a colored aqueous solution?
    (1) $CaCl_2$ (2) $CrCl_3$ (3) NaOH   (4) KBr

74 When a metal atom combines with a nonmetal atom, the nonmetal atom will
    (1) lose electrons and decrease in size
    (2) lose electrons and increase in size
    (3) gain electrons and decrease in size
    (4) gain electrons and increase in size

75 According to Reference Table *P*, which of the following elements has the smallest covalent radius?
    (1) nickel   (2) cobalt   (3) calcium   (4) potassium

76 Which element's ionic radius is smaller than its atomic radius?
    (1) neon   (2) nitrogen   (3) sodium   (4) sulfur

Group 5 – Mathematics of Chemistry
If you choose this group, be sure to answer questions 77–81.

77 What is the total number of moles of hydrogen gas contained in 9.03×10^{23} molecules?
    (1) 1.50 moles   (2) 2.00 moles   (3) 6.02 moles   (4) 9.03 moles

78 At the same temperature and pressure, which gas will diffuse through air at the fastest rate?
    (1) H_2 (2) O_2 (3) CO   (4) CO_2

79 How are the boiling and freezing points of a sample of water affected when a salt is dissolved in the water?
    (1) The boiling point decreases and the freezing point decreases.
    (2) The boiling point decreases and the freezing point increases.
    (3) The boiling point increases and the freezing point decreases.
    (4) The boiling point increases and the freezing point increases.

80 A sample of an unknown gas at STP has a density of 0.630 gram per liter. What is the gram molecular mass of this gas?
    (1) 2.81 g   (2) 14.1 g   (3) 22.4 g   (4) 63.0 g

81 A compound is 86% carbon and 14% hydrogen by mass. What is the empirical formula for this compound?
    (1) CH   (2) CH_2 (3) CH_3 (4) CH_4

Group 6 – Kinetics and Equilibrium
If you choose this group, be sure to answer questions 82–86.

82 In a chemical reaction, a catalyst changes the
    (1) potential energy of the products
    (2) potential energy of the reactants
    (3) heat of reaction
    (4) activation energy

83 Which statement describes characteristics of an endothermic reaction?
    (1) The sign of ΔH is positive, and the products have less potential energy than the reactants.
    (2) The sign of ΔH is positive, and the products have more potential energy than the reactants.
    (3) The sign of ΔH is negative, and the products have less potential energy than the reactants.
    (4) The sign of ΔH is negative, and the products have more potential energy than the reactants.

84 What is the K_{sp} expression for the salt PbI_2?
    (1) $[Pb^{2+}][I-]^2$ (2) $[Pb^{2+}][2I-]$ (3) $[Pb^{2+}][I_2]^2$ (4) $[Pb^{2+}][2I-]^2$

85 Given the equilibrium system: $PbCO_3(s) \leftrightarrow Pb^{2+}(aq) + CO_3{}^{2-}(aq)$
    Which changes occur as $Pb(NO_3)_2(s)$ is added to the system at equilibrium?
    (1) The amount of $PbCO_3(s)$ decreases, and the concentration of $CO_3{}^{2-}(aq)$ decreases.
    (2) The amount of $PbCO_3(s)$ decreases, and the concentration of $CO_3{}^{2-}(aq)$ increases.
    (3) The amount of $PbCO_3(s)$ increases, and the concentration of $CO_3{}^{2-}(aq)$ decreases.
    (1) The amount of $PbCO_3(s)$ increases, and the concentration of $CO_3{}^{2-}(aq)$ increases.

86 A chemical reaction will always occur spontaneously if the reaction has a negative
    (1) ΔG (2) ΔH (3) ΔS (4) T

Group 7 - Acids and Bases
If you choose this group, be sure to answer questions 87-91.

87 An acidic solution could have a pH of
    (1) 7   (2) 10   (3) 3   (4) 14

88 What is the pH of a 0.00001 molar HCl solution?
    (1) 1   (2) 9   (3) 5   (4) 4

89 According to the Brönsted-Lowry theory, an acid is any species that can
    (1) donate a proton   (3) accept a proton
    (2) donate an electron   (4) accept an electron

90 When the salt Na_2CO_3 undergoes hydrolysis, the resulting solution will be
    (1) acidic with a pH less than 7
    (2) acidic with a pH greater than 7
    (3) basic with a pH less than 7
    (4) basic with a pH greater than 7

91 In an aqueous solution, which substance yields hydrogen ions as the only positive ions?
(1) C_2H_5OH (2) CH_3COOH (3) KH   (4) KOH

Group 8 – Redox and Electrochemistry
If you choose this group, be sure to answer questions 92–96.

92 In which kind of cell are the redox reactions made to occur by an externally applied electrical current?
(1) galvanic cell   (3) electrochemical cell
(2) chemical cell   (4) electrolytic cell

93 According to Reference Table N, which metal will react spontaneously with Ag^+ ions, but not with Zn^{2+} ions?
(1) Cu   (2) Au   (3) Al   (4) Mg

94 Which atom forms an ion that would migrate toward the cathode in an electrolytic cell?
(1) F   (2) I   (3) Na   (4) Cl

95 Given the equations A, B, C, and D:
$$(A) \; AgNO_3 + NaCl \rightarrow AgCl + NaNO_3$$
$$(B) \; Cl_2 + H_2O \rightarrow HClO + HCl$$
$$(C) \; CuO + CO \rightarrow CO_2 + Cu$$
$$(D) \; NaOH + HCl \rightarrow NaCl + H_2O$$
Which two equations represent redox reactions?
(1) A and B (2) B and C (3) C and A (4) D and B

96 Given the unbalanced equation:
$$\_NO_3^- + 4H^+ \_Pb \rightarrow \_Pb^{2+} + \_NO_2 + 2H_2O$$
What is the coefficient of NO_2 when the equation is correctly balanced?
(1) 1   (2) 2   (3) 3   (4) 4

Group 9 – Organic Chemistry
If you choose this group, be sure to answer questions 97–101.

97 Which polymers occur naturally?
(1) starch and nylon   (3) protein and nylon
(2) starch and cellulose   (4) protein and plastic

98 Which statement explains why the element carbon forms so many compounds?
(1) Carbon atoms combine readily with oxygen.
(2) Carbon atoms have very high electronegativity.
(3) Carbon readily forms ionic bonds with other carbon atoms.
(4) Carbon readily forms covalent bonds with other carbon atoms.

99 Which structural formula represents a primary alcohol?

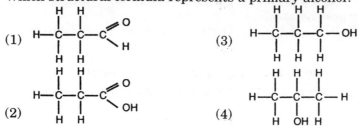

100 Which compounds are isomers?
(1) 1-propanol and 2-propanol
(2) methanoic acid and ethanoic acid
(3) methanol and methanal
(4) ethane and ethanol

101 Compared to the rate of inorganic reactions, the rate of organic reactions generally is
(1) slower because organic particles are ions
(2) slower because organic particles contain covalent bonds
(3) faster because organic particles are ions
(4) faster because organic particles contain covalent bonds

Group 10 – Applications of Chemical Principles
If you choose this group, be sure to answer questions 102–106.

102 Which products are obtained from the fractional distillation of petroleum?
(1) esters and acids   (3) soaps and starches
(2) alcohols and aldehydes   (4) kerosene and gasoline

103 Given the lead-acid battery reaction:
$$Pb + PbO_2 + 2H_2SO_4 \underset{\text{Charge}}{\overset{\text{Discharge}}{\rightleftharpoons}} 2\,PbSO_4 + 2H_2O$$
Which species is oxidized during battery discharge?
(1) Pb   (2) PbO_2 (3) $SO_4{}^{2-}$ (4) H_2O

104 Given the reaction: $ZnO + X + heat \rightarrow Zn + XO$
Which element, represented by X, is used industrially to reduce the ZnO to Zn?
(1) Cu   (2) C   (3) Sn   (4) Pb

105 Which metal is obtained commercially by the electrolysis of its salt?
(1) Zn   (2) K   (3) Fe   (4) Ag

106 The corrosion of aluminum (Al) is a less serious problem than the corrosion of iron (Fe) because
(1) Al does not oxidize
(2) Fe does not oxidize
(3) Al oxidizes to form a protective layer
(4) Fe oxidizes to form a protective layer

Group 11 – Nuclear Chemistry
If you choose this group, be sure to answer questions 107–111.

107 Fissionable uranium-233, uranium-235, and plutonium-239 are used in a nuclear reactor as
    (1) coolants   (3) moderators
    (2) control rods   (4) fuels

108 Which reaction illustrates fusion?

    (1) $^{2}_{1}H + ^{2}_{1}H \rightarrow ^{4}_{2}He$ (3) $^{27}_{13}Al + ^{4}_{2}He \rightarrow ^{30}_{15}P + ^{1}_{0}n$

    (2) $^{1}_{0}n + ^{27}_{13}Al \rightarrow ^{24}_{11}Na + ^{4}_{2}He$ (4) $^{14}_{7}N + ^{4}_{2}He \rightarrow ^{1}_{1}H + ^{17}_{8}O$

109 An accelerator can *not* be used to speed up
    (1) alpha particles   (3) protons
    (2) beta particles   (4) neutrons

110 Brain tumors can be located by using an isotope of
    (1) carbon-14   (3) technetium-99
    (2) iodine-131   (4) uranium-238

111 In the reaction $^{9}_{4}Be + X \rightarrow ^{12}_{6}C + ^{1}_{0}n$, the X represents

    (1) an alpha particle   (3) an electron
    (2) a beta particle   (4) a proton

Group 12 – Laboratory Activities
If you choose this group, be sure to answer questions 112–116.

112 Which piece of laboratory equipment should be used to remove a heated crucible from a ring stand?

    (1)    (3)

    (2)    (4)

113 The following set of procedures was used by a student to determine the heat of solution of NaOH.
    (A) Read the original temperature of the water.
    (B) Read the final temperature of the solution.
    (C) Pour the water into a beaker.
    (D) Stir the mixture.
    (E) Add the sodium hydroxide.
What is the correct order of procedures for making this determination?

    (1) $A \rightarrow C \rightarrow E \rightarrow B \rightarrow D$ (3) $C \rightarrow A \rightarrow E \rightarrow D \rightarrow B$

    (2) $E \rightarrow D \rightarrow C \rightarrow A \rightarrow B$ (4) $C \rightarrow E \rightarrow D \rightarrow A \rightarrow B$

114 In an experiment, a student found 18.6% by mass of water in a sample of $BaCl_2 \cdot 2H_2O$. The accepted value is 14.8%. What was the student's experimental percent error?

(1) $\dfrac{3.8}{18.6}$ x 100   (2) $\dfrac{3.8}{14.8}$ x 100   (3) $\dfrac{14.8}{18.6}$ x 100   (4) $\dfrac{18.6}{14.8}$ x 100

115 A student obtained the following data in a chemistry laboratory. Based on Reference Table *D*, which of the trials seems to be in error?

(1) 1   (3) 3
(2) 2   (4) 4

Trial	Temperature (°C)	Solubility (grams of KNO_3/100 g of H_2O)
1	25	40
2	32	50
3	43	70
4	48	60

116 A student using a Styrofoam cup as a calorimeter added a piece of metal to distilled water and stirred the mixture as shown in the diagram. The student data is shown in the table.

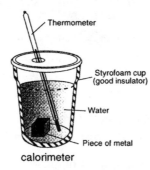

Thermometer

Styrofoam cup (good insulator)

Water

Piece of metal

calorimeter

DATA TABLE

Mass of H_2O ....50.0 g
Initial temperature of H_2O ....25.0°C
Mass of metal.......................................20.0 g
Initial temperature of metal.............100.°C
Final temperature of H_2O + metal.....32.0°C

Which statement correctly describes the heat flow in calories? [Ignore heat gained or lost by the calorimeter.]

(1) The water lost 1360 calories of heat and the metal gained 140. calories of heat.

(2) The water lost 350. calories of heat and the metal gained 350. calories of heat.

(3) The water gained 1360 calories of heat and the metal lost 140. calories of heat.

(4) The water gained 350. calories of heat and the metal lost 350. calories of heat.

698   **Chemistry Practice Examination 3**
Part 1
Answer all 56 questions in this part.
Directions (1-56): For *each* statement or question, select the word or expression that, of those given, best completes the statement or answers the question.

Page 297

1   The diagrams at the right represent two solids and the temperature of each. What occurs when the two solids are placed in contact with each other?

Solid A   Solid B

Temperature 50°C   Temperature 80°C

   (1) Heat energy flows from solid A to solid B. Solid A decreases in temperature.
   (2) Heat energy flows from solid A to solid B. Solid A increases in temperature.
   (3) Heat energy flows from solid B to solid A. Solid B decreases in temperature.
   (4) Heat energy flows from solid B to solid A. Solid B increases in temperature.

2   The particles of a substance are arranged in a definite geometric pattern and are constantly vibrating. This substance an be in
   (1) the solid phase, only   (3) either the liquid or the solid phase
   (2) the liquid phase, only   (4) neither the liquid nor the solid phase

3   What is the pressure of a mixture of CO_2, SO_2, and H_2O gases, if each gas has a partial pressure of 250 torr?
   (1) 250 torr   (2) 500 torr   (3) 750 torr   (4) 1000 torr

4   Which substances can be decomposed chemically?
   (1) CaO and Ca   (3) CO and Co
   (2) MgO and Mg   (4) CaO and MgO

5   A gas sample has a volume of 25.0 milliliters at a pressure of 1.00 atmosphere. If the volume increases to 50.0 milliliters and the temperature remains constant, the new pressure will be
   (1) 1.00 atm   (2) 2.00 atm   (3) 0.250 atm   (4) 0.500 atm

6   An atom with the electron configuration $1s^22s^22p^63s^213p^613d^54s^2$ has an incomplete
   (1) $2p$ sublevel   (3) third principal energy level
   (2) second principal energy level (4) $4s$ sublevel

7   Which orbital notation represents a boron atom in the ground state?

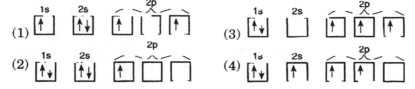

8   In the equation $^{234}_{90}\text{Th} \rightarrow {}^{234}_{91}\text{Pa} + X$, the symbol X represents

(1) $^{0}_{+1}\text{e}$ (2) $^{0}_{-1}\text{e}$ (3) $^{1}_{0}\text{n}$ (4) $^{1}_{1}\text{H}$

9   Which subatomic particle is found in the nucleus of all isotopes of hydrogen?
(1) proton   (2) neutron   (3) electron   (4) positron

10   What is the highest principal quantum number (n) for an electron in an atom of sulfur in the ground state?
(1) 1   (2) 2   (3) 3   (4) 4

11   What is the total number of electrons in a completely filled fourth principal energy level?
(1) 8   (2) 10   (3) 16   (4) 32

12   What is the total number of hydrogen atoms required to form 1 molecule of $C_3H_5(OH)_3$?
(1) 1   (2) 5   (3) 3   (4) 8

13   Which element is found in both potassium chlorate and zinc nitrate?
(1) Hydrogen   (2) oxygen   (3) potassium   (4) zinc

14   Which formula represents lead (II) phosphate?
(1) $PbPO_4$ (2) Pb_4PO_4 (3) $Pb_3(PO_4)_2$ (4) $Pb_2(PO_4)_3$

15   Atoms of which element have the *weakest* attraction for electrons?
(1) Na   (2) P   (3) Si   (4) S

16   The ability to conduct electricity in the solid state is a characteristic of metallic bonding. This characteristic is best explained by the presence of
(1) high ionization energies   (3) mobile electrons
(2) high electronegativities   (4) mobile protons

17   When ionic bonds are formed, metallic atoms tend to
(1) lose electrons and become negative ions
(2) lose electrons and become positive ions
(3) gain electrons and become negative ions
(4) gain electrons and become positive ions

18   The bond between hydrogen and oxygen in a water molecule is classified as
(1) ionic and nonpolar   (3) covalent and nonpolar
(2) ionic and polar   (4) covalent and polar

19   According to the Periodic Table, which element has more than one positive oxidation state?
(1) cadmium   (2) iron   (3) silver   (4) zinc

20   Which group contains a liquid that is a nonmetal at STP?
(1) 14   (2) 15   (3) 16   (4) 17

21   Which of these Group 14 elements has the most metallic properties
(1) C   (2) Ge   (3) Si   (4) Sn

22 As the elements in Group 2 are considered in order of increasing atomic number, the atomic radius of each successive element increases. This increase is primarily due to an increase in the number of
(1) occupied principal energy levels   (3) neutrons in the nucleus
(2) electrons in the outermost shell   (4) unpaired electrons

23 Which element is classified as a metalloid (semi-metal)?
(1) sulfur   (2) silicon   (3) barium   (4) bromine

24 Which element in Group 1 has the greatest tendency to lose an electron?
(1) cesium   (2) rubidium   (3) potassium   (4) sodium

25 The table at the right shows some properties of elements *A*, *B*, *C*, and *D*.
Which element is most likely a nonmetal?
(1) *A*   (3) *C*
(2) *B*   (4) *D*

Element	Ionization Energy	Electro-negativity	Conductivity of Heat and Electricity
A	low	low	low
B	low	low	high
C	high	high	low
D	high	high	high

26 What is the gram formula mass of $Ca_3(PO_4)_2$?
(1) 135 g/mol   (2) 215 g/mol   (3) 278 g/mol   (4) 310. g/mol

27 The gram atomic mass of oxygen is 16.0 grams per mole. How many atoms of oxygen does this mass represent?
(1) 16.0   (2) 32.0   (3) 6.02×10^{23} (4) $2(6.02 \times 10^{23})$

28 Given the *unbalanced* equation: $N_2(g) + H_2(g) \rightarrow NH_3(g)$
When the equation is balanced using smallest whole-number coefficients, the ratio of moles of hydrogen consumed to moles of ammonia produced is
(1) 1:3   (2) 2:3   (3) 3:1   (4) 3:2

29 What is the concentration of a solution of 10. moles of copper (II) nitrate in 5.0 liters of solution?
(1) 0.50 M   (2) 2.0 M   (3) 5.0 M   (4) 10. M

30 Given the balanced equation: $Mg(s) + 2HCl(aq) \rightarrow MgCl_2(aq) + H_2(g)$
At STP, what is the total number of liters of hydrogen gas produced when 3.00 moles of hydrochloric acid solution is completely consumed?
(1) 11.2 L   (2) 22.4 L   (3) 33.6 L   (4) 44.8 L

31 According to Reference Table *D*, which compound's solubility decreases most rapidly when the temperature increases from 50°C to 70°C?
(1) NH_3 (2) HCl   (3) SO_2 (4) KNO_3

32 Given the reaction at equilibrium: $2CO(g) + O_2(g) \leftrightarrow 2CO_2(g)$
When the reaction is subjected to stress, a change will occur in the concentration of
(1) reactants, only   (3) both reactants and products
(2) products, only   (4) neither reactants nor products

33   An increase in the temperature of a system at equilibrium favors the
     (1) endothermic reaction and decreases its rate
     (2) endothermic reaction and increases its rate
     (3) exothermic reaction and decreases its rate
     (4) exothermic reaction and increases its rate

34   Based on Reference Table *L*, which compound, when in aqueous
     solution, is the best conductor of electricity?
     (1) HF   (2) H_2S (3) H_2O (4) H_2SO_4

35   A compound that can act as an acid or a base is referred to as
     (1) a neutral substance   (3) a monomer
     (2) an amphoteric substance   (4) an isomer

Base your answers to questions 36 and 37
on the potential energy diagram of a chemical reaction shown at the right.

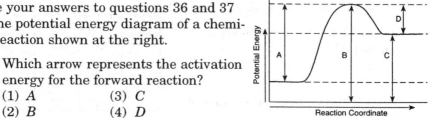

36   Which arrow represents the activation
     energy for the forward reaction?
     (1) *A*   (3) *C*
     (2) *B*   (4) *D*

37   The forward reaction is best described as an
     (1) exothermic reaction in which energy is released
     (2) exothermic reaction in which energy is absorbed
     (3) endothermic reaction in which energy is released
     (4) endothermic reaction in which energy is absorbed

38   In the reaction $HNO_3 + H_2O \leftrightarrow H_3O^+ + NO_3^-$, the two Brönsted acids are
     (1) H_2O and HNO_3 (3) H_2O and H_3O^+
     (2) H_2O and NO_3^- (4) HNO_3 and H_3O^+

39   Which substance can be classified as an Arrhenius acid?
     (1) HCl   (2) NaCl   (3) LiOH   (4) KOH

40   How many milliliters of 0.20 M HCl are needed to exactly
     neutralize 40. milliliters of 0.40 M KOH?
     (1) 20. mL   (2) 40. mL   (3) 80. mL   (4) 160 mL

41   Which 0.1 M solution will turn phenolphthalein pink?
     (1) HBr(aq)   (2) CO_2(aq)   (3) LiOH(aq)   (4) CH_2OH(aq)

42   Which compound is an electrolyte?
     (1) CH_3OH (2) CH_3COOH (3) $C_3H_5(OH)_3$ (4) $C_{12}H_{22}O_{11}$

43   What is the oxidation number of chlorine in $HClO_4$?
     (1) +1   (2) +5   (3) +3   (4) +7

44   Given the redox reaction: Fe^{2+}(aq) + Zn(s) → Zn^{2+}(aq) + Fe(s)
     Which species acts as a reducing agent?
     (1) Fe(*s*)   (2) Fe^{2+}(aq)   (3) Zn(*s*)   (4) Zn^{2+}(aq)

45 Given the redox reaction: $2I^-(aq) + Br_2(l) \rightarrow 2Br^-(aq) + I_2(s)$
    What occurs during this reaction?
    (1) The I^- ion is oxidized, and its oxidation number increases.
    (2) The I^- ion is oxidized, and its oxidation number decreases.
    (3) The I^- ion is reduced, and its oxidation number increases.
    (4) The I^- ion is reduced, and its oxidation number decreases.

46 Given the reaction: $Zn(s) + 2HCl(aq) \rightarrow ZnCl_2(aq) + H_2(g)$
    Which equation represents the correct oxidation half-reaction?
    (1) $Zn(s) \rightarrow Zn^{2+} + 2e^-$ (3) $Zn^{2+} + 2e \rightarrow Zn(s)$
    (2) $2H^+ + 2e^- \rightarrow H_2(g)$ (4) $2Cl^- \rightarrow Cl_2(g) + 2e^-$

47 According to Reference Table N, which redox reaction occurs
    spontaneously?
    (1) $Cu(s) + 2H^+ \rightarrow Cu^{2+} + H_2(g)$ (3) $2Ag(s) + 2H^+ \rightarrow 2Ag^+ + H_2(g)$
    (2) $Mg(s) + 2H^+ \rightarrow Mg^{2+} + H_2(g)$ (4) $Hg(l) + 2H^+ \rightarrow Hg^{2+} + H_2(g)$

48 Which quantities are conserved in all oxidation-reduction reactions?
    (1) charge, only   (3) both charge and mass
    (2) mass, only   (4) neither charge nor mass

49 The reaction $CH_2CH_2 + H_2 \rightarrow CH_3CH_3$ is an example of
    (1) substitution   (3) esterification
    (2) addition   (4) fermentation

50 Given the compound (at the right):
    Which structural formula represents
    an isomer?

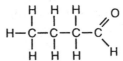

    (1)

    (2)

    (3)

    (4)

51 In which pair of hydrocarbons does each compound contain only one
    double bond per molecule?
    (1) C_2H_2 and C_2H_6 (3) C_4H_8 and C_2H_4
    (2) C_2H_2 and C_3H_6 (4) C_6H_6 and C_7H_8

52 Which organic compound is classified as an acid?
    (1) CH_3CH_2COOH (3) $C_{12}CH_{22}O_{11}$
    (2) CH_3CH_2OH (4) $C_6H_{12}O_6$

53 The products of the fermentation of a sugar are ethanol and
    (1) water   (3) carbon dioxide
    (2) oxygen   (4) sulfur dioxide

Note that questions 54 through 56 have only three choices.

54 As the temperature of $H_2O(l)$ in a closed system decreases, the vapor pressure of the $H_2O(l)$

(1) decreases   (2) increases   (3) remains the same

55 As the number of neutrons in the nucleus of a given atom of an element increases, the atomic number of that element

(1) decreases   (2) increases   (3) remains the same

56 Given the closed system at equilibrium: $CO_2(g) \leftrightarrow CO_2(aq)$
As the pressure on the system increases, the solubility of the $CO_2(g)$

(1) decreases   (2) increases   (3) remains the same

Part II

This part consists of twelve groups, each containing five questions. Each group tests a major area of the course. Choose seven of these twelve groups. Be sure that you answer all five questions in each group chosen.

Group 1 – Matter and Energy

If you choose this group, be sure to answer questions 57–61.

57 The table at the right shows the temperature, pressure, and volume of five samples. Which sample contains the same number of molecules as sample *A*?

Sample	Substance	Temperature (K)	Pressure (atm)	Volume (L)
A	He	273	1	22.4
B	O_2	273	1	22.4
C	Ne	273	2	22.4
D	N_2	546	2	44.8
E	Ar	546	2	44.8

(1) *E*   (2) *B*   (3) *C*   (4) *D*

58 The energy absorbed when ammonium chloride dissolves in water can be measured in

(1) degrees   (3) moles per liter
(2) kilocalories   (4) liters per mole

59 At 1 atmosphere of pressure, the steam-water equilibrium occurs at a temperature of

(1) 0 K   (2) 100 K   (3) 273 K   (4) 373 K

60 The graph at the right represents the uniform cooling of a substance, starting with the substance as a gas above its boiling point. During which interval is the substance completely in the liquid phase?

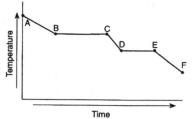

(1) *AB*   (3) *CD*
(2) *BC*   (4) *DE*

61 Which two compounds readily sublime at room temperature (25°C)?

(1) $CO_2(s)$ and $I_2(s)$ (3) $NaCl(s)$ and $I_2(s)$
(2) $CO_2(s)$ and $C_6H_{12}O_6(s)$ (4) $NaCl(s)$ and $C_6H_{12}O_6(s)$

Group 2 - Atomic Structure
If you choose this group, be sure to answer questions 62–66.

62 Which electron configuration is possible for a nitrogen atom in the excited state?
(1) $1s^2 2s^2 2p^3$ (2) $1s^2 2s^2 2p^2 3s^1$ (3) $1s^2 2s^2 2p^4$ (4) $1s^2 2s^2 2p^2$

63 What is the total amount of energy required to remove the most loosely bound electron from each atom in a mole of gaseous Ca?
(1) 100 kcal/mol (2) 119 kcal/mol (3) 141 kcal/mol (4) 176 kcal/mol

64 What is the total number of unpaired electrons in an atom of nickel in the ground state?
(1) 0   (2) 2   (3) 3   (4) 4

65 The characteristic bright-line spectrum of an element is produced when its electrons
(1) form a covalent bond   (3) move to a higher energy state
(2) form an ionic bond   (4) return to a lower energy state

66 Which emanation has *no* mass and *no* charge?
(1) alpha   (2) beta   (3) gamma   (4) neutron

Group 3 - Bonding
If you choose this group, be sure to answer questions 67–71.

67 Given the incomplete equation: $2N_2O_5(g) \rightarrow$ Which set of products completes and balances the incomplete equation?
(1) $2N_2(g) + 3H_2(g)$ (3) $4NO_2(g) + O_2(g)$
(2) $2N_2(g) + 2O_2(g)$ (4) $4NO(g) + 5O_2(g)$

68 Which structural formula represents a nonpolar molecule?

(1) H Cl   (2) H O / H   (3) H—H   (4) H—N—H / H

69 Compared to the boiling point of H_2S, the boiling point of H_2O is relatively high. Which type of bonding causes this difference?
(1) covalent   (2) hydrogen   (3) ionic   (4) network

70 In which system do molecule-ion attractions exist?
(1) NaCl(aq)   (2) NaCl(s)   (3) $C_6H_{12}O_6$(aq) (4) $C_6H_{12}O_6$(s)

71 An example of an empirical formula is
(1) C_4H_{10} (2) $C_6H_{12}O_6$ (3) $HC_2H_3O_2$ (4) CH_2O

Group 4 - Periodic Table
If you choose this group, be sure to answer questions 72–76.

72 the elements from which two groups of the Periodic Table are most similar in their chemical properties?
(1) 1 and 2   (2) 1 and 17   (3) 2 and 17   (4) 17 and 18

73 Which metal is most likely obtained by the electrolysis of its fused salt?
     (1) Au   (2) Ag   (3) Li   (4) An

74 Which aqueous solution is colored?
     (1) $CuSO_4$(aq)   (2) $BaCl_2$(aq)   (3) KCl(aq)   (4) $MgSO_4$(aq)

75 Because of its high reactivity, which element is *never* found free in nature?
     (1) O   (2) F   (3) N   (4) Ne

76 Which Group 18 element is most likely to form a compound with the element fluorine?
     (1) He   (2) Ne   (3) Ar   (4) Kr

Group 5 – Mathematics of Chemistry
If you choose this group, be sure to answer questions 77–81.

77 At STP, which gas will diffuse more rapidly than Ne?
     (1) He   (2) Ar   (3) Kr   (4) Xe

78 The heat of fusion of a compound is 30.0 calories per gram. What is the total number of calories of heat that must be absorbed by a 15.0-gram sample to change the compound from solid to liquid at its melting point?
     (1) 15.0 cal   (2) 45.0 cal   (3) 150. cal   (4) 450. cal

79 Which gas has a density of 1.70 grams per liter at STP?
     (1) F_2(g)   (2) He(g)   (3) N_2(g)   (4) SO_2(g)

80 Given the reaction: $2C_2H_2(g) + 5O_2(g) \rightarrow 4CO_2(g) + 2H_2O(g)$
     What is the total number of grams of O_2(g) needed to react completely with 0.50 mole of C_2H_2(g)?
     (1) 10.g   (2) 40. g   (3) 80. g   (4) 160 g

81 Which statement describes KCl(aq)?
     (1) KCl is the solute in a homogeneous mixture.
     (2) KCl is the solute in a heterogeneous mixture.
     (3) KCl is the solvent in a homogeneous mixture.
     (4) KCl is the solvent in a heterogeneous mixture.

Group 6 – Kinetics and Equilibrium
If you choose this group, be sure to answer questions 82–86.

82 According to Reference Table *G*, which compound will for spontaneously from its elements?
     (1) ethene   (3) nitrogen (II) oxide
     (2) hydrogen iodide   (4) magnesium oxide

83 Given the equilibrium reaction: $AgCl(s) \leftrightarrow Ag^+(aq) + Cl^-(aq)$
     At 25°C, the K_{sp} is equal to
     (1) 6.0×10^{-23} (2) 1.8×10^{-10} (3) 1.0×10^{-7} (4) 9.6×10^{-4}

84   Given the reaction: $2N_2(g) + O_2(g) \leftrightarrow 2N_2O(g)$ Which statement is true when this closed system reaches equilibrium?
(1) All of the $N_2(g)$ has been consumed.
(2) All of the $O_2(g)$ has been consumed.
(3) Pressure changes no longer occur.
(4) The forward reaction no longer occurs.

85   Which equation is used to determine the free energy change during a chemical reaction?
(1) $\Delta G = \Delta H - \Delta S$ (3) $\Delta G = \Delta H - T\Delta S$
(2) $\Delta G = \Delta H + \Delta S$ (4) $\Delta G = \Delta H + T\Delta S$

86   Which is the correct equilibrium expression for the reaction $2A(g) + 3B(g) \leftrightarrow C(g) + 3D(g)$?

(1) $K = \dfrac{[2A] + [3B]}{[C] + [3D]}$ (3) $K = \dfrac{[A]^2\,[B]^3}{[C]\,[D]^3}$

(2) $K = \dfrac{[C] + [3D]}{[2A] + [3B]}$ (4) $K = \dfrac{[C]\,[D]^3}{[A]^2\,[B]^3}$

Group 7 - Acids and Bases
If you choose this group, be sure to answer questions 87–91.

87   According to Reference Table *L*, what is the conjugate acid of the hydroxide ion (OH-)?
(1) O^{2-} (2) H^+ (3) H_2O (4) H_3O^+

88   Which of the following is the *weakest* Brönsted acid?
(1) NH_4^+ (2) HSO_4^- (3) H_2SO_4 (4) HNO_3

89   Which compound is a salt?
(1) CH_3OH (2) $C_6H_{12}O_6$ (3) $H_2C_2O_4$ (4) $KC_2H_3O_2$

90   What is the pH of a solution with a hydronium ion concentration of 0.01 mole per liter?
(1) 1   (2) 2   (3) 10   (4) 14

91   Given the equation: $H^+ + OH^- \rightarrow H_2O$
Which type of reaction does the equation represent?
(1) esterification   (3) hydrolysis
(2) decomposition   (4) neutralization

Group 8 - Redox and Electrochemistry
If you choose this group, be sure to answer questions 92–96.

92   Which reduction half-reaction has a standard electrode potential (E^0) of 1.50 volts?
(1) $Au^{3+} + 3e^- \rightarrow Au(s)$ (3) $Co^{2+} + 2e^- \rightarrow Co(s)$
(2) $Al^{3+} + 3e^- \rightarrow Al(s)$ (4) $Ca^{2+} + 2e^- \rightarrow Ca(s)$

93   Given the reaction: $2Li(s) + Cl_2(g) \rightarrow 2LiCl(s)$
As the reaction takes place, the $Cl_2(g)$ will
(1) gain electrons   (3) gain protons
(2) lose electrons   (4) lose protons

94 The diagram at the right shows the
    electrolysis of fused KCl. What occurs
    when the switch is closed?

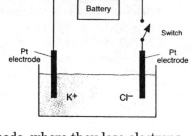

    (1) Positive ions migrate toward the
        anode, where they lose electrons.
    (2) Positive ions migrate toward the
        anode, where they lose electrons.
    (3) Positive ions migrate toward the
        anode, where they lose electrons.
    (4) Positive ions migrate toward the anode, where they lose electrons.

95 The diagram at the right represents an
    electrochemical cell at 298 K and 1
    atmosphere. What is the maximum cell
    voltage (E^0) when the switch is closed?
    (1) +1.61 V   (3) +3.13 V
    (2) -1.61 V   (4) -3.13 V

96 Given the balanced equation:
    $2AL(s) + 6H^+(aq) \rightarrow 2Al^{3+}(aq) + 3H_2(g)$

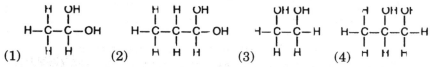

    When 2 moles of Al(s) completely reacts, what is the total number
    of moles of electrons transferred from Al(s) to H^+(aq)?
    (1) 5   (2) 6   (3) 3   (4) 4

Group 9 - Organic Chemistry
If you choose this group, be sure to answer questions 97-101.

97 A condensation polymerization reaction produces a polymer and
    (1) H_2 (2) O_2 (3) CO_2 (4) H_2O

98 Which organic compound is classified as a primary alcohol?
    (1) ethylene glycol   (3) glycerol
    (2) ethanol   (4) 2-butanol

99 What is the structural formula for 1,2-ethanediol?

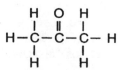

    (1)   (2)   (3)   (4)

100 Given the structural formula for ehtyne: $H- C \equiv C \quad H$
    What is the total number of electrons shared between the carbon
    atoms?
    (1) 6   (2) 2   (3) 3   (4) 4

101 What is the name of the compound with the
    formula at the right?
    (1) propanone   (3) propanal
    (2) propanol   (4) propanoic acid

Group 10 - Applications of Chemical Principles
If you choose this group, be sure to answer questions 102–106.

102 Given the overall reaction for the lead-acid battery:

$$Pb + PbO_2 + 2H_2SO_4 \underset{\text{Charge}}{\overset{\text{Discharge}}{\rightleftharpoons}} 1PbSO_4 + 2H_2O$$

Which element changes oxidation state when electric energy is produced?
(1) hydrogen　　(2) oxygen　　(3) sulfur　　(4) lead

103 Which substance is produced by the Haber process?
(1) aluminum　(2) ammonia　(3) nitric acid　(4) sulfuric acid

104 Iron corrodes more easily than aluminum and zinc because aluminum and zink both
(1) are reduced　　　　　　　(3) form oxides that are self-protective
(2) are oxidizing agents　　　(4) form oxides that are very reactive

105 Which balanced equation represents a cracking reaction?
(1) $2C_3H_6 + 9O_2 \rightarrow 6H_2O + 6CO_2$ (3) $C_{14}H_{28} + Cl_2 \rightarrow C_{14}H_{28}Cl_2$
(2) $C_{14}H_{30} \rightarrow C_7H_{16} + C_7H_{14}$　　　(4) $C_2H_6 + Cl_2 \rightarrow C_2H_5Cl + HCl$

106 During fractional distillation, hydrocarbons are separated according to their
(1) boiling points　　　　　　(3) triple points
(2) melting points　　　　　　(4) saturation points

Group 11 - Nuclear Chemistry
If you choose this group, be sure to answer questions 107–111.

107 In a fusion reaction, reacting nuclei must collide. Collisions between two nuclei are difficult to achieve because the nuclei are
(1) both negatively charged and repel each other
(2) both positively charged and repel each other
(3) oppositely charged and attract each other
(4) oppositely charged and repel each other

108 A particle accelerator can increase the kinetic energy of
(1) an alpha particle and a beta particle
(2) an alpha particle and a neutron
(3) a gamma ray and a beta particle
(4) a neutron and a gamma ray

109 Which nuclide is a radioisotope used in the study of organic reaction mechanisms?
(1) carbon-12　(2) carbon-14　(3) uranium-235 (4) uranium-238

110 To make nuclear fission more efficient, which device is used in a nuclear reactor to slow the speed of neutrons?
(1) internal shield　　　　　　(3) control rod
(2) external shield　　　　　　(4) moderator

111 Which equation is an example of artificial transmutation?

(1) $^{238}_{92}U \rightarrow {}^{4}_{2}He + {}^{234}_{90}Th$ (3) $^{14}_{6}C \rightarrow {}^{14}_{7}N + {}^{0}_{-1}e$

(2) $^{27}_{13}Al + {}^{4}_{2}He \rightarrow {}^{30}_{15}P + {}^{1}_{0}n$ (4) $^{226}_{88}Ra \rightarrow {}^{4}_{2}He + {}^{222}_{86}Rn$

Group 12 – Laboratory Activities
If you choose this group, be sure to answer questions 112–116.

112 Which measurement contains three significant figures?

    (1) 0.08 cm   (2) 0.080 cm   (3) 800 cm   (4) 8.08 cm

113 A student investigated the physical and chemical properties of a sample of an unknown gas and then identified the gas. Which statement represents a conclusion rather than an experimental observation?

    (1) The gas is colorless.
    (2) The gas is carbon dioxide.
    (3) When the gas is bubbled into limewater, the liquid becomes cloudy.
    (4) When placed in the gas, a flaming splint stops burning.

114 The table at the right shows properties of four solids, *A, B, C,* and *D*. Which substance could represent diamond, a network solid?

    (1) *A*   (3) *C*
    (2) *B*   (4) *D*

Substance	Melting Point	Conductivity in Solid State	Solubility in Water
A	high	no	soluble
B	high	yes	insoluble
C	high	no	insoluble
D	low	no	insoluble

115 A student obtained the following data to determine the percent by mass of water in a hydrate. What is the approximate percent by mass of the water in the hydrated salt?

    (1) 2.5%   (3) 88%
    (2) 12%   (4) 98%

Mass of empty crucible + cover	11.70 g
Mass of crucible + cover + hydrated salt before heating	14.90 g
Mass of crucible + cover + anhydrous salt after thorough heating	14.53 g

116 A student wishes to prepare approximately 100 milliliters of an aqueous solution of 6 M HCl using 12 M HCl. Which procedure is correct?

    (1) adding 50 mL of 12 M HCl to 50 mL of water while stirring the mixture steadily
    (2) adding 50 mL of 12 M HCl to 50 mL of water then stirring the mixture steadily
    (3) adding 50 mL of water to 50 mL of 12 M HCl while stirring the mixture steadily
    (4) adding 50 mL of water to 50 mL of 12 M HCl then stirring the mixture steadily

699
Chemistry Practice Examination 4
Part 1
Answer all 56 questions in this part.
Directions (1-56): For *each* statement or question, select the word or expression that, of those given, best completes the statement or answers the question.

Page 309

1   Solid A at 80°C is immersed in liquid B at 60°C. Which statement correctly describes the energy changes between A and B?
    (1) A releases heat and B absorbs heat.
    (2) A absorbs heat and B releases heat.
    (3) Both A and B absorb heat.
    (4) Both A and B release heat.

2   Given the phase equilibrium at a pressure of 1 atmosphere:
$$H_2O(s) \leftrightarrow H_2O(l)$$
    What is the temperature of the equilibrium mixture?
    (1) 273°C   (2) 273 K   (3) 373°C   (4) 373 K

3   Which statement is an identifying characteristic of a mixture?
    (1) A mixture can consist of a single element.
    (2) A mixture can be separated by physical means.
    (3) A mixture must have a definite composition by weight.
    (4) A mixture must be homogeneous.

4   Which phase change is endothermic?
    (1) gas $\rightarrow$ solid   (2) gas $\rightarrow$ liquid   (3) liquid $\rightarrow$ solid   (4) liquid $\rightarrow$ gas

5   What volume will a 300.-milliliter sample of a gas at STP occupy when the pressure is doubled at constant temperature?
    (1) 150. mL   (2) 450. mL   (3) 300. mL   (4) 600. mL

6   Each stoppered flask at the right contains 2 liters of a gas at STP. Each gas sample has the same
    (1) density
    (2) mass
    (3) number of molecules
    (4) number of atoms

7   Which electron configuration represents an atom in the excited state?
    (1) $1s^22s^22p^63s^2$ (2) $1s^22s^22p^63s^1$ (2) $1s^22s^22p^6$ (4) $1s^22s^22p^53s^2$

8   Which electron-dot symbol represents an atom of chlorine in the ground state?

    (1) C l⁝   (2) ·Ċl·   (3) ⁚Ċl·   (4) ⁚Ċl⁚

9   What is the total number of electrons in an atom of an element with an atomic number of 18 and a mass number of 40?
    (1) 18   (2) 22   (3) 40   (4) 58

10 After bombarding a gold foil sheet with alpha particles, scientists concluded that atoms consist mainly of
    (1) electrons   (2) empty space   (3) protons   (4) neutrons

11 Which element has atoms in the ground state with a sublevel that is only half filled?
    (1) helium   (2) beryllium   (3) nitrogen   (4) neon

12   Which sublevel contains a total of 5 orbitals?
      (1) *s*   (2) *p*   (3) *d*   (4) *f*

13   In how many days will a 12-gram sample of $^{131}_{53}$ I decay, leaving a total of
      1.5 grams of the original isotope?
      (1) 8.0   (2) 16   (3) 20.   (4) 24

14   Electronegativity is a measure of an atom's ability to
      (1) attract the electrons in the bond between the atom and another atom
      (2) repel the electrons in the bond between the atom and another atom
      (3) attract the protons of another atom
      (4) repel the protons of another atom

15   The molecular formula of a compound is represented by X_3Y_6. What is the
      empirical formula of this compound?
      (1) X_3Y (2) X_2Y (3) XY_2 (4) XY_3

16   Given the unbalanced equation:
                $\_CaSO_4+\_AlCl_3 \rightarrow \_Al_2(SO_4)_3+\_CaCl_2$
      What is the coefficient of $Al_2(SO_4)_3$ when the equation is completely
      balanced using the smallest whole-number coefficients?
      (1) 1   (2) 2   (3) 3   (4) 4

17   What is the total number of moles of sulfur atoms in 1 mole of $Fe_2(SO_4)_3$?
      (1) 1   (2) 15   (3) 3   (4) 17

18   The elements Li and F combine to form an ionic compound. The electron
      configurations in this compound are the same as the electron
      configurations of atoms in Group
      (1) 1   (2) 14   (3) 17   (4) 18

19   An element with an electronegativity of 0.9 bonds with an element with an
      electronegativity of 3.1. Which phrase best describes bond between these
      elements?
      (1) mostly ionic in character and formed between two nonmetals
      (2) mostly ionic in character and formed between a metal and a nonmetal
      (3) mostly covalent in character and formed between two nonmetals
      (4) mostly covalent in character and formed between a metal and a nonmetal

20   The symmetrical structure of the CH_4 molecule is due to the fact that the
      four single bonds between carbon and hydrogen atoms are directed toward
      the corners of a
      (1) triangle   (2) tetrahedron   (3) square   (4) rectangle

21   As the elements in Group 15 are considered in order of increasing atomic
      number, which sequence in properties occurs?
      (1) nonmetal → metalloid → metal
      (2) metalloid → metal → nonmetal
      (3) metal → metalloid → nonmetal
      (4) metal → nonmetal → metalloid

22   In which set do the elements exhibit the most similar chemical properties?
      (1) N, O, and F   (3) Li, Na, and K
      (2) Hg, Br, and Rn   (4) Al, Si, and P

23   Which reactant is most likely to have *d* electrons involved in a chemical
      reaction?
      (1) a halogen   (3) a transition element
      (2) a noble gas   (4) an alkali metal

24   Elements in a given period of the Periodic Table contain the same number of
      (1) protons in the nucleus   (3) electrons in the outermost level
      (2) neutrons in the nucleus   (4) occupied principal energy levels

25   The elements known as the alkaline earth metals are found in Group
      (1) 1   (2) 2   (3) 16   (4) 17

26   The properties of silicon are characteristic of
      (1) a metal, only   (3) both a metal and a nonmetal
      (2) a nonmetal, only   (4) neither a metal nor a nonmetal

27   Which substance has the greatest molecular mass?
      (1) H_2O_2 (2) NO   (3) CF_4 (4) I_2

28   According to Reference Table *D*, which of the following substances is *least*
      soluble in 100 grams of water at 50°C?
      (1) NaCl   (2) KCl   (3) NH_4Cl (4) HCl

29   Which sample contains a total of 6.0 x 10^{23} atoms?
      (1) 23 g Na   (2) 24 g C   (3) 42 g Kr   (4) 78 g K

30   A 20.-milliliter sample of 0.60 M HCl is diluted with water to a volume of
      40. milliliters. What is the new concentration of the solution?
      (1) 0.15 M   (2) 0.60 M   (3) 0.30 M   (4) 1.2 M

31   Given the reaction:

$$4Al(s) + 3O_2(g) \rightarrow 2Al_2O_3(s)$$

      What is the minimum number of grams of oxygen gas required to produce
      1.00 mole of aluminum oxide?
      (1) 32.0 g   (2) 48.0 g   (3) 96.0 g   (4) 192 g

32   The minimum amount of energy required to start a chemical reaction is called
      (1) entropy   (3) free energy
      (2) enthalpy   (4) activation energy

33   Beaker *A* contains a 1-gram piece of zinc and beaker *B* contains 1 gram of
      powdered zinc. If 100 milliliters of 0.1 M HCl is added to each of the
      beakers, how does the rate of reaction in beaker *A* compare to the rate of
      reaction in beaker *B*?
      (1) The rate in *A* is greater due to the smaller surface area of the zinc.
      (2) The rate in *A* is greater due to the larger surface area of the zinc.
      (3) The rate in *B* is greater due to the smaller surface area of zinc.
      (4) The rate in *B* is greater due to the larger surface area of the zinc.

34   Based on Reference Table *D*, which amount of a compound dissolved in 100
      grams of water at the stated temperature represents a system at
      equilibrium?
      (1) 20 g $KClO_3$ at 80°C   (3) 40 g KCl at 60°C
      (2) 40 g KNO_3 at 25°C   (4) 60 g $NaNO_3$ at 40°C

35   Given the reaction at equilibrium:

$$N_2(g) + 3H_2(g) \leftrightarrow 2NH_3(g) + 22 \text{ kcal}$$

      Which stress would cause the equilibrium to shift to the left?
      (1) increasing the temperature   (3) adding $N_2(g)$ to the system
      (2) increasing the pressure   (4) adding $H_2(g)$ to the system

36 The potential energy diagram of a
    chemical reaction is shown at the right.
    Which arrow represents the part of the
    reaction most likely to be affected by the
    addition of a catalyst?
    (1) *A*   (3) *C*
    (2) *B*   (4) *D*

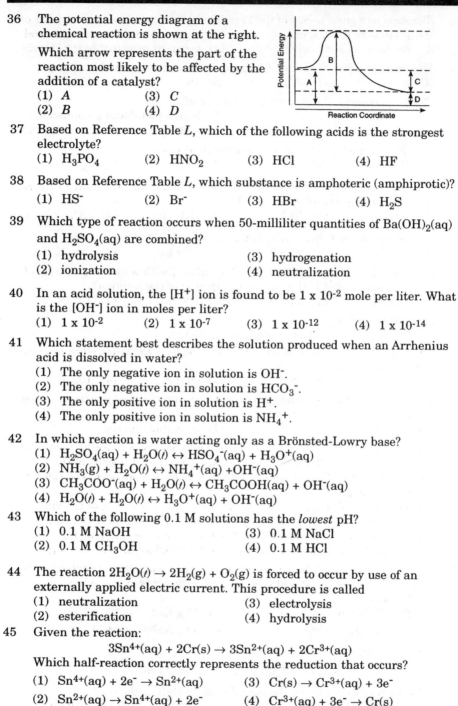

37 Based on Reference Table *L*, which of the following acids is the strongest
    electrolyte?
    (1) H_3PO_4 (2) HNO_2 (3) HCl (4) HF

38 Based on Reference Table *L*, which substance is amphoteric (amphiprotic)?
    (1) HS^- (2) Br^- (3) HBr (4) H_2S

39 Which type of reaction occurs when 50-milliliter quantities of $Ba(OH)_2(aq)$
    and $H_2SO_4(aq)$ are combined?
    (1) hydrolysis   (3) hydrogenation
    (2) ionization   (4) neutralization

40 In an acid solution, the $[H^+]$ ion is found to be 1×10^{-2} mole per liter. What
    is the $[OH^-]$ ion in moles per liter?
    (1) 1×10^{-2} (2) 1×10^{-7} (3) 1×10^{-12} (4) 1×10^{-14}

41 Which statement best describes the solution produced when an Arrhenius
    acid is dissolved in water?
    (1) The only negative ion in solution is OH^-.
    (2) The only negative ion in solution is HCO_3^-.
    (3) The only positive ion in solution is H^+.
    (4) The only positive ion in solution is NH_4^+.

42 In which reaction is water acting only as a Brönsted-Lowry base?
    (1) $H_2SO_4(aq) + H_2O(l) \leftrightarrow HSO_4^-(aq) + H_3O^+(aq)$
    (2) $NH_3(g) + H_2O(l) \leftrightarrow NH_4^+(aq) + OH^-(aq)$
    (3) $CH_3COO^-(aq) + H_2O(l) \leftrightarrow CH_3COOH(aq) + OH^-(aq)$
    (4) $H_2O(l) + H_2O(l) \leftrightarrow H_3O^+(aq) + OH^-(aq)$

43 Which of the following 0.1 M solutions has the *lowest* pH?
    (1) 0.1 M NaOH   (3) 0.1 M NaCl
    (2) 0.1 M CH_3OH (4) 0.1 M HCl

44 The reaction $2H_2O(l) \rightarrow 2H_2(g) + O_2(g)$ is forced to occur by use of an
    externally applied electric current. This procedure is called
    (1) neutralization   (3) electrolysis
    (2) esterification   (4) hydrolysis

45 Given the reaction:
              $3Sn^{4+}(aq) + 2Cr(s) \rightarrow 3Sn^{2+}(aq) + 2Cr^{3+}(aq)$
    Which half-reaction correctly represents the reduction that occurs?
    (1) $Sn^{4+}(aq) + 2e^- \rightarrow Sn^{2+}(aq)$ (3) $Cr(s) \rightarrow Cr^{3+}(aq) + 3e^-$
    (2) $Sn^{2+}(aq) \rightarrow Sn^{4+}(aq) + 2e^-$ (4) $Cr^{3+}(aq) + 3e^- \rightarrow Cr(s)$

46 The oxidation number of nitrogen in N_2 is
    (1) +1   (2) 0   (3) +3   (4) -3

47   Which reaction is an organic reaction?
     (1) $C_3H_8(g) + 5O_2(g) \rightarrow 3CO_2(g) + 4H_2O(g)$
     (2) $2H_2(g) + O_2(g) \rightarrow 2H_2O(g)$
     (3) $3CU^{2+}(aq) + 2Fe(s) \rightarrow 3Cu(s) + 2Fe^{3+}(aq)$
     (4) $NaOH(aq) + HCl(aq) \rightarrow NaCl(aq) + H_2O(l)$

48   Given the reaction:
$$Zn(s) + 2HCl(aq) \rightarrow ZnCl_2(aq) + H_2(g)$$
     Which substance is oxidized?
     (1) $Zn(s)$ (2) $HCl(aq)$ (3) $Cl^-(aq)$ (4) $H^+(aq)$

49   Which organic compound will dissolve in water to produce a solution that
     will turn blue litmus red?

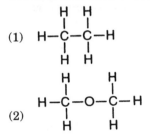

50   A redox reaction always demonstrates the conservation of
     (1) mass, only   (3) both mass and charge
     (2) charge, only   (4) neither mass nor charge

51   In which organic reaction is sugar converted to an alcohol and carbon
     dioxide?
     (1) esterification   (3) substitution
     (2) addition   (4) fermentation

52   Which three compounds belong to the same homologous series?
     (1) CH_4, C_2H_6, C_3H_4 (3) $C_4H_{10}, C_5H_{10}, C_6H_6$
     (2) $C_3H_6, C_4H_8, C_5H_{10}$ (4) C_2H_2, C_3H_4, C_4H_8

53   Which formula represents a saturated compound?
     (1) C_2H_4 (2) C_2H_2 (3) C_3H_6 (4) C_3H_8

Note that questions 54 through 56 have only three choices.

54   Given the reaction: $2Na(s) + Cl_2(g) \rightarrow 2NaCl(s)$
     As the reactants form products, the stability of the chemical system will
     (1) decrease   (2) increase   (3) remain the same

55   As the elements of Group 1 are considered in order from top to bottom, the
     first ionization energy of each successive element will
     (1) decrease   (2) increase   (3) remain the same

56   Given the reaction at equilibrium:
$$4HCl(g) + O_2(g) \leftrightarrow 2Cl_2(g) + 2H_2O(g)$$
     If the pressure on the system is increased, the concentration of $Cl_2(g)$ will
     (1) decrease   (2) increase   (3) remain the same

Part II

This part consists of twelve groups, each containing five questions. Each group tests a major area of the course. Choose seven of these twelve groups. Be sure that you answer all five questions in each group chosen.

Group 1 - Matter and Energy
If you choose this group, be sure to answer questions 57-61.

57   Which substance can be decomposed by a chemical change?
     (1) ammonia   (2) aluminum   (3) magnesium   (4) manganese

58   The heat required to change 1 gram of a solid at its normal melting point to a liquid at the same temperature is called the heat of
     (1) vaporization   (2) fusion   (3) reaction   (4) formation

59   A real gas would behave most like an ideal gas under conditions of
     (1) low pressure and low temperature
     (2) low pressure and high temperature
     (3) high pressure and low temperature
     (4) high pressure and high temperature

60   The volume of a sample of a gas at 273°C is 200. liters. If the volume is decreased to 100. liters at constant pressure, what will be the new temperature of the gas?
     (1) 0 K   (2) 100. K   (3) 273 K   (4) 546 K

61   The graph at the right represents the relationship between pressure and volume of a given mass of a gas at constant temperature.

The product of pressure and volume is constant. According to the graph, what is the product in atm•mL?
     (1) 20.   (3) 60.
     (2) 40.   (4) 80.

Group 2 - Atomic Structure
If you choose this group, be sure to answer questions 62-66.

62   The mass of a calcium atom is due primarily to the mass of its
     (1) protons, only   (3) protons and neutrons
     (2) neutrons, only   (4) protons and electrons

63   What is the maximum number of electrons that can occupy the fourth principal energy level of an atom?
     (1) 6   (2) 8   (3) 18   (4) 32

64   Which element has no known stable isotope?
     (1) Hg   (2) Po   (3) Se   (4) Zn

65   The characteristic spectral lines of elements are caused when electrons in an excited atom move from
     (1) lower to higher energy levels, releasing energy
     (2) lower to higher energy levels, absorbing energy
     (3) higher to lower energy levels, releasing energy
     (4) higher to lower energy levels, absorbing energy

66 What is the total number of protons contained in the nucleus of a carbon-14 atom?
    (1) 6   (2) 8   (3) 12   (4) 14

Group 3 – Bonding
If you choose this group, be sure to answer questions 67–71.

67 At 298 K, the vapor pressure of H_2O is less than the vapor pressure of CH_3OH because H_2O has
    (1) larger molecules   (3) stronger ionic bonds
    (2) a larger molecular mass   (4) stronger hydrogen bonds

68 The chemical formula $CaCO_3$ is an example of an expression that is
    (1) quantitive, only
    (2) qualitative, only
    (3) both quantitative and qualitative
    (4) neither quantitative nor qualitative

69 When NaCl(s) is dissolved in $H_2O(l)$, the sodium ion is attracted to the water molecule's
    (1) negative end, which is hydrogen
    (2) negative end, which is oxygen
    (3) positive end, which is hydrogen
    (4) positive end, which is oxygen

70 Which electron-dot formula represents a substance that contains a nonpolar covalent bond?

    (1) $[Na]^+ [\overset{x\ x}{\underset{x\ x}{\vdots C \vert x}}]^-$ (3) $H \overset{x\ x}{\underset{x\ x}{\vdots C \vert x}}$

    (2) $\overset{x\ x}{\underset{x\ x}{x C \vert}} \vdots \overset{\bullet\bullet}{\underset{\bullet\bullet}{C \vert}}$ (4) $\overset{\bullet\bullet}{\underset{\bullet\ x}{\vdots O \vdots}} H$ \\ H

71 What is the correct formula for iron (II) sulfide?
    (1) FeS   (2) $FeSO_3$ (3) Fe_2S_3 (4) $Fe_2(SO_4)_3$

Group 4 – Periodic Table
If you choose this group, be sure to answer questions 72–76.

72 Which of the following groups in the Periodic Table contain elements so highly reactive they are never found in the free state?
    (1) 1 and 2   (2) 1 and 11   (3) 2 and 15   (4) 11 and 15

73 The presence of which ion usually produces a colored solution?
    (1) K^+ (2) F^- (3) Fe^{2+} (4) S^{2-}

74 How does the size of a barium ion compare to the size of a barium atom?
    (1) The ion is smaller because it has fewer electrons.
    (2) The ion is smaller because it has more electrons.
    (3) The ion is larger because it has fewer electrons.
    (4) The ion is larger because it has more electrons.

75 Which element is brittle in the solid phase and is a *poor* conductor of heat and electricity?
    (1) calcium   (2) sulfur   (3) strontium   (4) copper

76 Which halogen can only be prepared by the electrolysis of its fused compounds?
    (1) I_2 (2) Cl_2 (3) Br_2 (4) F_2

Group 5 – Mathematics of Chemistry
If you choose this group, be sure to answer questions 77–81.

77   What is the mass of 1 mole of a gas that has a density of 2.00 grams per liter at STP?
(1) 11.2 g   (2) 22.4 g   (3) 33.6 g   (4) 44.8 g

78   Dissolving 1 mole of KCl in 1,000 grams of H_2O affects
(1) the boiling point of the H_2O, only
(2) the freezing point of the H_2O, only
(3) both the boiling point and the freezing point of the H_2O
(4) neither the boiling point nor the freezing point of the H_2O

79   The heat of vaporization of a liquid is 320. calories per gram. What is the minimum number of calories needed to change 40.0 grams of the liquid to vapor at the boiling point?
(1) 8.00   (2) 320.   (3) 3,280   (4) 12,800

80   A compound was analyzed and found to contain 75% carbon and 25% hydrogen by mass. What is the compound's empirical formula?
(1) CH   (2) CH_2 (3) CH_3 (4) CH_4

81   Which gas diffuses most rapidly at STP?
(1) Ne   (2) Ar   (3) Cl_2 (4) F_2

Group 6 – Kinetics and Equilibrium
If you choose this group, be sure to answer questions 82–86.

82   According to Reference Table *G*, which compound is spontaneously formed even though the reaction is endothermic?
(1) ICl(g)   (2) $CO_2(g)$ (3) $H_2O(l)$ (4) $Al_2O_3(s)$

83   Given the reaction at equilibrium
$$BaCrO_4(s) \leftrightarrow Ba^{2+}(aq) + CrO_4^{2-}(aq)$$
Which substance, when added to the mixture, will cause an increase in the amount of $BaCrO_4(s)$?
(1) K_2CO_3 (2) $CaCO_3$ (3) $BaCl_2$ (4) $CaCl_2$

84   At 1 atmosphere and 298 K, which saturated salt solution is most concentrated? [Refer to Reference Table *M*.]
(1) $PbCO_3$ (2) $PbCrO_4$ (3) AgBr   (4) AgCl

85   Given the reaction: $A_2B(s) \leftrightarrow 2A^+(aq) + B^{2-}(aq)$
What is the solubility product constant expression (K_{sp}) for this reaction?
(1) $2[A^+] [B^{2-}]$ (3) $2[A^+]^2 [B^{2-}]$
(2) $2[A^+] + [B^{2-}]$ (4) $2[A^+]^2 + [B^{2-}]$

86   Which factors must be equal when a reversible chemical process reaches equilibrium?
(1) mass of the reactants and mass of the products
(2) rate of the forward reaction and rate of the reverse reaction
(3) concentration of the reactants and concentration of the products
(4) activation energy of the forward reaction and activation energy of the reverse reaction

Group 7 - Acids and Bases
If you choose this group, be sure to answer questions 87–91.

87 The diagram at the right shows an apparatus used to test the conductivity of various materials. Which aqueous solution will cause the bulb to light?

Bulb

Electrodes

Aqueous solution

Source of power

(1) $C_6H_{12}O_6$(aq)   (3) CH_3OH(aq)
(2) $C_{12}H_{22}O_{11}$(aq)   (4) LiOH(aq)

88 If 50.0 milliliters of 3.0 M HNO_3 completely neutralized 150.0 milliliters of KOH, what was the molarity of the KOH solution?
(1) 1.0 M   (2) 4.5 M   (3) 3.0 M   (4) 6.0 M

89 According to the Arrheius theory, which list of compounds includes only bases?
(1) KOH, $Ca(OH)_2$, and CH_3OH (3) LiOH, $Ca(OH)_2$, and $C_2H_4(OH)_2$
(2) KOH, NaOH, and LiOH   (4) NaOH, $Ca(OH)_2$, and CH_3COOH

90 Given the reaction:

$$NH_3(g) + H_2O(l) \leftrightarrow NH_4^+(aq) + OH^-(aq)$$

Which is a conjugate acid-base pair?
(1) $H_2O(l)$ and $NH_4^+(aq)$ (3) $NH_3(g)$ and $OH^-(aq)$
(2) $H_2O(l)$ and $NH_3(g)$ (4) $NH_3(g)$ and $NH_4^+(aq)$

91 Given the reaction:
$$HC_2H_3O_2(aq) + KOH(aq) \rightarrow KC_2H_3O_2(aq) + H_2O(l)$$
The products of this reaction form a salt solution that is
(1) acidic and turns litmus blue   (3) basic and turns litmus blue
(2) acidic and turns litmus red   (4) basic and turns litmus red

Group 8 - Redox and Electrochemistry
If you choose this group, be sure to answer questions 92–96.

92 What is the E^0 for the half-reaction $Cu^+ + e^- \rightarrow Cu(s)$?
(1) -0.52 V   (2) -0.34 V   (3) +0.34 V   (4) +0.52 V

93 Given the reaction: $2Al(s) + 6H^+(aq) \rightarrow 2Al^{3+}(aq) + 3H_2(g)$

What is the total number of moles of electrons gained by $H^+(aq)$ when 2 moles of Al(s) is completely reacted?
(1) 6   (2) 2   (3) 3   (4) 12

94 Given the redox reaction:
$Mg(s) + CuSO_4(aq) \rightarrow MgSO_4(aq) + Cu(s)$
Which species acts as the oxidizing agent?
(1) Cu(s)   (2) $Cu^{2+}(aq)$ (3) Mg(s)   (4) $Mg^{2+}(aq)$

95 In an electrolytic cell, the negative electrode is called the
(1) anode, at which oxidation occurs
(2) anode, at which reduction occurs
(3) cathode, at which oxidation occurs
(4) cathode, at which reduction occurs

96   The diagram at the right represents an
     electrochemical cell.

     What occurs when the switch is closed?

     (1) Zn is reduced.
     (2) Cu is oxidized.
     (3) Electrons flow from Cu to Zn.
     (4) Electrons flow from Zn to Cu.

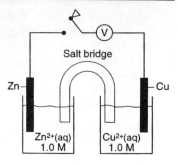

Group 9 – Organic Chemistry
If you choose this group, be sure to answer questions 97–101.

97   Which structural formula represents a primary alcohol?

(1)

(3)

(2)

(4)

98   Which materials are naturally occurring polymers?
     (1) nylon and cellulose   (3) starch and cellulose
     (2) nylon and polyethylene   (4) starch and polyethylene

99   Which formula represents an isomer of the compound propanoic acid,
     CH_3CH_2COOH?
     (1) $CH_3CH_2CH_2OH$ (3) $CH_3CH(OH)CH_2OH$
     (2) $CH_3CH_2CH_2COOH$ (4) CH_3COOCH_3

100 The compound 1,2-ethanediol is a
     (1) monohydroxy alcohol   (3) primary alcohol
     (2) dihydroxy alcohol   (4) secondary alcohol

101 Which reaction best represents the complete combustion of ethene?
     (1) $C_2H_4 + HCl \rightarrow C_2H_5Cl$ (3) $C_2H_4 + 3O_2 \rightarrow 2CO_2 + 2H_2O$
     (2) $C_2H_4 + Cl_2 \rightarrow C_2H_4Cl_2$ (4) $C_2H_4 + H_2O \rightarrow C_2H_5OH$

Group 10 – Applications of Chemical Principles
If you choose this group, be sure to answer questions 102–106.

102 During the contact process, the ores of which kind of compounds are
     burned to produce SO_2?
     (1) bromides   (2) carbides   (3) phosphides   (4) sulfides

102 Given the equation for the overall reaction in a lead-acid storage battery:

$$Pb(s) + PbO_2(s) + 2H_2SO_4(aq) \underset{\text{Charge}}{\overset{\text{Discharge}}{\rightleftharpoons}} 2PbSO_4(s) + 2H_2O(l)$$

Which occurs during the charging of the battery?
(1) The concentration of H_2SO_4 decreases and the number of moles of Pb(s) increases.
(2) The concentration of H_2SO_4 decreases and the number of moles of $H_2O(l)$ increases.
(3) The concentration of H_2SO_4 increases and the number of moles of Pb(s) decreases.
(4) The concentration of H_2SO_4 increases and the number of moles of $H_2O(l)$ decreases.

104 The corrosion of iron is an example of
(1) an oxidation-reduction reaction   (3) a substitution reaction
(2) an addition reaction   (4) a neutralization reaction

105 The separation of petroleum into components based on their boiling points is accomplished by
(1) cracking   (3) fractional distillation
(2) melting   (4) addition polymerization

106 Petroleum is a complex mixture of many
(1) hydrocarbons (2) aldehydes   (3) organic halides (4) ketones

Group 11 – Nuclear Chemistry
If you choose this group, be sure to answer questions 107–111.

107 The diagram at the right represents a nuclear reaction in which a neutron bombards a heavy nucleus.

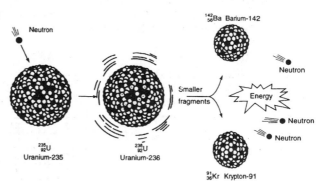

Which type of reaction does the diagram illustrate?
(1) fission   (2) fusion   (3) alpha decay   (4) beta decay

108 Within a nuclear reactor, the purpose of the moderator is to
(1) absorb neutrons in the reactor core
(2) absorb neutrons in the outer containment structure
(3) slow down neutrons in the reactor core
(4) slow down neutrons in the outer containment structure

109 The radioisotope I-131 is used to
(1) control nuclear reactors   (3) diagnose thyroid disorders
(2) determine the age of fossils   (4) trigger fusion reactors

110 In which list can all particles be accelerated by an electric field?
    (1) alpha particles, beta particles, and neutrons
    (2) alpha particles, beta particles, and protons
    (3) alpha particles, protons, and neutrons
    (4) beta particles, protons, and neutrons

111 Given the nuclear reaction: $\quad {}^{9}_{4}Be + X \rightarrow {}^{12}_{6}C + {}^{1}_{0}n$

    What is the identity of particle X?
    (1) alpha particle (2) beta particle (3) proton   (4) neutron

Group 12 - Laboratory Activities
If you choose this group, be sure to answer questions 112-116.

112 Which set of laboratory equipment would most likely be used with a crucible?

    (1)   (2)   (3)   (4)

113 A student calculated the percent by mass of water in a sample of $BaCl_2\bullet 2H_2O$ to be 16.4%, but the accepted value is 14.8%. What was the student's percent error?

    (1) $\dfrac{14.8}{16.4}$ x 100%   (3) $\dfrac{1.6}{14.8}$ x 100%

    (2) $\dfrac{16.4}{14.8}$ x 100%   (4) $\dfrac{14.8}{1.6}$ x 100%

114 A student observed the following reaction:
    $AlCl_3(aq) + 3NaOH(aq) \rightarrow Al(OH)_3(s) + 3NaCl(aq)$
    After the products were filtered, which substance remained on the filter paper?
    (1) NaCl   (2) NaOH   (3) $AlCl_3$ (4) $Al(OH)_3$

115 The table at the right shows the data collected by a student as heat was applied at a constant rate to a solid below its freezing point.

What is the boiling point of this substance?

    (1) 32°C
    (2) 54°C
    (3) 62°C
    (4) 100°C

Time (min)	Temperature (°C)	Time (min)	Temperature (°C)
0	20	18	44
2	24	20	47
4	28	22	51
6	32	24	54
8	32	26	54
10	32	28	54
12	35	30	54
14	38	32	58
16	41	34	62

116 Which quantity expresses the sum of the given masses to the correct number of significant figures?

$$\begin{array}{r} 22.1 \text{ g} \\ 375.66 \text{ g} \\ + \underline{5400.132} \text{ g} \end{array}$$

    (1) 5800 g   (2) 5798 g   (3) 5797.9 g   (4) 5797.892 g